T0311452

*Mathematical Methods
in Biology*

Mathematical Methods in Biology

J. DAVID LOGAN
University of Nebraska Lincoln

WILLIAM R. WOLESENSKY
Doane College

A JOHN WILEY & SONS, INC., PUBLICATION

Published by John Wiley & Sons, Inc., Hoboken, New Jersey.
Published simultaneously in Canada.

For general information on our other products and services or for technical support, please contact our Customer Care Department within the United States at (800) 762-2974, outside the United States at (317) 572-3993 or fax (317) 572-4002.

Wiley also publishes its books in a variety of electronic formats. Some content that appears in print may not be available in electronic format. For information about Wiley products, visit our web site at www.wiley.com.

Library of Congress Cataloging-in-Publication Data:

Logan, J. David (John David)
  Mathematical methods in biology / J. David Logan, William R. Wolesensky.
      p. cm.
  Includes bibliographical references and index.
  ISBN 978-0-470-52587-6 (pbk.)
  1. Biomathematics. I. Wolesensky, William R., 1962– II. Title.
  QH323.5.L64 2009
  570.15'1—dc22   2009013339

Printed in the United States of America.

10 9 8 7 6 5 4 3 2 1

*To Tess
and to Kris, Danielle, and Billy*

Contents

Preface

This textbook illustrates mathematical techniques used in many areas in the biological sciences. The last several years have witnessed a revolution in the connections between mathematics and biology. This interest in quantitative methods in biology has spawned a large number of outstanding textbooks in mathematical biology. (Many of these are included in the Reference Notes section at the end of each chapter.) This book differs from most others in that it covers *both deterministic* and *probabilistic* models. The first chapter is a long introduction and review of ideas about biological modeling, calculus, differential equations, dimensionless variables, and descriptive statistics. In the next three chapters we examine standard discrete and continuous models using difference and differential equations, and matrix algebra (there is a long appendix in Chapter 3 on matrices). The final three chapters cover probability, statistics, and stochastic processes, including bootstrap methods and stochastic differential equations.

The book is for students in mathematics or biology who have studied at least two semesters of calculus and have the ability to reason quantitatively. Topics such as differential equations, matrix algebra, and probability theory are introduced. The sophistication, or level of exposition, requires increasing maturity in thinking about quantitative issues as the chapters progress. In the first chapter some calculus ideas are reviewed, and by the final chapter concepts involving stochastic processes are introduced. The best way to see what is covered in this book is to peruse the table of contents. Portions of all this material have been taught numerous times by both authors to undergraduate students in mathematics, undergraduate students jointly majoring in mathematics and biology, and graduate students in mathematics and biology.

An instructor has considerable flexibility in choosing topics for a one- or two-semester course. At the end of the preface are suggestions for the books's use. One needs to be highly selective when choosing topics for an elementary course for students who have studied only a calculus sequence. Overall, it would be difficult to cover the entire book in two semesters.

One feature of the text not shared by many others is that there are a large number of exercises at the end of each section that vary in difficulty from routine to more intermediate problems that build technique. Over fifty pages of solutions and hints to odd-numbered exercises are provided in Appendix A at the end of the book.

The text uses MATLAB® to illustrate many of the algorithms, although it does not rely on any single computer algebra package. An instructor can easily redirect and adapt this material to another computer algebra system, such as R, Mathematica, or Maple. or even a spreadsheet.

Of course, the life sciences is an enormous landscape. What we discuss herein focuses mostly in one area: theoretical ecology. Actually, ecology has become extremely quantitative, and the mathematical techniques used in ecology are applicable to most other areas in the life sciences. For example, some virus–immune system models closely resemble ecosystem dynamics. Ecology provides an especially accessible context for study by mathematics majors. Moreover, we choose ecology for our motivations and examples because our own interests and research are in that area.

Our connection to ecological theory was stimulated almost a decade ago by our biology mentor, Professor Tony Joern, an eco-physiologist formerly at the University of Nebraska and now at Kansas State University. Our years of exciting discussions on the ecology of grasslands are high points in both of our careers, and they represent the ideal of interdisciplinary interactions.

We are also grateful to our editor at Wiley, Susanne Steitz-Filler, for her enthusiasm about this project and her skill in making it a painless, efficient process. The reviewers of the book deserve a special acknowledgement for their participation and insightful comments. These include: Myron Allen, University of Wyoming; Fwu-Ranq Chang, Indiana University; David Edwards, University of Delaware; Ellina Grigorieva, Texas Women's University; Andrew Nevai, University of Central Florida; Robert O'Malley, University of Washington; Nori Tarui, University of Hawai'i at Manoa; Dan Van Peursem, The University of South Dakota; and Thomas Witelski, Duke University.

Suggestions for the use of this book

1. A one-semester introductory course for students who have had calculus 1 and 2 (Selected topics from Chapters 1, 2, 3)

2. A one-semester course for students who have had differential equations and linear algebra (Chapters 1, 2, 3, 4)

3. A one-semester course focusing on stochastic modeling (Chapters 5, 6, 7)

4. A one-semester general course for students who have had calculus 1 and 2 (Selected topics from Chapters 1, 2, 5, 6)

5. A two-semester course (Selected topics from Chapters 1 through 8)

Finally, we welcome suggestions, comments, and corrections. Contact information is on the web site: `http://www.math.unl.edu/~dlogan`. Also at this site there is supplementary material, including, for example, an additional chapter on natural selection and population genetics. A solution manual, with solutions to even-numbered problems, can be obtained by instructors from the Publisher. Contact details for obtaining the manual are available on the web site above.

David Logan, Lincoln, Nebraska
Bill Wolesensky, Crete, Nebraska
June, 2009

1

Introduction to Ecological Modeling

In ecology we want to quantify theories about population growth, animal behavior, life histories, environmental influences, and so on. A mathematical model is an analytical statement that quantifies and explains a certain phenomenon or observation.

1.1 Mathematical Models

By a **mathematical model** we usually mean an equation, or set of equations, or some other relationships that describe some phenomenon that we observe in science, engineering, economics, or some other area, that provides a quantitative prediction of observations. By **mathematical modeling** we mean the process by which we formulate and analyze model equations and compare observations to the predictions that the models makes. This process includes introducing the important and relevant quantities or variables involved in the model; making model-specific assumptions about those quantities; solving the model equations, if possible; comparing the solutions to real data, and interpreting the results. Often, the solution method involves computer simulation or approximation. The comparison to data may lead to revision and refinement until we are satisfied that the model describes the phenomenon accurately and is predictive of similar observations. This process is depicted schematically in Fig. 1.1. In summary, mathematical modeling involves physical intuition, formula-

Mathematical Methods in Biology,
By J. David Logan and William R. Wolesensky
Copyright © 2009 John Wiley & Sons, Inc.

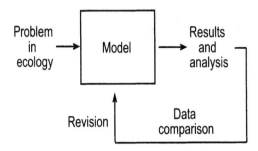

Figure 1.1 Schematic of the modeling process.

tion of equations, solution techniques, analysis, and tests of validity. A good mathematical model is simple, applies to many situations, and is predictive. Stated in a different way, in the modeling process the overarching objective is to make sense of the natural world as we observe it, by inventing caricatures of reality. Scientific exactness is sometimes sacrificed for mathematical tractability. Models help us clarify verbal descriptions of nature and the mechanisms that make up natural laws, and they help us determine which processes are important and which are unimportant.

One issue is the level of complexity of a model. With modern computer technology it is tempting to build complicated models that include every possible effect we can think of, with large numbers of parameters and variables. *Simulation models* (also called *agent-based models*) like these have their place, but computer output does not always allow us to discern which are the important processes and which are not. In building a model, it is usually a good idea to err on the side of simplicity and then build in complexity as needed or desired.

Mathematical models are classified in several ways: stochastic vs. deterministic, continuous vs. discrete, static vs. dynamic, quantitative vs. qualitative, descriptive vs. explanatory, and so on. In this book we are interested in modeling the underlying reasons for the phenomena we observe (explanatory) rather than fitting the data with formulas (descriptive) as is often done in statistics. For example, fitting measurements of the size of an animal over its lifetime by a regression curve is descriptive, and it gives some information. But describing the dynamics of growth by a differential equation relating growth rates, food assimilation rates, and energy maintenance requirements tells more about the underlying processes involved.

The reader is already familiar with many mathematical models. For example, in an elementary science course we learn that Newton's second law, $F = ma$ (force equals mass times acceleration), governs mechanical systems. The law of mass action in chemistry describes how fast chemical reactions occur, and the logistics equation models growth and competition in a population.

The first step in the modeling process is to select relevant variables (independent and dependent) and parameters that describe the problem. Biological quantities have *dimensions* such as time, mass, and degrees, or corresponding *units* such as seconds, kilograms, and degrees Celsius. The equations we write down as models must be dimensionally correct. Apples cannot equal oranges. Verifying that each term in a model equation has the same dimensions is the first task in obtaining a correct model. We should always be aware of the dimensions of the quantities, both variables and parameters, and we should always try to identify the biological meaning of the terms in the equations we obtain. Another general rule is always to allow the biological issues to drive the mathematics, not vice versa.

Example 1.1

Suppose that we want to know how fast a certain species of fish grow. One strategy is to go to the lab and every few days measure the length of each member of a cohort of newly hatched eggs. We can record at each time the average length L of the fish and plot the data points on a set of time–length, tL, axes. These data points are shown as pluses in Fig. 1.2. Then we can fit a curve to the data that has the same shape as the data. For example, we might determine constants a and b for which the curve

$$L = b(1 - e^{-at}) \tag{1.1}$$

fits the data. Equation (1.1) is a satisfactory model for the growth. It is a *descriptive model* that explains *how* the growth occurs, but it gives no clue as to *why* it occurs in this way. An *explanatory model*, on the other hand, is a model that provides reasons for the shape of the growth curve. That is, it gives mechanisms that explain the growth. In this case, as we observe later, the explanatory model is a differential equation which states that the growth rate is equal to the rate at which nutrients are assimilated minus the rate at which they are used for body maintenance and respiration. Out of this explanatory model comes equation (1.1) automatically; and there is an underlying mechanism based on energetics that tells us why the growth occurs in the way it does.   □

Example 1.2

MATLAB has simple commands to plot curves. Let us consider the two growth formulas $L_1 = 5(1 - e^{-0.5t})$ and $L_2 = 8(1 - e^{-0.5t})$ for times $0 \le t \le 10$. We

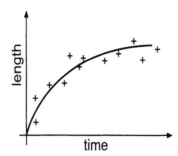

Figure 1.2 A set of data points $(+)$ and a curve of the form $L = b(1 - e^{-at})$, $a, b > 0$, that fits the data. Such a curve, which is a descriptive model, can be found by regression, which is discussed in a later section.

can use MATLAB to plot these curves on the same axes using the commands

```
t=0:0.05:10;
L1=5*(1-exp(-0.5*t));
L2=8*(1-exp(-0.5*t));
plot(t,L1,t,L2)
```

The command plot(t,L1) graphs just a single curve. To make two plots side by side, use the subplot command:

```
subplot(1,2,1), plot(t,L1)
subplot(1,2,2), plot(t,L2)
```

The command subplot(n,m,p) creates an $n \times m$ array of plots, putting the current plot in the pth position. □

EXERCISES

1. Suppose that a marine animal essentially has a spherical shape with radius r. If it assimilates nutrients at a rate proportional to its surface area, and if it uses nutrients at a rate proportional to its volume, what would be its size (radius) if its intake and use rates are in balance?

2. *Darlington's rule* is a descriptive model that relates the number of species S of birds on an island of area A (mi^2) near a mainland. The rule is $S = cA^b$, where c and b are constants found by fitting the curve to data. For example, in the West Indies, $c = 8.76$ and $b = 0.113$.

   (a) Plot the West Indies model on the AS axes (with A the abscissa, or horizontal axis). Then plot the model on a set of $\log A$, $\log S$ axes, where the logarithm is base 10.

(b) Can you give reasons why the curve has the shape it has?

3. The immigration rate of bird species (species per time) from a mainland to an offshore island is $I_m(1 - S/P)$, where I_m is the maximum immigration rate, P is the size of the source pool of species on the mainland, and S is the number of species already occupying the island. Further, the extinction rate is ES/P, where E is the maximum extinction rate. The growth rate of the number of species on the island is the immigration rate minus the extinction rate.

   (a) Plot the immigration and extinction rates vs. S, and determine the number of species for which the net rate of growth is zero, or the number of species is in equilibrium. (This exercise requires generic plots of functions. A *generic plot* is a plot of the *form* of the equation, regardless of what the constants may be. For example, we know that $y = mx + b$ plots as a straight line regardless of m and b, so a generic plot is just a straight line. We could indicate on the plot, if possible, what the constants mean; for example, b is the y intercept.)

   (b) Suppose that two islands of the same size are at different distances from the mainland. Birds arrive from the source pool, and they have the same extinction rate on each island. However, the maximum immigration rate is larger for the island farther away. Which island will have the larger number of species at equilibrium?

4. If a herbivore enters a patch where food has density F (items per area), one can model the rate of consumption C (items per time), the rate at which food items are eaten, by the equation

$$C = \frac{adF}{1 + a\tau dF},\tag{1.2}$$

where a is the search rate of the herbivore (area per time), d is the fraction of the food items discovered of those present, and τ is the time it takes to consume a single item. This equation, called a **type II functional response**, will be derived later.

   (a) Make a table of the quantities in model (1.2) and indicate the dimensions of each as well as a suitable set of units for a field mouse hunting for small edible food items.

   (b) What is the shape of the graph of C vs. F for F taking on all positive values? At what value of C does the rate *saturate* as F gets very large? What is the ecological meaning of this saturation value? Why is this response an example of a law of diminishing returns?

(c) At what food density is the consumption rate half its saturation value?

(d) The **risk** of any one food item being consumed is $R = C/F$. What are the dimensions of R? What is the shape of the graph of the risk R vs. F?

(e) If the fraction d of items discovered is a linear function of food density F (i.e., $d = \gamma F$,) find C as a function of F and plot the shape of the curve. This model for C is called a **type III functional response**. What are the dimensions of the proportionality constant γ?

(f) A simple consumption rate where there is no handling time is the linear rate, $C = adF$, which is called a **type I functional response**. Sketch this consumption rate and the risk vs. F.

(g) Contrast the shapes of the three consumption curves as well as the shapes of the associated curves for risk.

(h) Discuss situations where these responses might be appropriate or inappropriate.

1.2 Rates of Change

There are two fundamental concepts in calculus, the derivative and the integral. In this section we review key ideas about the derivative. Derivatives allow us to calculate how fast certain quantities, such as population, temperature, and growth, are changing. The derivative is also the slope of the tangent line to a curve at a point, and it permits us to approximate the curve near that point with a straight line. Going one step further, we discuss the higher-order approximations given by Taylor polynomials.

The **derivative** of a quantity measures how fast the quantity is changing with respect to another quantity, which is usually time. If t is time and the quantity y is a function of t, we write

$$y = f(t), \tag{1.3}$$

where f is the name of the function. For definiteness, think of y as being a population, or numbers of animals in a given region. A word about notation—when there is no confusion, we sometimes write (1.3) as $y = y(t)$, using the same letter to denote the quantity, or dependent variable, and the name of the function. Often, the right side of (1.3) is given by a specific formula.

At a certain *fixed* value of time, $t = t_0$, we can ask how fast the population y is changing at that instant. The answer is given by the derivative, which we

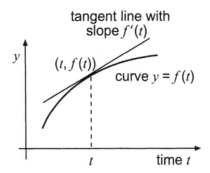

Figure 1.3 Generic plot of a function $y = f(t)$ and a tangent line at a time t. The slope of the tangent line at the point $(t, f(t))$ is the derivative, and its value measures how fast the function is changing at that instant.

denote by $f'(t_0)$. Graphically, the derivative measures the slope of the tangent line to the graph of $y = f(t_0)$ at the point $(t_0, f(t_0))$, as shown in Fig. 1.3. The slope measures how fast the quantity, in this case the population y, is changing. Clearly, we can consider the slope of the tangent line at any arbitrary value t, which is $f'(t)$. Thus, we can regard the derivative itself as a function of t. If the derivative is large, the slope is steep and y is changing rapidly, and if the derivative is small, the slope is shallow and y is changing slowly. If the derivative is positive at a value $t = a$, $f'(a) > 0$ and the graph of f must be rising at that instant; and if the derivative of f is negative at a time $t = c$, $f'(c) < 0$ and the graph is decreasing, or falling, at that time. If $f'(b) = 0$, the graph of f is flat at $(b, f(b))$, possibly signaling a maximum or a minimum value of the population. See Fig. 1.4.

Another common notation for the derivative $f'(t)$ is

$$f'(t) = \frac{dy}{dt}. \tag{1.4}$$

This notation is suggestive that the derivative is a ratio of dy and dt, where dy is an infinitesimally small change in population and dt is an infinitesimally small change in time. This is clarified below. Thinking in this way, the derivative has dimensions of population per time. The derivative is called the **rate of change**, measuring how fast the quantity y, in this case, population, is changing instantaneously at a fixed time t. Many of the laws of biology are expressed in terms of rates, or derivatives. These often arise as differential equations, which are equations that relate an unknown function to its rates.

It is important to understand how the derivative is defined. Let us fix an instant of time t_0 and ask how much the population changes from t_0 to $t_0 + h$,

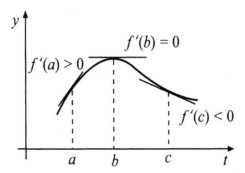

Figure 1.4 A function $y = f(t)$ will have a positive derivative at times where the function is increasing, and a negative derivative at times where the function is decreasing. At a point were the function has a local maximum (or a local minimum), the derivative, or slope, is zero.

where h is a small amount of time. The actual change is $f(t_0 + h) - f(t)$, so the average change in the population *over this interval* is

$$\text{average change} = \frac{f(t_0 + h) - f(t_0)}{h}. \tag{1.5}$$

Geometrically, on the graph of $y = f(t)$, this average change is the slope of the line connecting the two points $(t_0, f(t_0))$ and $(t_0 + h, f(t_0 + h))$. The derivative is the limiting value of the average change as the interval of time h approaches zero. In symbols,

$$f'(t_0) = \lim_{h \to 0} \frac{f(t_0 + h) - f(t_0)}{h}. \tag{1.6}$$

Notice that if h is very small,

$$f'(t_0) \approx \frac{f(t_0 + h) - f(t_0)}{h},$$

which means that the derivative is approximated by the average change over a very small interval of time. Rewriting this expression with $t = t_0 + h$ gives

$$f(t) \approx f(t_0) + f'(t_0)h. \tag{1.7}$$

This important equation states that the actual population $f(t)$ at time $t = t_0 + h$ can be approximated by the population at time t_0, provided that we know the derivative, or how fast the population is changing, at time t_0. The right side of expression (1.7) is called the **linear approximation** or **linearization**, and it implies that the curve $y = f(t)$ can be approximated near a fixed time t_0 by the tangent line through the point $(t_0, f(t_0))$. The right side of (1.7), using $h = t - t_0$, is $f(t_0) + f'(t_0)(t - t_0)$, which is the equation of a straight line.

If $y = N(t)$ represents a population, we say that $N'(t)$ is the **growth rate**. It measures how fast the population is changing. The *relative growth rate*, or *per capita growth rate*, is defined by the ratio

$$\text{per capita growth rate } = \frac{N'(t)}{N(t)}. \tag{1.8}$$

The per capita growth rate, which has units of time^{-1}, measures how fast the population is changing relative to the current value of the population. To say that the growth rate of a population is 3% and the per capita growth rate is 3% is to say two very different things. Really, saying that a population grew 3% actually says very little. Is it a large change in the population? Is it a small change?

1.2.1 Taylor Polynomials

The approximation of $f(t)$ at a value $t = a$ by its linearization (a straight line) can be improved by including higher-order derivatives. The linear approximation, which we denote by

$$P_1(t) = f(a) + f'(a)(t - a),$$

agrees with both $f(a)$ and $f'(a)$ at the value $t = a$, and it approximates $f(t)$ near $t = a$. To get more accuracy, we can also require the approximation to have the same concavity at $t = a$. Recall that the concavity is measured by the second derivative. Therefore, let us try a parabolic approximation and define

$$P_2(t) = f(a) + f'(a)(t - a) + c_2(t - a)^2,$$

where c_2 is to be determined so that $P_2''(a) = f''(a)$. By direct differentiation we get $P_2''(a) = 2c_2$. So we choose $c_2 = \frac{1}{2}f''(a)$. Then

$$P_2(t) = f(a) + f'(a)(t - a) + \frac{1}{2}f''(a)(t - a)^2,$$

which is a **quadratic**, or parabolic, **approximation**. To do even better, we can try to add a cubic term to make the third derivatives equal also. So, let

$$P_3(t) = f(a) + f'(a)(t - a) + \frac{1}{2}f''(a)(t - a)^2 + c_3(t - a)^3.$$

Requiring that $P_3'''(a) = f'''(a)$ forces $P_3'''(a) = 3 \cdot 2c_3 = f'''(a)$, which means that $c_3 = 1/(2 \cdot 3)f'''(a)$. Therefore,

$$P_3(t) = f(a) + f'(a)(t - a) + \frac{1}{2}f''(a)(t - a)^2 + \frac{1}{2 \cdot 3}f'''(a)(t - a)^3,$$

which is called the **cubic approximation**. We can continue this process indefinitely, assuming that f has the required derivatives, to obtain

$$P_n(t) = f(a) + f'(a)(t - a) + \frac{1}{2}f''(a)(t - a)^2 + \frac{1}{2 \cdot 3}f'''(a)(t - a)^3$$
$$+ \cdots + \frac{1}{n!}f^{[n]}(a)(t - a)^n,$$

where $f^{[n]}(a)$ denotes the nth derivative of f at $t = a$. These approximating polynomials, $P_1(t)$, $P_2(t)$, $P_3(t)$, ..., $P_n(t)$, are called the **Taylor polynomials**. These polynomials approximate $f(t)$ for values t near $t = a$; generally, the higher the degree of polynomial, the more accurate the approximation and the larger the interval over which the approximation is good. As an aside, we define $P_0(t) = f(a)$, which is a constant function.

Example 1.3

Let $f(t) = \ln t$. Find a cubic approximation to $\ln t$ at $t = 1$. We need the derivatives of f at $t = 1$:

$$f(t) = \ln t, \ f'(t) = \frac{1}{t}, \ f''(t) = -\frac{1}{t^2}, \ f'''(t) = \frac{2}{t^3}.$$

Here $a = 1$, and therefore

$$f(1) = \ln 1 = 0, \ f'(1) = 1, \ f''(1) = -1, \ f'''(1) = 2.$$

Therefore, the cubic approximation is

$$P_3(t) = f(1) + f'(1)(t - 1) + \frac{1}{2}f''(1)(t - 1)^2 + \frac{1}{2 \cdot 3}f'''(1)(t - 1)^3$$
$$= (t - 1) - \frac{1}{2}(t - 1)^2 + \frac{1}{3}(t - 1)^3.$$

We urge the reader to use a calculator and plot both $f(t) = \ln t$ and $P_3(t)$ on the same set of axes. For example, we can approximate $\ln 1.5$ by $P_3(1.5) = 0.4167$. The actual value is $\ln 1.5 = 0.4055$, so we make an error of magnitude 0.0112. □

One can show (see most calculus texts) that the exact error in the Taylor approximation $P_n(t)$ at t involves the next-higher derivative of f and is given by

$$\text{error} = \frac{1}{(n + 1)!}f^{[n+1]}(c)(t - a)^{n+1},$$

where the $(n+1)$st derivative of f is evaluated at some point c *between* a and t; we don't know the value of c, but we can use this formula to estimate an upper bound on the error in an interval $[a - \delta, a + \delta]$. We have

$$|\text{error}| \leq \frac{M}{(n+1)!}(2\delta)^{n+1},$$

where M is the maximum value of $|f^{[n+1]}(t)|$ over $[a - \delta, a + \delta]$.

Example 1.4

Returning to Example 1.3, the maximum error in $P_3(t)$ over the interval $[0.5, 1.5]$ (so $\delta = 0.5$) is bounded by

$$\frac{M}{(4)!},$$

where $M = \max |f^{[4]}(t)| = \max \left|-6t^{-4}\right| \leq 6(0.5)^{-4} = 16$. Therefore, the absolute error is bounded by $16/24 = 0.667$. □

Example 1.5

The logistic growth law for a population x is given by

$$g(x) = rx\left(1 - \frac{x}{K}\right),$$

where r is the growth rate and K is the carrying capacity. The populations $x = 0$ and $x = K$ have zero growth rate and are therefore called **equilibrium populations**. We find the linearization of the growth rate at the carrying capacity K. First,

$$g'(x) = r - 2\frac{r}{K}x; \quad g'(K) = -r.$$

Therefore, the linearization is

$$P_1(x) = g(K) + g'(K)(x - K) = -r(x - K).$$

Later, we show that we can use this simple linear approximation for $g(x)$ to examine the behavior of the population at values near the carrying capacity. □

If a function f has infinitely many continuous derivatives at $t = a$, we can form the Taylor polynomial $P_n(t)$ for arbitrarily large n. In the limit as $n \to \infty$, we have the infinite sum

$$\sum_{k=0}^{\infty} \frac{1}{k!} f^{[k]}(a)(t - a)^k.$$

Here, to get the first term, when $k = 0$ we use $0! = 1$ and $f^{[0]}(a) = f(a)$, both by convention. This series is called the **Taylor series** for $f(t)$ centered about $t = a$. One can show that it converges *either* for all real numbers t, for only $t = a$, or for t in a symmetric interval $|t - a| < r$ about a; the radius r is called the *radius of convergence*. When the series converges for a value t, it converges to $f(t)$, and we write

$$f(t) = \sum_{k=0}^{\infty} \frac{1}{k!} f^{[k]}(a)(t - a)^k.$$

Example 1.6

The Taylor series for the functions e^{rt} and $\sin at$ about $t = 0$ are, respectively,

$$e^{rt} = \sum_{k=0}^{\infty} \frac{1}{k!}(rt)^k, \tag{1.9}$$

$$\sin at = \sum_{k=0}^{\infty} \frac{(-1)^k}{(2k + 1)!}(at)^{2k+1}. \tag{1.10}$$

Both converge for all values of t. The function $f(t) = 1/(1 - t)$ has a Taylor series about $t = 0$ given by

$$\frac{1}{1 - t} = \sum_{k=0}^{\infty} t^k = 1 + t + t^2 + t^3 + \cdots, \tag{1.11}$$

which is called the **geometric series**. The geometric series converges in the interval $|t| < 1$. Taylor series are used extensively in mathematical biology and in other sciences to make approximations. □

We end this section with two interesting models that require only calculus techniques for their analysis.

1.2.2 Foraging Theory

There are different theories about animals' foraging strategies. For a broad introduction, the reader should consult Stephens & Krebs (1986), which has become a standard reference on the topic. Here we investigate which is the best of two strategies. The idea is that a forager is consuming prey items from a single patch that has two types of prey items, 1 and 2. Each type has a total energy associated with it (E_1 and E_2), a time that it takes for the forager to handle an item (h_1 and h_2), and a rate that the particular type of item is encountered (r_1 and r_2), the latter measured in items per unit of time. The question is: What is the rate at which energy is gained by a generalist who consumes each item as it is encountered vs. a specialist who consumes only the item with the higher-energy payoff? Without loss of generality, we assume that eating item 1 has a higher-energy payoff, or $E_1 > E_2$.

To compare these strategies, we first calculate the *rate* R_s at which energy is gained by the specialist from consuming *only* item 1 over a time period T, which is the total time available to the forager. Clearly, the total time T must be the sum of the time S spent searching and the time H it takes to handle the food items encountered. So $T = S + H$. But during the time searching S, the forager encounters $r_1 S$ food items and gains total energy $E_1 r_1 S$; the total time it takes to handle those items is $H = h_1 r_1 S$. Therefore,

$$T = S + h_1 r_1 S,$$

or

$$S = \frac{T}{1 + h_1 r_1}.$$

Thus, the rate that energy is gained by selecting only item 1 is

$$R_s = \frac{E_1 r_1 S}{T} = \frac{E_1 r_1}{1 + h_1 r_1}.$$

For the generalist who consumes both items, the total time T is

$$T = S + H_1 + H_2,$$

where H_1 and H_2 are the total times to handle all items of type's 1 and 2, respectively. Then

$$T = S + h_1 r_1 S + h_2 r_2 S,$$

which gives

$$S = \frac{T}{1 + h_1 r_1 + h_2 r_2}.$$

The total energy gained is $E_1 r_1 S + E_2 r_2 S$. It follows that the rate at which energy is gained by the generalist in selecting both items is

$$R_g = \frac{E_1 r_1 S + E_2 r_2 S}{T} = \frac{E_1 r_1 + E_2 r_2}{1 + h_1 r_1 + h_2 r_2}.$$

It is interesting to compare R_g and R_s. By direct algebra it is easy to show that $R_s > R_g$ implies that

$$r_1 > \frac{E_2}{h_2 E_1 - h_1 E_2},$$

which occurs when the encounter rate for the most profitable item is sufficiently high. The right side can be thought of as a cut off value when a forager should switch from being a generalist to a specialist.

Optimum Time in a Patch

Now let's look at a different foraging situation. Suppose that there are several patches, all the same, each with the same food items, and the energy gain $G(t)$ for a forager in a given patch is a function of the *residence time* t spent in that patch, but with diminishing returns. In other words, the energy gain curve has the form shown in Fig. 1.5 with $G(0) = 0$ and $G(t) > 0$, $G'(t) > 0$, and $G''(t) < 0$ for $t > 0$. Therefore, if the forager remains in the patch too long, it gets less and less energy gain per unit time. Therefore, when should the forager leave the patch and go to another patch? To complicate matters, let us impose the condition that it takes T units of time to move from one patch to another, so moving is costly because of the time lost in eating.

Again, the forager wants to maximize its rate of energy gain. Over one cycle (residing in a patch, then moving) the time is $t+T$ and the energy gain is $G(t)$. Hence, the rate of energy gain over that cycle is

$$R(t) = \frac{G(t)}{t+T}.$$

See Fig. 1.5 for a typical plot of R. We can maximize R using elementary calculus. Taking the derivative and setting it to zero gives

$$R'(t) = \frac{(t+T)G'(t) - G(t)}{(t+T)^2} = 0,$$

or the condition on the residence time t:

$$G'(t) = \frac{G(t)}{t+T}.$$

We can determine geometrically the value of t that solves this equation. Let us write this condition as

$$G'(t) = \frac{G(t)}{t+T} = \frac{G(t) - 0}{t - (-T)}.$$

The right side is the slope of the straight-line segment from the point $(-T, 0)$ to $(t, G(t))$. So the value of the optimum residence time t is given by the value of t where the straight-line segment is tangent to the graph of $G(t)$. See Fig. 1.5. This construction, adapted by Charnov (1976) from economics, is often called the **marginal value theorem**.

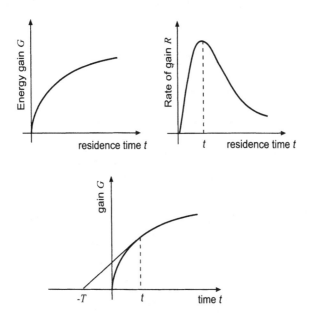

Figure 1.5 Plots of the energy gain and the rate of energy gain vs. residence time, and the graphical interpretation of the marginal value theorem.

EXERCISES

1. A population y is modeled by the equation $y = f(t) = \sqrt{t}$, where t is measured in days and y is measured in hundreds of animals. The derivative is $f'(t) = 1/(2\sqrt{t})$.

   (a) Graph $f(t)$ and $f'(t)$ for $t > 0$.

   (b) What are the units of $f'(t)$?

   (c) What is the slope of the tangent line to the graph of $f(t)$ when $t = 4$? Illustrate this on the plot of $f(t)$, and find the equation of this tangent line.

   (d) Use a calculator to compute the average change of f over the interval $[4, 4.2]$. Indicate this quantity on the plot of $f(t)$.

   (e) Approximate the average change in the population over the interval $[4, 4.2]$ using the derivative.

   (f) Find the growth rate and the per capita growth rate at time $t = 4$.

2. A descriptive model of population growth is a logistic growth curve, which

is given by the formula

$$y = f(t) = \frac{y_0 K}{y_0 + (K - y_0)e^{-rt}},$$

where r, K, and y_0 are positive parameters, and y is the population.

(a) Use MATLAB to plot the logistic curve when $r = 1$, $K = 10$, and $y_0 = 2$.

(b) Use a calculator and estimate the growth rate and the per capita growth rate when $t = 3$. (*Hint:* You will have to approximate the derivative.)

3. A population has a constant growth rate of r per day. If the population at time $t = 0$ is y_0, find a specific formula $y = f(t)$ for the population as a function of time. (*Hint:* What expressions have a constant derivative?)

4. A population $y = f(t)$ has a per capita growth rate of r per day. Show that a population law of the form $y = Ce^{rt}$, where C is any constant, satisfies the condition of constant per capita growth. If the population at time $t = 0$ is y_0, what is the formula for the population as a function of time?

5. The length L (cm) of an organism is changing according to $L(t) = 3(1 - e^{-0.2t})$. One finds that $L'(t) = 0.6e^{-0.2t}$. Plot L and L' vs. t on the same axes. How fast is the length changing at time $t = 10$ days?

6. Consider the exponential function $f(t) = e^{rt}$, where r is a fixed constant. Near $t = 0$, find the Taylor polynomial approximations $P_1(t)$, $P_2(t)$, and $P_3(t)$. What is $P_n(t)$?

7. The **Ricker growth law** with mortality, which models some fish populations, is

$$g(x) = bxe^{-cx} - mx, \quad b > 1,$$

where x is the population, b is the growth rate, m is the mortality rate, and c is a predation rate. Find the nonzero equilibrium population and determine the linearization of $g(x)$ about that equilibrium. Simplify completely.

8. The **Gompertz growth law** for some tumors depends on the tumor's radius R and is given by

$$G(R) = aR(\ln R_m - \ln R),$$

where a and R_m are positive constants. Find the equilibrium radius, and determine the linearization of $G(R)$ about that equilibrium.

9. The growth rate of a plant is dependent on the nitrogen concentration N in the soil, and it is given by **Tilman's law**:

$$f(N) = \frac{aN}{1 + bN},$$

where a and b are positive constants. What is the limiting (or, saturating) growth rate for very large nitrogen concentrations? Find a quadratic approximation for $f(N)$ valid near $N = 0$.

10. Verify the Taylor series formulas (1.9), (1.10), and (1.11).

11. Find Taylor series about $t = 0$ for the following functions: $\cos at$, $\ln(1 + t)$, and $1/(1 + t^2)$.

12. The survival of a fish egg through its critical period is a function of its mass x. The larger the egg, the more nutrients are present and the more likely it is to hatch successfully. This survivorship is often modeled by a function of the form

$$s(x) = 1 - cx^{-b}, \quad x \geq c,$$

where c and b are positive parameters. If G is the total gonadal mass of the female, the number of eggs laid by the female is G/x, and the number of eggs that survive through the critical period is

$$E(x) = \frac{G}{x}(1 - cx^{-b}).$$

Show that the egg size that optimizes the female's number of eggs is

$$x^* = (c + bc)^{1/b}.$$

Sketch generic plots of $s(x)$ and $E(x)$.

13. In the optimum residence-time problem (Section 1.2.2), take the energy gain function to be

$$G(t) = \frac{t}{t + 3}$$

and the travel time between patches to be $T = 3$. Use calculus to find the optimum residence time and illustrate your result by the Charnov construction.

1.3 Balance Laws

Many models in ecology come from a simple bookkeeping or accounting for a given quantity–where it comes from and where it goes. This bookkeeping, or balancing, gives a law which is often expressed as a differential equation.

In terms of setting up models, this section may be the most important in the book! It is based on a very simple idea that is used in all of science and engineering: that of a balance law. To explain what we mean, let Q be any quantity whatsoever in a fixed, well-defined domain. To list a few examples, Q could be:

- The number of animals in a fixed area

- The number of milligrams of a medicine in a person's blood

- The mass (kilograms) of a toxic chemical pollutant in a lake

- The amount of heat energy (calories) in a small animal's body

- The number of individuals in a community infected with a communicable disease

To determine how fast a quantity Q changes in a fixed domain all we have to do is keep track of where it comes from and where it goes. There are four possibilities. Q can enter the domain from outside; it can leave the domain and go outside; it can be created inside the domain; or it can be destroyed inside. Figure 1.6 shows how we represent these four notions pictorially in the case of animals in a fixed region. They can immigrate into the region or emigrate out of the region. Inside the region they can be born or they can die, which we represent by appropriate arrows. For the scenario of a toxin in a lake, the chemical can be pumped into the lake by a factory, and it can flow out of the lake through estuaries; while in the lake, the chemical can degrade or be consumed by reactions.

We can write the balance, or accounting law, in symbols. The rate of change of the total quantity Q in the domain is, by definition, the derivative dQ/dt. We give each of the four cases a name:

$$I = \text{ rate } Q \text{ flows into the domain from outside}$$
$$E = \text{ rate } Q \text{ flows out of the domain}$$
$$G = \text{ rate that } Q \text{ is created inside the domain}$$
$$D = \text{ rate that } Q \text{ is destroyed inside the domain}$$

Notice that I, E, G, and D are rates, measured in quantity per time. For the population scenario, I is the immigration rate, E is the emigration rate, G is

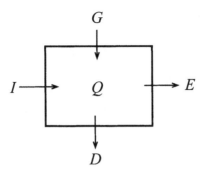

Figure 1.6 Net rate of change of Q is $I - E + G - D$. The rates are added or subtracted according to whether they increase or decrease Q.

the birth rate (gain), and D is the death rate (loss). For the lake, I is the rate at which toxins are pumped into the lake, E is the rate at which they flow out into the estuaries, and D is the degradation rate; in this example, $G = 0$. The **balance law** states that

$$\frac{dQ}{dt} = I - E + G - D. \tag{1.12}$$

In other words, all the rates have to balance; the right side of (1.12) accounts for how fast Q can change. This is a fundamental law of science.

Now, here is how this helps us. Once we determine expressions for each of the rates on the right side of (1.12), we will have a specific equation for Q. Because some of these rates may depend on Q itself (e.g., the emigration rate E of a population may depend on the population Q), the balance law (1.12) becomes a differential equation for the *unknown* function $Q = Q(t)$. Often, our goal is to solve the differential equation: that is, to find $Q(t)$ or to understand qualitatively how $Q(t)$ behaves.

Example 1.7

If all the rates on the right side of (1.12) are constant, $r = I - E + G - D$ is constant and the balance law is

$$\frac{dQ}{dt} = r.$$

This is a simple differential equation: It says that the derivative of Q is constant. So Q must be changing as a linear function of t, or $Q = rt + c$, where c is *any* constant. In different words, Q is the antiderivative of r. If we have an initial

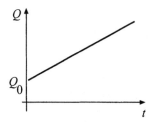

Figure 1.7 The linear function $Q = rt + Q_0$ with slope $r > 0$ and intercept Q_0.

condition on Q, say $Q(0) = Q_0$, then $c = Q_0$ and

$$Q = rt + Q_0.$$

A generic plot is shown in Fig. 1.7. $\square$

Example 1.8

The equation

$$\frac{dy}{dt} = ry,$$

where r is a fixed, constant parameter, is a differential equation for an unknown function $y = y(t)$; it relates the rate that y is changing to the quantity y itself. We want to find a $y = y(t)$ that works to make the equation true; such a y is called a *solution*. It is easily checked that a solution is

$$y(t) = Ce^{rt},$$

where C is any constant, called an *arbitrary constant*. If $y(0) = y_0$ is imposed, where y_0 is a given, fixed initial value, then $C = y_0$ and the solution is

$$y(t) = y_0 e^{rt}.$$

Typically, a differential equation has infinitely many solutions containing an arbitrary constant (here, C). But specifying an initial condition at $t = 0$ picks out one of these many solutions by fixing a value of C. Then we get a unique solution, as we expect in scientific problems. The differential equation in this example models exponential growth when $r > 0$ and exponential decay when $r < 0$. $\square$

Example 1.9

Here is a more difficult example. Let V be the volume of blood in the body and let $C = C(t)$ be the concentration (milligrams per deciliter) of a cancer-fighting chemical in the blood. Through chemotherapy, a patient is injected with the chemical at the rate of I milligrams per hour. At the same time, the body tissues absorb the chemical at a rate proportional to its concentration, or at rate kC, where k is the constant of proportionality, given in units of hours^{-1}. If the initial concentration is zero, or $C(0) = 0$, what is the concentration $C(t)$ in the blood at any time t? We set up a differential equation for the *mass* of the chemical. (A rule: *Masses are always balanced; concentrations never.*) Let $M = M(t)$ be the mass of the chemical in the blood. Mass and concentration are related by

$$\text{mass} = \text{concentration} \times \text{volume},$$

so that $M = CV$. Then, by the balance law,

$$\frac{dM}{dt} = \text{ rate mass flows in } - \text{ rate mass is absorbed.}$$

We always have to be careful about units; each term must be in mass per time. The rate at which mass flows in is given to be I; the rate at which mass is absorbed is kC, which is mass per volume per unit of time. So the rate that mass is absorbed is $kVC = kM$, which is mass per unit of time. Thus, the balance law is

$$\frac{dM}{dt} = I - kM,$$

which is a differential equation for $M = M(t)$. We can write the balance equation in terms of concentration (which is what we would measure) as

$$\frac{d(VC)}{dt} = I - kVC,$$

or, using the fact that $d(VC)/dt = V(dC/dt)$, we have

$$\frac{dC}{dt} = \frac{I}{V} - kC. \tag{1.13}$$

We know, initially, that $C(0) = 0$, which is the *initial condition*. We can now use a computer software program, for example, to find a formula for the solution $C(t)$ to this equation. But here is how we can do it ourselves. We choose a new dependent variable $y = y(t)$, defined by

$$y = \frac{I}{V} - kC.$$

Then $dy/dt = -k(dC/dt)$, and (1.13) can be written

$$-\frac{1}{k}\frac{dy}{dt} = y$$

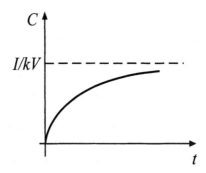

Figure 1.8 Concentration of the medicine in the blood.

or

$$\frac{dy}{dt} = -ky.$$

This is the exponential decay equation, and from the last example we know that its solution is

$$y = Ae^{-kt},$$

where A is an arbitrary constant. But this means going back to the variable C,

$$\frac{I}{V} - kC = Ae^{-kt}.$$

Solving for C then yields

$$C(t) = \frac{I}{kV} - \frac{A}{k}e^{-kt}.$$

We can find the value of the arbitrary constant A using the initial condition. We have $C(0) = I/(kV) - A/k = 0$, which gives

$$A = \frac{I}{V}.$$

Therefore, the concentration of the medicine in the blood is

$$C(t) = \frac{I}{kV} - \frac{I}{kV}e^{-kt}.$$

This formula gives the solution to the model (1.13) with initial condition $C(0) = 0$. We can draw a generic graph of C vs. t. Notice that $C(0) = 0$, and, from calculus, the limiting value of the concentration C as $t \to \infty$ is $I/(kV)$, which means that the graph is approaching the line $C = I/(kV)$. The plot is shown in Fig. 1.8.   □

Remark 1.10

The method we used in Example 1.9 is applicable to any differential equation of the form

$$\frac{dQ}{dt} = a + bQ,$$

where $Q = Q(t)$ is the unknown quantity and a and b are fixed parameters. This equation occurs very frequently in biological applications. The change of dependent variable

$$y = a + bQ$$

transforms the differential equation for Q into

$$\frac{dy}{dt} = by,$$

which is solved by $y = Ae^{bt}$, where A is an arbitrary constant. Then $a + bQ = Ae^{bt}$, which can be solved algebraically for Q. The value of A is determined by an initial condition $Q(0) = Q_0$, where Q_0 is a fixed number. $\square$

Example 1.11

A population has N individuals, and initially I_0 of them are infected with a communicable illness, while the remaining S_0 are susceptible to the illness. Let us set up a model that tracks the number $I = I(t)$ of infective individuals over time. Let $S(t) = N - I(t)$ be the total number susceptible to the illness. By the balance law, the rate of change of the number of infectives, dI/dt, is equal to the rate that susceptible individuals become infected. We will assume no deaths, no births, and that no individuals get over the illness. If there are I infectives and S suspectibles, we can argue that the rate that individuals become infected is proportional to the number of encounters between susceptibles and infectives. For example, if there were 200 susceptibles and 8 infectives, there would be 1600 possible encounters; a fraction of those, say a, will result in an infection. The constant a is the **transmission rate**. In general, the rate of infection is

$$\text{infection rate} = aSI = a(N - I)I.$$

Therefore the balance law is

$$\frac{dI}{dt} = a(N - I)I,$$

which is a differential equation for $I = I(t)$. This is called an *SI model*. The solution to this equation (which is derived in an exercise) is

$$I(t) = \frac{N}{1 - Ce^{-aNt}},$$

where C is an arbitrary constant. It may determined by the condition $I(0) = I_0$, to get $C = 1 - N/I_0$.   □

EXERCISES

1. Find and plot a function $y = y(t)$ that solves the problem $dy/dt = -2$, $y(0) = 5$.

2. Find and plot the function $y = y(t)$ that solves the problem $dy/dt = -2y$, $y(0) = 5$.

3. Find and plot the function $y = y(t)$ that solves the problem $dy/dt = 20-2y$, $y(0) = 5$.

4. This exercise leads to the solution $I = I(t)$ of the disease model in Example 1.11:

$$\frac{dI}{dt} = a(N - I)I.$$

   (a) Rewrite the equation in terms of the dependent variable w defined by $w = 1/I$, or $I = 1/w$. You should get

$$\frac{dw}{dt} = a - aNw.$$

   [*Hint:* The chain rule for derivatives requires that $dI/dt = -(1/w^2)(dw/dt)$. Why?]

   (b) Next observe that the equation in part (a) has the form of that in Remark 1.10. Make the appropriate transformation and solve, and then rewrite the solution in terms of the original variable I.

   (c) Sketch a generic graph of I vs. t for different initial conditions.

5. A nonreactive chemical toxin of concentration $C = C(t)$ grams per volume is dissolved uniformly in a pond of volume V gallons. Initially, the concentration is C_0. The toxin flows into the pond from a stream at a volumetric flow rate of q gallons per day at concentration γ. It is perfectly mixed and flows out in another stream at the same rate q.

   (a) Write down a differential equation and an initial condition whose solution would give the concentration in the pond at any time t.

   (b) Find a formula for the concentration and show a generic plot. What is the eventual concentration of the toxin in the pond?

   (c) Set up the differential equation for $C(t)$ if the volumetric flow rate is q_i and the flow rate out is q_o, with $q_i > q_o$. (Note that the volume changes in this case.)

1.4 Temperature in the Environment

How does an animal's body temperature depend on its environment? In this section we present a model for calculating an animal's equilibrium temperature based on the environmental temperature, solar radiation, and the heating characteristics of the animal.

All life forms have to deal with their environment. That is partly what ecology is about. The environment consists of **biotic** influences (other organisms, competition, availability of food resources, etc.) and **abiotic** influences (weather, temperature, etc.). Here we discuss temperature effects and how we can model those effects on certain organisms. Questions such as this are crucial in times of global climate change.

1.4.1 Heat Transfer

For survival, an animal typically has to maintain its body temperature within a certain range. Either it generates it own heat, as a mammal does, or it does not, as in the case of an insect or reptile. Animals that generate their own heat are called **homeothermic**, and those that do not are **poikilothermic**. To fix the idea, we consider a small reptile, such as a lizard with body temperature θ. (We always measure temperature in degrees Celsius.) In the interest of formulating a simple model of how an animal's temperature may be determined, we consider only two effects: heat transfer with its microhabitat and radiative heating from the sun. We let the microhabitat temperature be T, and we assume that the solar radiation has value q, measured in calories per hour. See Fig. 1.9. We are avoiding the effects of both convective cooling caused by the wind and the complex behavior of many poikilothermic animals, who thermoregulate by exposing themselves to direct or indirect sunlight, or orient their bodies to receive more, or less, sunlight.

First we ignore solar heating. If the ambient temperature T is larger than the body temperature θ (i.e., $\theta - T < 0$), the body temperature increases, and if T is smaller than θ (i.e., $\theta - T > 0$), the body temperature decreases. We are assuming that heat flows from hotter objects to colder objects, which is the second law of thermodynamics; so the reptile exchanges heat with its environment. It is a common assumption in science and engineering to assume that the rate heat flows between two adjacent regions of different temperatures (here, the body and the air) is proportional to the difference between the two temperatures, $\theta - T$. Therefore, the rate at which heat is exchanged is

$$k(\theta - T),$$

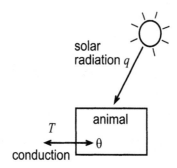

Figure 1.9 An animal at temperature θ receiving solar energy q and exchanging heat energy with its environment at a rate proportional to $\theta - T$, where T is the environmental temperature.

where k is the constant of proportionality. We call k the **heat transfer coefficient** and its units are calories/(time·deg). This constant is characteristic of the animal, and it measures how fast the animal conducts and transfers heat.

Therefore, the net rate f at which the animal receives heat, in calories per hour, is

$$f = q - k(\theta - T).$$

The quantity f is called the **heat flux**; the right side of this equation is the rate at which heat flows in minus the rate at which heat flows out. To fix the idea, reasonable values for k and q might be 1000 cal/(hours·deg) and 50 cal/h, respectively.

The animal's body temperature is in equilibrium when $f = 0$, or $q = k(\theta - T)$. Equilibrium occurs when the rate at which heat flows in equals the rate at which heat flows out. Consequently, the animal's equilibrium temperature is given by the formula

$$\theta_e = T + \frac{q}{k},$$

which occurs when the solar radiation balances the heat loss due to conduction. For example, if the environmental temperature were 20° C, then $\theta_e = 40°$ C, which may be near a lethal temperature for the animal. In this case the animal would have to seek shade or thermoregulate to lower its temperature.

We can plot the equilibrium temperature θ_e vs. q, which is a linear relationship, using MATLAB commands in the command window (see Fig. 1.10):

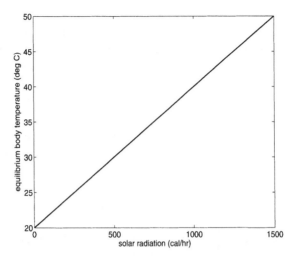

Figure 1.10 MATLAB plot of the equilibrium body temperature as as function of solar radiation when $k = 50$ and $T = 20$.

```
k=50; T=20;
q=0:1500;
theta=T+q/k;
plot(q,theta)
xlabel('solar radiation (cal/h)')
ylabel('equilibrium body temperature (deg C)')
```

1.4.2 Dynamic Temperatures

To understand how changing environmental temperatures affect the body temperature of an animal, we must develop a dynamic model, or a differential equation. In this section we develop such models and present two MATLAB procedures to calculate how an animal heats up.

In the preceding section we considered only steady states (equilibria). All the quantities were constant. We ignored the fact that the solar radiation q and the environmental temperature T may change during the day. If we include time-dependent parameters [e.g., $q = q(t)$ and $T = T(t)$], we have a *dynamic* problem, and the body temperature θ changes with time, or $\theta = \theta(t)$. If we knew formulas for $q(t)$ and $T(t)$, how can we find $\theta(t)$? In these cases we cannot use the equilibrium formulas from Section 1.4.1; they were derived under constant conditions, and now we are considering dynamical conditions.

Even if the parameters q and T are constant, we can still have a dynamical problem if the animal begins the day at a temperature different from its equilibrium temperature. For example, if its temperature is lower than the equilibrium value, we expect its temperature to increase in time up to that value. The reader should compare this heating problem to the problem of putting a turkey at room temperature in a hot oven and asking how fast it will heat up; that is, what is the temperature history?

To create a dynamical model, we first need to determine how time progresses. Does it progress continuously, or do we want it to tick off in discrete steps, say hourly? This question arises in every dynamic model. The answer often depends on when we take data, census a population, and so on. In the present problem, temperature is changing continuously in time.

To fix the idea, let us ask how we can obtain an expression for the rate of change of body temperature, $d\theta/dt$. From elementary science we know that the total amount of heat energy E, in calories, in a object of mass m is $E = mc\theta$, where c is the **specific heat** (cal/(g·deg)) of the object, or the amount of heat energy required to raise the temperature of a 1-g mass exactly 1 degree.[1] Therefore, the time rate of change of energy (cal/h) is

$$\frac{dE}{dt} = \frac{d}{dt}(mc\theta) = mc\frac{d\theta}{dt}. \tag{1.14}$$

We reason that this rate of change must equal the net energy flux into the animal, $q - k(\theta - T)$, because of the balance law. Hence,

$$mc\frac{d\theta}{dt} = q - k(\theta - T),$$

or

$$\frac{d\theta}{dt} = \frac{q}{mc} - \frac{k}{mc}(\theta - T). \tag{1.15}$$

This dynamical equation, which is called **Newton's law of cooling**, provides a relation between the unknown temperature $\theta = \theta(t)$ and its derivative $d\theta/dt$, which is also unknown. Equation (1.15) is another example of a **differential equation**. To review our earlier comments, a differential equation is an equation that relates an unknown function to some of its derivatives. If we know the initial temperature $\theta(0) = \theta_0$ of the animal, we fully expect that there should be a temperature function $\theta(t)$ that describes its temperature at any time $t > 0$. The problem of finding the specific function $\theta(t)$ that solves (1.15) with the initial condition $\theta(0) = \theta_0$ is called in mathematics an **initial value problem**. By *solve* we mean that (1.15) is satisfied identically for all

---

[1] Note that the mass m of an object is related to it volume V and its density ρ by the formula $\rho = m/V$. So $E = \rho V c\theta$.

times t when the formulas for $\theta = \theta(t)$ and its derivative $d\theta/dt$ are substituted into the equation.

How can we discover the formula for the temperature function $\theta = \theta(t)$? Well, there are techniques, taught in elementary courses in differential equations, for determining the unknown function that solves a simple differential equation. In Section 1.3 we learned a general method to solve this equation. Software packages (e.g., Maple, Mathematica, and the Symbolic Toolbox in MATLAB), and even calculators (e.g., the TI–89, or the TI Voyage 200), can find solution formulas for simple equations with an initial condition. In the next example we present a script MATLAB m-file that uses commands from the symbolic toolbox to solve and and plot the solution to (1.15) with $\theta(0) = \theta_0$. The reader should enter this code in MATLAB and run it. The solution to (1.15) with $\theta(0) = \theta_0$, which is found by the first line of the code below, is

$$\theta(t) = \frac{q}{k} + T - e^{-kt/(mc)}\left(q + kT - k\theta_0\right)k^{-1}.$$

Example 1.12

```
theta=dsolve('Dtheta=q/(m*c)-(k/(m*c))*(theta-T)','theta(0)=theta0');
theta=vectorize(theta);
k=50; q=1000; m=1; c=1;T=38; theta0=15;
t=0:.005:0.2;
theta=eval(theta);
plot(t,theta,'r')
ylim([0 60])
title('How an Animal Heats Up','FontSize',14)
xlabel('time (hrs)','FontSize',14)
ylabel('body temp (deg C)','FontSize',14)   □
```

Now comes a big caveat! For the most part, differential equations in ecology cannot be solved with a formula; another type of solution must be determined. One type is a numerical solution, which is only an approximate solution. We explain this next.

A **numerical solution** is an approximation to the actual solution $\theta = \theta(t)$. A numerical solution leads to approximate values of $\theta = \theta(t)$ at a set of discrete times as well as an approximate graph of the actual solution. The idea is to select a discrete set of equally spaced times $t_0 = 0$, $t_1 = h$, $t_2 = 2h$, $t_3 = 3h, ...$, where h is the **step size**. Note that $t_{n+1} = t_n + h$ for $n = 0, 1, 2, ...$ If we evaluate the differential equation at the discrete time t_n, we have

$$\theta'(t_n) = \frac{q}{mc} - \frac{k}{mc}\left(\theta(t_n) - T\right). \qquad (1.16)$$

Now, if h is reasonably small, the derivative on the left side can be approximated by its difference quotient, or

$$\frac{\theta(t_{n+1}) - \theta(t_n)}{h} \approx \frac{q}{mc} - \frac{k}{mc}\left(\theta(t_n) - T\right). \tag{1.17}$$

This equation is approximate, but it allows us to solve for $\theta(t_{n+1})$ in terms of $\theta(t_n)$. To this end, we have

$$\theta(t_{n+1}) \approx \theta(t_n) + h\left(\frac{q}{mc} - \frac{k}{mc}(\theta(t_n) - T)\right). \tag{1.18}$$

We can use this approximate equation as a recursive algorithm to obtain approximate values of $\theta(t)$ at the discrete times $t_1, t_2, t_3, \dots$. We simply take $\theta(t_0)$, which is given, and find $\theta(t_1)$ from the formula. Then we take this computed value of $\theta(t_1)$ and use the formula to compute $\theta(t_2)$, and so on, until we cover the range of times that we desire. This is easily accomplished using a recursive *for–end* loop on a computer. This numerical method is called the **Euler algorithm** for solving a differential equation and initial condition.

Example 1.13

Let us fix the parameter values $k = 50$, $T = 35$, $q = 800$, $m = 10$, and $c = 1$, and assume that the initial temperature of the animal is $\theta(0) = 15$. Next we create in MATLAB an m-file titled NewtonHeating.m, and we simulate time for 2 hours with 100 time steps, giving $h = 0.02$. Using a for–end loop we recursively calculate the temperatures theta, saving them in a list thetahistory, which we then plot. The program listing is given below and the temperature history is given in Fig. 1.11.

```
function NewtonHeating
clear all
theta=15; thetahistory=15;
k=50; T=35; c=1; q=800; m=10;
Time=2; N=100; h=Time/N;
for n=1:N
theta=theta+(h/(m*c))*(q-k*(theta-T));
thetahistory=[thetahistory theta];
end
t=0:h:Time;
plot(t,thetahistory)
xlabel('time'),
ylabel('body temperature')
axis([0 Time])   □
```

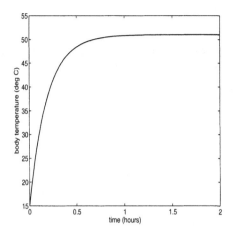

Figure 1.11 Temperature vs. time of an animal.

Example 1.14

Next we ask how to include variable environmental temperatures and variable sunlight. The same initial value problem holds, but now T and q are functions of t. Let us specify

$$T(t) = 26 - 11\cos(\pi \times t/12)$$

$$q(t) = \frac{1000}{64}(1 - \cos(\pi \times t/12))^6.$$

The preceding m-file is modified by including two function definitions (the last four lines) as follows:

```
function variabletemp
clear all
theta=15; thetahistory=15; k=50;
T=35; c=1; q=800; m=10; Time=24; N=10000; h=Time/N;
t=0;
for n=1:N
theta=theta+(h/(m*c))*(sunlight(t)-k*(theta-ambient(t)));
thetahistory=[thetahistory,theta];
t=t+h;
end
hrs=0:h:Time;
plot(hrs,thetahistory,hrs,26-11*cos(pi*hrs/12),hrs,35)
xlabel('time (hours)'), ylabel('temperature (deg C)')
```

```
axis([0 Time0 75])
function que=sunlight(t)
que=(1000/64)*(1-cos(pi*t/12))^6;
function tee=ambient(t)
tee=26-11*cos(pi*t/12);
```

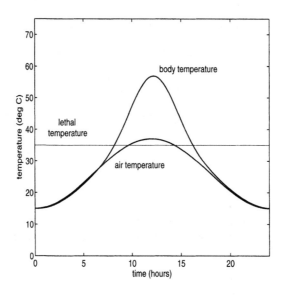

Figure 1.12 Hourly body temperature of an animal.

Figure 1.12 shows the body temperature, the air temperature, and an upper lethal temperature for the animal over the period of a day. We can infer that the animal is active only when the body temperature is under the lethal value; during the period of higher body temperatures the animal would seek refuge in a cooler environment. Activity times are those times when a predator may be seeking prey, or when prey may be foraging. Changing temperature levels, such as those associated with global climate change, can affect the interactions of predators and prey; we discuss this issue in a later section.   □

1.4.3 Development Rate

Temperature is a key factor in the rate at which many plants and poikilothermic animals develop and mature. Development is usually measured in degree-days, and the development rate is measured in degree-days per day. An organism accumulates 1 degree-day if its body temperature $\theta(t)$ is maintained at exactly

1 degree above some minimal, threshold temperature θ_h for one 24-hour period, 1 day. Of course, body temperature fluctuates throughout the day, so the total degree-days $D(t)$ accumulated from time $t = 0$ to time t is the integrated value

$$D(t) = \int_0^t [\theta(\tau) - \theta_h]^+ d\tau,$$

where $\theta(t)$ is the body temperature and

$$[\theta(t) - \theta_h]^+ = \left\{ \begin{array}{ll} 0, & \theta(t) \leq \theta_h, \\ \theta(t) - \theta_h, & \theta(t) > \theta_h, \end{array} \right.$$

denotes the positive part of the difference between $\theta(t)$ and θ_h. Geometrically, the number of degree-days accumulated is the area from $t = 0$ to t under the temperature curve $\theta(t)$ and above the constant threshold temperature θ_h. The development rate r is the derivative of $D(t)$ and is given by

$$r(\theta(t)) = \frac{dD}{dt} = [\theta(t) - \theta_h]^+. \tag{1.19}$$

Full development, or maturity, occurs at a time T when the organism accumulates a certain number of degree-days D_0. Thus,

$$D_0 = D(T) = \int_0^T [\theta(\tau) - \theta_h]^+ d\tau = \int_0^T r(\theta(t)) \, dt.$$

Therefore, $D(t)$ satisfies the initial value problem

$$\frac{dD}{dt} = r(\theta(t)), \quad t > 0; \quad D(0) = 0. \tag{1.20}$$

The reader should take care going through these definitions and formulas, which are just restatements of the fundamental theorems of calculus.

We observe that the body temperature of a poikilothermic animal is typically a function of its microhabitat temperature, which in turn is related to the ambient air temperature. However, because poikilothermic animals thermoregulate, their actual body temperature remains in a fairly narrow range.

Equation (1.19) states that the development rate is a linear function of temperature. This, however, is just a commonly used approximation that holds for limited temperature ranges. In fact, most plants and animals have a nonlinear development rate curve that has the shape shown in Fig. 1.13. There is a minimal temperature below which no development occurs (θ_h), and there is an optimum temperature value θ_m where the maximum rate occurs. Beyond θ_m temperatures are often lethal and there is a rapid drop-off in development rate. Equation (1.20) also holds for any rate function $r = r(\theta)$.

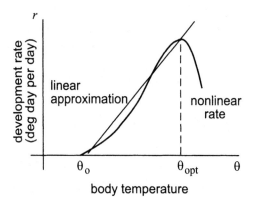

Figure 1.13 Nonlinear development rate (degree-days per day) as a function of body temperature θ. Also shown is a linear approximation that many researchers use.

Often, the development is normalized such that full development occurs when $D = 1$. We can carry out this normalization by defining a new development variable x by

$$x = \frac{D}{D_0}.$$

Then x is dimensionless and

$$\frac{dx}{dt} = R(\theta(t)), \quad x(0) = 0,$$

where $R(\theta) = (1/D_0)r(\theta)$ is the normalized development rate. Then the development time T is defined by $x(T) = 1$.

Example 1.15

A grasshopper beginning its third instar (stage) has a development rate $R(\theta) = 0.004[\theta - 15]^+$ during that instar. If its body temperature is a constant $35°$ C degrees throughout, how many days will it take the insect to reach the end of the instar? We have $R(\theta) = 0.004[35 - 15] = 0.08$. Then

$$1 = \int_0^T 0.08dt = 0.08T.$$

Therefore, $T = 12.5$ days. $\square$

Example 1.16

Suppose that the body temperature varies periodically over a day from $23°$ C to $33°$ C deg via

$$\theta(t) = 28 + 5\cos 2\pi t,$$

where t is given in days, and suppose that its development rate $R(\theta)$ is given as in Example 1.15. To determine the time of development T, we must solve the differential equation

$$\frac{dx}{dt} = 28 + 5\cos 2\pi t - 15, \quad x(0) = 0,$$

and stop the calculation when $x = 1$. Although this differential equation can be solved by direct integration, it is easier to write MATLAB code using the Euler method to solve the equation.

```
function degreeday
clear all
x=0; maxtime=30; numsteps=100000; h=0.01;
for n=1:numsteps
    if x<1
        x=x+h*0.004*(28+5*cos(2*pi*(n-1)*h)-15);
    else
        break
    end
end
x
days=n*h
```

The output is "days $= 19.2$."   □

EXERCISES

1. Suppose that the nonlethal body temperature range for an animal is $22 \leq \theta_e \leq 38$. If the animal's heat conduction coefficient is $k = 50$, find and sketch the region in *climate space*, the qT plane, where the animal's body temperature is in the nonlethal range.

2. Use the MATLAB program NewtonHeating.m to examine how animals of different sizes heat up. For example, use $m = 5$, 10, 25, and 50. Can you make a general conclusion? Make similar calculations and conclusions for different values of k.

3. Suppose that the upper lethal temperature of the animal studied in Example 1.13 is 35° C. How long can the animal safely survive before having to seek shade? Draw the plot in Fig. 1.12 with the lethal temperature $\theta = 35$ superimposed on the graph.

4. Sketch plots of the air temperature function $T = T(t)$ and the solar flux $q = q(t)$ in Example 1.14.

5. Referring to Exercise 3 in Section 1.1, state why changes in the number of species S on an island satisfies the differential equation

$$\frac{dS}{dt} = I\left(1 - \frac{S}{P}\right) - \frac{ES}{P}.$$

   (a) Find a formula for the solution $S = S(t)$ of the equation if $S(0) = S_0$.

   (b) Plot the solution to the problem if the parameters are given by $I = 8$ per year, $E = 3$ per year, $P = 48$ species, and $S_0 = 11$ species.

   (c) On your plot in part (b), graph the number of species vs. time when the island has an equilibrium number of species.

   (d) Referring to part (b), graph the solution to the differential equation if the immigration rate is a periodic function given by $I(t) = 8 - 5\cos 4t$.

6. In Exercise 2 in Section 1.1 we assumed that the growth rate of a spherical marine animal of radius r is the rate at which it consumes nutrients minus the rate at which it uses the nutrients. These two rates are proportional to the animal's surface area $(4\pi r^2)$ and to its volume $(\frac{4}{3}\pi r^3)$, respectively. Therefore, if we use mass m as a measure of growth, we can write the rate of increase in its mass as

$$\frac{dm}{dt} = 4a\pi r^2 - \frac{4b}{3}\pi r^3,$$

   where a and b are constants of proportionality. But mass can be written as density times volume, or $m = \rho\frac{4}{3}\pi r^3$.

   (a) Using the chain rule of calculus to calculate dm/dt in terms of dr/dt, show that the radius of the animal is governed by the differential equation

$$\frac{dr}{dt} = \frac{a}{\rho} - \frac{b}{3\rho}r. \tag{1.21}$$

   (b) Use the MATLAB symbolic toolbox commands as in Example 1.12 to find a formula for the solution $r(t)$ to equation (1.21) if $r(0) = 0$.

   (c) Take $a = 200$, $b = 100$, and $\rho = 10$, and plot r vs. t from part (b).

7. An insect with a temperature threshold of $12°$ C has a linear development rate function with a maximum development rate occurring at $40°$ C. If it develops fully in 8 days at a constant temperature of $40°$ C, what is its normalized development rate $R(\theta)$? How many days will it take to develop at a constant body temperature of $35°$ C? How many days will it take to develop if its body temperature is

$$\theta = 30 + 10\cos 2\pi t?$$

1.5 Dimensionless Variables

When we formulate models involving differential equations, the variables and the parameters have dimensions (such as time or distance), and they are given by specific units (such as minutes and meters). It is a often a good idea to re-formulate the model in a form where all the variables and parameters have no dimensions; that is, they are dimensionless. Usually, there is considerable economy associated with the dimensionless form of a problem because the number of parameters is reduced.

Consider the dynamic temperature model we derived in Section 1.4:

$$\frac{d\theta}{dt} = \frac{q}{mc} - \frac{k}{mc}(\theta - T), \tag{1.22}$$

$$\theta(0) = \theta_0. \tag{1.23}$$

The variables θ and t have dimensions (degrees and time), and all the parameters have dimensions as well—m is mass, q is energy per time, k is energy/(deg·time), and T and θ_0 are in degrees. Instead of measuring the temperature θ of the animal, we can opt to measure the temperature relative to the ambient air temperature T. This leads to the introduction of a new dependent variable y, defined by

$$y = \frac{\theta}{T}, \tag{1.24}$$

which is dimensionless (degrees $\div$ degrees). We say that y is a *dimensionless temperature*. Similarly, we can measure time t relative to the value mc/k, which has units of time. So we introduce a new *dimensionless* time τ by defining

$$\tau = \frac{t}{mc/k}. \tag{1.25}$$

Now we can rewrite the model (1.22)–(1.23) in terms of the new dimensionless quantities. First, note that the derivative becomes

$$\frac{d\theta}{dt} = \frac{d(Ty)}{d\left((mc/k)\tau\right)} = \frac{kT}{mc}\frac{dy}{d\tau}.$$

Then (1.22) and (1.23) become

$$\frac{kT}{mc}\frac{dy}{d\tau} = \frac{q}{mc} - \frac{k}{mc}(Ty - T),$$
$$Ty(0) = \theta_0.$$

This simplifies to

$$\frac{dy}{d\tau} = Q - (y - 1), \tag{1.26}$$
$$y(0) = B, \tag{1.27}$$

where Q and B are dimensionless constants given by

$$Q = \frac{q}{kT}, \quad B = \frac{\theta_0}{T}. \tag{1.28}$$

Let us review what we accomplished. By introducing dimensionless dependent and independent variables y and τ, respectively, the dimensioned problem (1.22)–(1.23) has been replaced by the dimensionless problem (1.26)–(1.27). The dimensioned problem has six parameters, whereas the dimensionless form of the problem has only two! Therefore, if we were going to perform a parameter study, it would be much simpler to work in a two-dimensional Q, B parameter space than in a six-dimensional parameter space. Our analysis actually shows that there are only two independent parameters in this problem.

The value T against which we measure temperature is called a *temperature scale* for the problem; we say that the temperature has been *scaled* by T. Similarly, mc/k is a *time scale*, and we say that we have *scaled* time by mc/k. The time scale, which is often not unique, is roughly the order in which time processes occur in the problem; time scales should be chosen so that the dimensionless time $\tau = t/T$ is neither large nor small. [A thorough discussion of dimensionless variables and how to select them may be found in Logan (2006); it is also shown that every consistent physical law can be reformulated in terms of dimensionless variables; the latter result is called the *Buckingham Pi Theorem.*]

In the last section we showed two ways in which an initial value problem for a differential equation could be solved using MATLAB: either with the symbolic toolbox, which gives an exact formula for the solution (when it works), or approximately, by discretizing time and writing a simple MATLAB m-file to perform the recursion required (the Euler algorithm). But there is another method that uses highly accurate, built-in MATLAB routines that solve differential equations automatically, with much greater accuracy than the Euler algorithm.

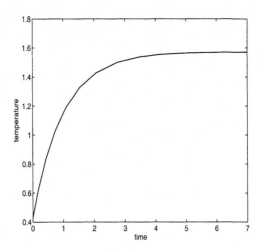

Figure 1.14 Plot of the dimensionless temperature vs. dimensionless time.

Example 1.17

Consider the dimensionless problem (1.26)–(1.27) with $B = 0.4286$ and $Q = 0.5714$. [The values of B and Q are computed from (1.28) with $q = 1000$, $k = 50$, $T = 35$, and $\theta_0 = 15$.] The third line in the code uses the package ode23, which calls the differential equation newton, defined in the last three lines.

```
function heateqn
global B Q
B=0.4286; Q=0.5714;
[time,temp]=ode23(@newton,0,2,B);
plot(time,temp)
function yprime=newton(t,y)
global B Q
yprime=Q-(y-1);
```

A plot of the dimensionless solution is shown in Fig. 1.14. □

EXERCISES

1. Use the MATLAB routine ode23 to find graphical solutions of the following initial value problems:

   (a) $dy/dt = -0.1y, \quad y(0) = 6.$

(b) $dY/dt = 1.5Y(1 - Y/20), \quad Y(0) = 2.$

(c) $dS/dt = 3(1 - S/30) - S/15, \quad S(0) = 6.$

(d) $dy/dt = -0.1y + 5e^{-t}, \quad y(0) = 6.$

2. In the biogeography model for the number of species on an island,

$$\frac{dS}{dt} = I\left(1 - \frac{S}{P}\right) - \frac{ES}{P},$$

find the dimensions of all the parameters and variables, and nondimensionalize the equation, scaling time by I^{-1} and species by P.

3. Nondimensionalize the problem (see Exercise 6 in Section 1.4)

$$\frac{dr}{dt} = \frac{a}{\rho} - \frac{b}{3\rho}r. \tag{1.29}$$

Use the MATLAB symbolic toolbox to find the solution to the dimensionless form of the differential equation with the initial condition $r(0) = 0$.

4. The logistic law for population growth is the differential equation

$$\frac{dP}{dt} = rP\left(1 - \frac{P}{K}\right),$$

where t is time in days, P is the number of animals, r is the growth rate in days^{-1}, and K is the carrying capacity in number of animals. Replace P and t by dimensionless variables and rewrite the logistics law in dimensionless form. (You will find that all of the parameters disappear!)

1.6 Descriptive Statistics

Researchers in all areas generate and collect data. Elementary techniques used to organize and understand the data sets are called *descriptive statistics*. In Chapter 6 we take a more advanced approach to data analysis.

Ecologists collect experimental data of all types. Statistics is the mathematical science of analyzing the data. Notwithstanding the classical quotes, "You can prove anything with statistics" (unknown) and "There are three kinds of lies: lies, damned lies, and statistics" (Disraeli), statistics offers well-defined mathematical methods to describe and understand the overall features of data and how the data might be used to predict future events and develop valid models. In this module we discuss how data can be described in simple terms that are familiar to most people.

There are three common characteristics of a data set that describe its nature: its *central tendency*, its *spread*, and its *shape*. The central tendency is a single value that is representative of the set. The usual measures of central tendency are the arithmetic mean, geometric mean, median, and mode. Each has advantages, depending on the intended purpose. (This is where statistics can lie!) If the data set is x_1, x_2, x_3,...,x_N, the **arithmetic mean** is

$$\overline{x} = \frac{x_1 + x_2 + x_3 + \cdots + x_N}{N}.$$

We usually call the arithmetic mean the *average*, or just the *mean*. The **geometric mean** is

$$x_{\text{geom}} = (x_1 x_2 x_3 \cdots x_N)^{1/N}.$$

One can show that $\overline{x} > x_{\text{geom}}$. The *median* is the value M for which half the values are lower and half the values are higher, and the *mode* is the value in the data set that occurs most often, or the most probable value; there may be more than one value of the mode.

Now consider the two data sets $2, 2, 2, 2, 2, 2, 2, 2$ and $0, 0, 0, 2, 2, 4, 4, 4$. Both have mean $\overline{x} = 2$, but they are clearly different. In the first there is no spread; in the second there is spread. To determine the spread of the data, or its dispersion, we use the variance, which measures how the data spread about the mean. The **sample variance** is defined by

$$s^2 = \frac{(x_1 - \overline{x})^2 + (x_2 - \overline{x})^2 + \cdots + (x_N - \overline{x})^2}{N - 1}.$$

The positive square root s of s^2 is the **sample standard deviation**. The numerator is the sum of the squares of the deviations from the mean. There are other ways to define deviations from the center (mean) rather than the square of the difference, but we do not consider them in this book. In our simple illustration, the variance of the first data set is zero, whereas in the second data set the variance is $24/7$. There are reasons that we divide by $N - 1$ rather than N.

Consider the two data sets $21, 28, 30, 33, 33, 35$ and $881, 888, 890, 893, 893, 895$. Most people would say that the first data set has more variability. The means are 30 and 890, but both have the same variance (5.0596) because the second data set is just the first data set with 860 added to each datum! Our intuitive answer is based on the size of the numbers involved. The **coefficient of variation** is defined by

$$\text{CV} = \frac{s_x}{\overline{x}}.$$

For these two data sets, $\text{CV} = 0.0750$ and $\text{CV} = 0.0025$. We are comparing the variation with the size of the numbers—the standard deviation relative to the mean. The coefficient of variation is a dimensionless quantity that serves our intuition in measuring variability.

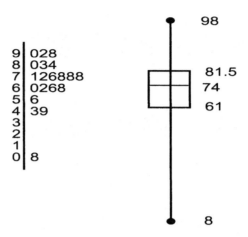

```
9 | 028
8 | 034
7 | 126888
6 | 0268
5 | 6
4 | 39
3 |
2 |
1 |
0 | 8
```

Figure 1.15 Stem plot of the data.

Example 1.18

During an 8-year period from 1950 through 1957 data were collected on the number of Canadian lynx furs sold in various provinces. Ecologists use this information to estimate population sizes. The data are given in the vector furs=[9592 6653 12636 10876 13876 9660 8397 8958]. Then MATLAB can compute the basic statistics as follows:

MATLAB command	MATLAB response
length(furs)	8
sum(furs)	80648
mean(furs)	10081
std(furs)	2324.6
max(furs)	13386
min(furs)	6653
median(furs)	9626

If year=[1950 1951 1952 1953 1954 1955 1956 1957], the commands bar(year,furs), plot(year,furs), and scatter(year,furs) plot a bar graph, a line graph, and a scatter diagram, respectively. The command sort(furs) sorts the data vector. The z-scores can be computed via z=(furs-mean(furs))/std. MATLAB responds z = [-0.2104 -1.4747 1.0991 0.3420 1.6325 -0.1811 -0.7244 -0.4831]. The coefficient of variation is CV=std(furs)/mean(furs)=0.2036, or about 20% variability. □

If the data set is large, we are interested in its shape, or distribution. Usually we draw a picture of the data, either a box plot or a stem-and-leaf plot.

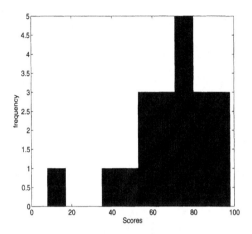

Figure 1.16 Histogram.

Example 1.19

Figure 1.15 (left) shows a stem plot of 20 scores on a recent examination. The scores are 98, 92, 90, and so on, down to 8. To obtain a box plot we find the minimum L and maximum H of the scores: $L = 8$ and $H = 98$. The median score is $M = 74$. The first quartile, Q_1, is the median of the numbers below the median M, and the third quartile, Q_3, is the median of the numbers above M. Here $Q_1 = 61$ and $Q_3 = 81.5$. A box plot displays the numbers L, Q_1, M, Q_1, H, as shown in Fig. 1.16 (right). A histogram is a frequency diagram that displays the number of data values in certain bins. The MATLAB command hist(scores,10) creates a histogram (Fig. 1.16) of the scores that has 10 bins, where scores is a vector of the scores, or scores=[98 92 90 84 83 $\cdots$ 49 43 8]. In place of the frequency on the vertical axis, we often plot the relative frequency, or percentage of the whole. $\square$

EXERCISE

1. In a medical study, 50 small rodents were infected with a virus, and the number of days they survived was recorded. The results are as follows:

40	47	53	57	57	60	68	68	75	76
80	80	87	89	95	95	97	98	99	101
108	110	120	121	123	123	125	128	129	130
133	135	136	139	140	144	150	154	165	171
190	202	221	251	280	305	330	362	380	402

(a) Find the mean, sum, product, standard deviation, median, and coefficient of variation.

(b) Sketch a stem plot with multiples of 10's in the left column.

(c) Sketch a box plot.

(d) Plot a histogram with eight bins.

1.7 Regression and Curve Fitting

Fitting a curve to a data set is a good way of extracting the main features of data and discovering important patterns and trends. This process, called *regression*, is carried out by the method of least squares, and it is fundamental in developing descriptive models of ecological phenomena.

In this section we show how to set up a descriptive model of a relationship between two variables for which data have been collected. The process of constructing the model is called **regression**, or curve fitting. We begin with an example that illustrates the process.

The table shows data collected that relate the photosynthetic rate P of a certain species of grass to the temperature T of the environment.

T	−1.5	0	2.5	5	7	10	12	15	17	20	22	25	27	30
P	33	46	55	80	87	93	95	91	89	77	72	54	46	34

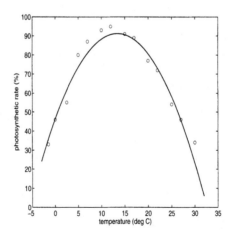

Figure 1.17 Data with a parabolic trend.

T is given in degrees Celsius and P is given as a percentage. Figure 1.17 [use scatter(T,P)] shows a scatter diagram of the data, and it appears that the shape, or trend, of the data is parabolic, as shown. This suggests a quadratic, descriptive model of the form

$$P = P(T) = a + bT + cT^2 \tag{1.30}$$

for some constants a, b, and c to be selected. In this context, we sometimes call the independent variable T the *explanatory variable* and the dependent variable P the *response variable*. We want to find the constants for which the parabolic curve best fits the data. So this brings up the issue of what we mean by *best fit*. Imagine for a moment that we have values of a, b, and c. There are many ways to measure the error we would make in approximating the data by (1.30), but the most successful way is to minimize the sum of squares of the errors at each data point. This is called the method of **least squares**. To be precise, at each temperature T_i the *error* in the approximation is $d_i = P(T_i) - P_i$. That is, the error at the ith data point is the computed value $P(T_i)$ (found using the model) minus the observed value P_i. See Fig. 1.18. Then we define the total error

$$S = \sum_{i=1}^{N}(P(T_i) - P_i)^2 = \sum_{i=1}^{N}(a + bT_i + cT_i^2 - P_i)^2, \tag{1.31}$$

which is the sum of the squares of the errors over the $N = 14$ data points. Notice that S depends on the choice of the constants a, b, and c, or $S = S(a, b, c)$. The method of least squares is, simply:

Find a, b, c such that $S = S(a, b, c)$ is minimum.

This is a calculus problem. We recall that to minimize a function we take the derivative and set it equal to zero. Here there are three derivatives of S, one with respect to each variable. So if the values a, b, and c provide a local minimum, then

$$S_a = 0, \quad S_b = 0, \quad S_c = 0,$$

where the subscripts on S mean to take the derivative with respect to that variable while holding the other variables fixed. Using the chain rule, and dropping the understood indices on the summation, we have

$$S_a = 2\sum(a + bT_i + cT_i^2 - P_i) = 0,$$

$$S_b = 2\sum(a + bT_i + cT_i^2 - P_i)T_i = 0,$$

$$S_c = 2\sum(a + bT_i + cT_i^2 - P_i)T_i^2 = 0.$$

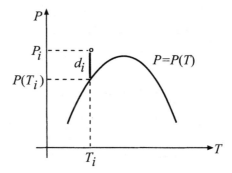

Figure 1.18 The error at T_i is $d_i = P(T_i) - P_i$. The quantity d_i^2 is the square of the error at that temperature. The total error is the sum of the d_i^2 over all the points.

These equations can be written in the form of three linear equations in three unknowns, a, b, and c. We write

$$Na + \left(\sum T\right) b + \left(\sum T^2\right) c = \left(\sum P\right),$$

$$\left(\sum T\right) a + \left(\sum T^2\right) b + \left(\sum T^3\right) c = \left(\sum PT\right), \qquad (1.32)$$

$$\left(\sum T^2\right) a + \left(\sum T^3\right) b + \left(\sum T^4\right) c = \left(\sum PT^2\right),$$

where we are using the simplified notation

$$\sum T_i = \sum T, \quad \sum T_i^2 = \sum T^2, \dots .$$

The system (1.32) can be solved by elementary methods to obtain a, b, and c, and therefore the quadratic (1.30) that best fits the data.

We can easily set up the system (1.32) in MATLAB by defining vectors T and P containing the 14 data points, and then compute the sums that give the coefficients. For example, $\sum T = \mathsf{sum(T)}, \sum T^2 = \mathsf{sum(T.\char`^2)}, \dots, \sum PT = \mathsf{sum(P.*T)}$, and so on. In matrix notation, (1.32) has the form

$$A\mathbf{x} = \mathbf{f},$$

where the matrix A and the vector $\mathbf{f}$ are given by

$$A = \begin{pmatrix} N & \sum T & \sum T^2 \\ \sum T & \sum T^2 & \sum T^3 \\ \sum T^2 & \sum T^3 & \sum T^4 \end{pmatrix}, \quad \mathbf{f} = \begin{pmatrix} \sum P \\ \sum PT \\ \sum PT^2 \end{pmatrix},$$

and **x** is the vector of unknowns

$$\mathbf{x} = \begin{pmatrix} a \\ b \\ c \end{pmatrix}.$$

Once A and $\mathbf{f}$ are entered into MATLAB as A and f, the vector **x** of coefficients can be found from the command A\f.

In practice, we do not always choose to solve the system (1.32). Rather, we use MATLAB to minimize (1.31) directly using a search method that is shown in the following m-file. We make a guess [1 1 1] and use the built-in command fminsearch.

```
function quadraticfit
global T P T=[-1.5 0 2.5 5 7 10 12 15 17 20 22 25 27 30];
P=[33 4655 80 87 93 95 91 89 77 72 54 46 34];
[x,fval]=fminsearch(quadSQ,[1 1 1])
function S=quadfitQ(x)
global T P
S=sum((x(1)+x(2)*T+x(3)*T. ^2 -P).^2);
```

MATLAB returns x =46.3705   6.7671   -0.2488, fval =292.1180. So the quadratic of best fit is

$$P(T) = 46.3705 + 6.7671T - 0.2488T^2.$$

Figure 1.17 is a plot of the data and the parabolic curve of best fit.

This process of least squares can be carried out to fit data with any function we choose—a linear function, a polynomial function, exponential functions, trigonometric functions, and so on. The idea is the same. Let $x_1, ..., x_N$ and $y_1, ..., y_N$ be two data sets. A scatter plot of all the points (x_i, y_i) usually suggests a function $y = f(x, a_1, a_2, ..., a_r)$, depending on r parameters $a_1, a_2, ..., a_r$, that may fit the data. The sum of the errors squared is a function S of the r parameters,

$$S(a_1, a_2, ..., a_r) = \sum_{i=1}^{N} (f(x_i, a_1, a_2, ..., a_r) - y_i)^2.$$

The least squares criterion is to determine the values $a_1, a_2, ..., a_r$ that minimize S.

If the r parameters $a_1, a_2, ..., a_r$ occur linearly in the model $y = f(x, a_1, a_2, ..., a_r)$, the system of r equations

$$S_{a_1} = 0, ..., S_{a_r} = 0$$

that we have to solve is a linear system. If the parameters occur nonlinearly, then the model is called a **nonlinear regression model** and the system of equations for the parameters is a nonlinear system. Typical nonlinear models are

$$y = a_1 + a_2 e^{a_3 x} \quad \text{(exponential)}$$

$$y = \frac{a_1 e^{a_2 x}}{1 + a_1 e^{a_2 x}} \quad \text{(logistics)}.$$

How do we measure the goodness of the fit when we obtain a regression formula? Let us answer this question for linear regression, where $y = f(x) = a_1 + a_2 x$. Then the least squares error is

$$S(a_1, a_2) = \sum_{i=1}^{N} (a_1 + a_2 x_i - y_i)^2.$$

We leave it to the reader (Exercise 1) to show that

$$a_1 = \frac{\sum X^2 \cdot \sum Y - \sum X \cdot \sum XY}{N \sum X^2 - (\sum X)^2}, \quad a_2 = \frac{N \sum XY - \sum X \cdot \sum Y}{N \sum X^2 - (\sum X)^2}, \quad (1.33)$$

with the obvious notation for the sums. Here a_1 and a_2 represent the y-intercept and the slope of the line of best fit, respectively. This straight line is called the **regression line**. The reader may be familiar with this idea from elementary algebra courses. At this point the calculation is strictly formal. What do the coefficients a_1 and a_2 have to do with the actual goodness of fit? With a little algebra we can show that

$$a_2 = r \frac{s_y}{s_x}, \quad a_1 = \overline{y} - a_2 \overline{x},$$

where $\overline{x}$ and $\overline{y}$ are the mean values of the x and y data sets and s_x and s_y are their sample standard deviations. The constant r is given by

$$r = \frac{1}{N-1} \sum \left(\frac{x_i - \overline{x}}{s_x} \right) \left(\frac{y_i - \overline{y}}{s_y} \right), \quad (1.34)$$

and it is a measure of how well the x and y data sets *correlate*; it is called the **correlation coefficient**. Note that r is an average of the product of z-scores for the two data sets. It is a dimensionless quantity with

$$-1 \le r \le 1.$$

(Can you show this?) We say that the data sets are *positively correlated* if the x_i values above the mean match up with the y_i values above the mean, and the values of x_i below the mean match up with the values of y_i below the mean. If they match up mostly in this way, the product of the z-scores will be positive and r will be near its maximum value of 1, and the line of best fit will have a positive slope. If the x_i values above the mean match up with the y_i values below the mean, and conversely, the products of the z-scores are mostly negative and we say that the data sets are *negatively correlated*. In this case, r is near its minimum value of -1 and the line of best fit has a negative slope. In the third case there may be little matching of the z-scores, and some terms in (1.34) will be positive and some will be negative. Thus, they will add and subtract and r will be near zero; in this case we that say there is no correlation. Figure 1.19 illustrates the three cases. If we find the regression line, we can ask how well it

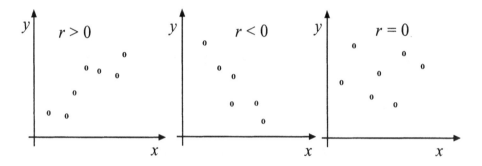

Figure 1.19 Three sets of data, showing positive, negative, and no correlation.

explains the data. The original variability of the response variable is measured with respect to the mean value of the y_i by $\sum (y_i - \bar{y})^2$. After the regression line is fit in, there is still unexplained variability measured by how much the regression line deviates from the data, or $\sum (y_i - Y_i)^2$, where $Y_i = a_1 + a_2 x_i$ is the estimated value on the regression line. The original variation minus the unexplained variability is the explained variability. See Fig. 1.20. Measured relative to the original amount of variability, we have

$$\text{explained variability} = R^2 = \frac{\sum (y_i - \bar{y})^2 - \sum (y_i - Y_i)^2}{\sum (y_i - \bar{y})^2}.$$

With algebra one can show that this value is equal to the square of the correlation coefficient, $R^2 = r^2$. Thus, $100 \times r^2$ is interpreted as the *percentage of the total variability that is explained by the linear fit*. So a value, for example, $r^2 = 0.91$, would mean that 91% of the data is explained by the linear model. Despite the high values of r^2, we have to be careful in remembering that *correlation does not imply causation*. Although there are other measures of fit, the value R^2 is common and it is in the output for most statistical packages. Moreover, the R^2 value computed in (1.7) does not require that the model be linear, so it may be extended to nonlinear regression.

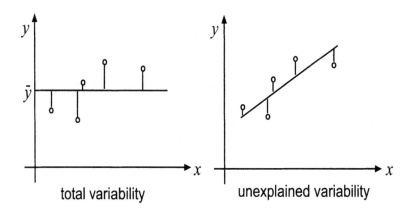

Figure 1.20 Total variability and unexplained variability.

Remark 1.20

There is another MATLAB routine that determines the best least squares *polynomial*

$$p(x) = b_0 x^n + b_1 x^{n-1} + \cdots + b_{n-1} x + b_n$$

that fits the data $x_1, ..., x_N$ and $y_1, ..., y_N$. If the data are typed in as vectors x and y, then polyfit(x,y,n) returns the coefficients $b_0, b_1, ..., b_n$. (Note the order of the coefficients.) This routine may be used to find the regression line when $n = 1$.

Example 1.21

The following m-file plots the data and the regression line. The result is shown in Fig. 1.21.

```
function polynomialfit
x=[1 2 3 4 5 6 7]; y=[2 6 12 19 30 26 40];
p=polyfit(x,y,1);
xx=0:.1:8;
plot(x,y,'.',xx,polyval(p,xx),'-','MarkerSize',25,'LineWidth',1.5)
```

Many calculators also perform simple regression. □

EXERCISES

1. (Linear regression) For linear regression, derive the formulas (1.33) for the constants a_1 and a_2. [Hint: Multivariable calculus is required for this problem; set the partial derivatives of $S(a_1, a_2)$ equal to zero and solve for a_1 and a_2.]

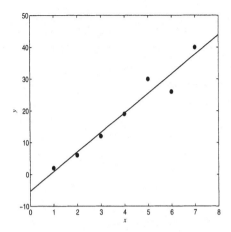

Figure 1.21 Regression line and data.

2. The average weights (in kilograms) of female black bears in age classes 1 through 15 years were found to be 35, 55, 68, 70, 71, 75, 79, 82, 81, 80, 78, 99, 99, 82. Fit the data with a straight line and determine the goodness of the fit.

3. In Exercise 2 fit the data with a logistics curve. Compute the R^2 value and compare to the linear fit.

4. The population of insects in a hostile environment over days $t = 0, 1, 3, 4$ was found to be $y = 200, 129, 58$, and 33, respectively. Find the curve of the form $y = ae^{-bt}$ that best fits the data in the least squares sense.

5. The pesticide DDT was one reason for the decline of the bald eagle population. In 1972 the pesticide was banned and eagle populations began to rise. From 1973 to 1979 the number of young eagles per unit area were counted to be 0.78, 0.86, 0.96, 0.82, 0.98, 1.12, and 0.82. Find the regression line and the R^2 value, and then use the regression line to predict the number of young eagles for 1982 through 1985.

1.8 Reference Notes

There is an extensive list of broadly based textbooks below, all covering aspects of mathematical ecology and biology. These are the texts on our shelves, and all of them have something to offer a reader; some are elementary, some are advanced, and some are more mathematically than biologically oriented. We recognize the omission of many other excellent books, but we wanted to list the books with which we are extremely familiar. We have found success using several of these in courses in mathematical biology (in both mathematics and biology departments) over the last seven years. We have not listed specialty books, which is an even longer list.

Three books are listed that focus on modeling issues. Mooney & Smith (1999) is an excellent elementary introduction. The classic text in applied mathematics is Lin & Segel (1974), and Logan (2006) has a beginning chapter devoted to modeling issues. An outstanding instruction manual for MATLAB, with examples, is Higham & Higham (2005).

Some of the texts fall into the elementary class in that they require limited mathematics skills at the beginning. Allman & Rhodes (2004), Vandermeer & Goldberg (2003), Gotelli (2008), Berryman (1999), Hastings (1997), and Otto & Day (2007) fit into this category, and are all excellent. Each requires grasping at the mathematics as the exposition proceeds. Neuhauser's beginning (2004) calculus text for biology students is an outstanding resource and it includes material on linear algebra, matrices, differential equations, and probability.

The remaining books on the list fit into the category of mathematical modeling in biology, and they introduce a substantial amount of mathematical ideas and symbolism. Some are penned by biologists, and some by mathematicians, but both lie in that intersection that brings the ideas of mathematical modeling and formalism to some of the central questions in biology.

A notable book is that of Roughgarden (1998), who incorporates MATLAB programs and techniques thoroughly into an exposition of ecological principles. Two excellent mathematical resources for optimal control in biology settings are Clark (2005) and Lenhart & Workman (2007). Wodarz (2007) is a very readable introduction to mathematical modeling of the virus–immune system dynamics and it shows the close similarities to ecosystem models.

References

Allen, L. J. S. 2007. *An Introduction to Mathematical Biology*, Prentice Hall, Upper Saddle River, NJ.

Allman, E. S. & Rhodes, J. A. 2004. *Mathematical Models in Biology*, Cambridge University Press, Cambridge, UK.

Berryman, A. A. 1999. *Principles of Population Dynamics and Their Applications*, Stanley Thornes, Cheltenham, UK.

Brauer, F. & Castillo-Chavez, C. 2001. *Mathematical Models in Population Biology and Epidemiology*, Springer-Verlag, New York.

Britton, N. F. 2003. *Essential Mathematical Biology*, Springer-Verlag, London.

Case, T. J. 2000. *An Illustrated Guide to Theoretical Ecology*, Oxford University Press, New York.

Charnov, E.L. 1976. Optimal foraging: The marginal value theorem. *Theoretical Population Biology* 9:129-136.

Clark, C. C. 2005. *Mathematical Bioeconomics*, 2nd ed., Wiley-Interscience, Hoboken, NJ.

de Vries, G., Hillen, T., Lewis, M., Müller, J., & Schnfisch, B. 2006. *A Course in Mathematical Biology: Quantitative Modeling with Mathematical and Computational Methods*, SIAM, Philadelphia.

Edelstein-Keshet, L. 2005. *Mathematical Models in Biology*, SIAM, Philadelphia (reprinted from the of 1988 Random House edition).

Ellner, S. P. & Guckenheimer, J. 2006. *Dynamic Models in Biology*, Princeton University Press, Princeton, NJ.

Gotelli, N. J. 2008. *A Primer of Ecology*, 4th ed., Sinauer Associates, Sunderland, MA.

Gurney, W. S. C. & Nisbet, R. M. 1998. *Ecological Dynamics*, Oxford University Press, Oxford, UK.

Hastings, A. 1997. *Population Biology: Concepts and Models*, Springer-Verlag, New York.

Higham, D. J. & Higham, N. J. 2005. *MATLAB Guide*, 2nd ed., SIAM, Philadelphia.

Kot, M. 2001. *Elements of Mathematical Ecology*, Cambridge University Press, Cambridge, UK.

Lenhart, S. & Workman, J. T. 2007. *Optimal Control Applied to Biological Models*, Chapman & Hall/CRC, London.

Lin, C. C. & Segel, L. A. 1974. *Mathematics Applied to Deterministic Problems in the Natural Sciences*, SIAM, Philadelphia (reprinted from the 1974 Macmillan edition).

Logan, J. D. 2006. *Applied Mathematics*, 3rd ed., Wiley-Interscience, Hoboken, NJ.

Mangel, M. 2006. *The Theoretical Biologist's Toolbox*, Cambridge University Press, Cambridge, UK.

Mazumdar, 1999. *An Introduction to Mathematical Physiology and Biology*, 2nd ed., Cambridge University Press, Cambridge, UK.

Mooney, D. & Smith, R. 1999. *A Course in Mathematical Modeling*, Mathematical Association of America, Washington, DC.

Murray, J. D. 2002. *Mathematical Biology I: An Introduction*, 3rd ed., Springer-Verlag, New York.

Neuhauser, C. 2004. *Calculus for Biology and Medicine*, 2nd ed., Prentice Hall, Upper Saddle River, NJ.

Otto, S. P. & Day, T. 2007. *A Biologist's Guide to Mathematical Modeling in Ecology and Evolution*, Princeton University Press, Princeton, NJ.

Pastor, J. 2008. *Mathematical Ecology of Populations and Ecosystems*, Wiley-Blackwell, Chichester, UK.

Roughgarden, J. 1998. *Primer of Ecological Theory*, Prentice Hall, Upper Saddle River, NJ.

Smith, J. M. 1974. *Models in Ecology*, Cambridge University Press, Cambridge, UK.

Stephens, D.W. & Krebs, J.R. 1986. *Foraging Theory*, Princeton University Press, Princeton, NJ.

Vandermeer, J. H. & Goldberg, D. E. 2003. *Population Ecology*, Princeton University Press, Princeton, NJ.

Wodarz, D. 2007. *Killer Cell Dynamics*, Springer-Verlag, New York.

2

Population Dynamics for Single Species

Perhaps the basic problem in ecology is to understand how populations of animals and plants are regulated. Are there laws in population ecology as there are in the other physical sciences, such as physics and chemistry? For example, in classical particle mechanics, Newton's second law of motion, $F = ma$ (force = mass × acceleration), governs the motion of particles and systems of particles. Are there analogs in population ecology?

Well, first of all, to say that Newton's second law of motion is the fundamental law of dynamics is a little deceptive. In the early 1900s, Albert Einstein showed that Newton's equation does not hold for particles moving at very high speeds. Newton's law was supplanted by relativistic mechanics, and later by quantum mechanics for particles on an atomic scale. So any law we write down is simply a model for some realities, but not for all realities. Newton's law works well for the motion of planets and other objects, but it not correct on small scales. The same can be said to be true for any population equation that we write. It is only a model of certain situations and cannot cover the entire scope of population regulation. Moreover, even if we believe Newton's law of motion, at some point we must specify the forces F on the particles and that almost always involves approximations or constitutive relations that define the forces. A mass balance equation in fluid flow that accounts for all the fluxes and sources or sinks is ideally correct, but when we model a specific system, we are forced into making assumptions about the form of the fluxes and the origins of the sources. The second law of thermodynamics correctly tracks all the energy in the system, but specific systems require approximate assumptions about the materials involved, and we often assume an equation of state

Mathematical Methods in Biology,
By J. David Logan and William R. Wolesensky
Copyright © 2009 John Wiley & Sons, Inc.

that relates pressure, density, and temperature (e.g., an ideal gas law). It is no different in ecology.

2.1 Laws of Population Dynamics

We can take a basic law of population dynamics to be a specification of the per capita growth rate of a population $x = x(t)$,

$$\frac{1}{x}x' = r. \tag{2.1}$$

But specific systems require defining what the growth rate r is and how it varies with exogeneous variables such as temperature or the presence of predators, or endogeneous factors witnin the population itself such as competition or cooperation with the members. Whatever reasoning we use to define those factors involves specifying constitutive relations, which make the law we obtain only a model for certain realities, and an approximate one at that. This is not unlike specifying the force on a particle. Our point of view is that (2.1) is a fundamental law of ecology, but it requires additional assumptions. These assumptions are often made on the basis of experiment, measurement, and data analysis.

How we measure time in population dynamics is an important issue as well. Equation (2.1) is a differential equation, where time runs continuously. In a discrete model we can formulate the fundamental principle by specifying the offspring/parent ratio R, or

$$\frac{X_{t+1}}{X_t} = R, \tag{2.2}$$

where X_t is the population at discrete times t, $t = 0, 1, 2, 3, ...$

If R is constant, the solution to (2.2) is

$$X_{t+1} = X_0 R^t,$$

which is geometric growth; the solution to the continuous-time model (2.1) is

$$x(t) = x(0)e^{rt}$$

which is exponential growth. To give the same solution, we must have

$$R = e^r,$$

or

$$r = \ln R.$$

Therefore, we can write the discrete law alternatively as

$$\frac{X_{t+1}}{X_t} = e^r$$

or

$$\ln \frac{X_{t+1}}{X_t} = r.$$

In this context, Berryman (1999), in his classification of population models, refers to r as the r-function, and to write a discrete model he imposes constitutive relations on the per capita growth rate r.

Which law is more natural, (2.2) or (2.1), depends on how we census the population, whether or not there are distinct generations, whether births and deaths occur continuously, and so on. But in either case we want to measure the per capita growth rate r to formulate the basic model.

Moreover, in a single-species model, whether continuous or discrete in time, the population is assumed to be homogeneous and there are no differences in size, age, gender, or other structure.

Example 2.1

(**Logistic model**) The classic logistic population law can be understood as follows. Suppose that the per capita growth rate of a population is the birth rate minus the death rate,

$$r_0 - c_i x,$$

where r_0 is the birth rate and c_i is the coefficient of intraspecific (internal, within the population) competition. As the population increases there is greater competition for the existing resources, which decreases the growth rate and limits growth. We can define c_i as

$$c_i = \frac{\text{demand for resources}}{\text{total resources}} = \frac{D}{H}.$$

The dimension of D is resources/time per animal, and H is given in resources. Then the per capita growth rate is

$$r_0 - \frac{D}{H}x = r_0\left(1 - \frac{x}{K}\right),$$

where $K = r_0 H/D$, given in animals, is the carrying capacity. Therefore, we have the continuous-time logistic population law

$$\frac{1}{x}x' = r_0\left(1 - \frac{x}{K}\right),$$

where the per capita growth rate is on the right side. The corresponding discrete model is therefore

$$\ln \frac{X_{t+1}}{X_t} = r_0 \left(1 - \frac{X_t}{K} \right),$$

or, in standard form,

$$X_{t+1} = X_t e^{r_0(1 - X_t/K)}$$
$$= RX_t e^{-r_0(X_t/K)}, \qquad R = e^{r_0}, \qquad (2.3)$$

which is the **Ricker law**. If we expand the right side of (2.3) using the approximation (Taylor polynomial) $e^z = 1 + z$, we have

$$X_{t+1} = X_t \left(1 + r_0 \left(1 - \frac{X_t}{K} \right) \right), \qquad (2.4)$$

which is the often quoted form of the discrete logistic law [e.g., see Neuhauser 2004, p 104]. The value of (2.3) over (2.4) is that the former always gives nonnegative population values, whereas the latter can become unrealistically negative. □

Example 2.2

In a given region of Africa, let W_t denote the population of wildebeest at year t, and let L be the number of lions. Then, if a is the number of wildebeest eaten per lion per year (e.g., $a = 10$), a single wildebeest's risk of being killed is

$$\frac{aL}{W_t}.$$

Because lions consume other prey, they have an alternative food source. If F denotes an alternative food supply for lions measured in an equivalent number of wildebeest, a wildebeest's risk is reduced to

$$\frac{aL}{F + W_t}.$$

Consequently, if λ is the per capita growth rate of wildebeest, the total per capita mortality rate is

$$r = \lambda - \frac{aL}{F + W_t}.$$

We conclude that the discrete population law is

$$\frac{W_{t+1}}{W_t} = e^{\lambda - aL/(F + W_t)} = Re^{-aL/(F + W_t)},$$

with $R = \exp(\lambda)$. A corresponding continuous law is

$$\frac{1}{w}w' = \lambda - \frac{aL}{F+w},$$

or

$$w' = \lambda w - \frac{aLw}{F+w}.$$

The latter equation is a familiar form of a standard growth–predation law with a type II, or hyperbolic, predation rate. A detailed discussion of predation comes later. □

EXERCISES

1. The discrete **Beverton–Holt recruitment curve** for population growth is

$$X_{t+1} = \frac{RX_t}{1 + [(R-1)/K]X_t}.$$

What is a reasonable corresponding continuous population law? [*Hint:* Use a Taylor expansion on $\ln(1+(R-1)x/K)$ and retain the lowest-order term.]

2. This exercise leads to another derivation of the Ricker law. Let A_n be the number of adult fish at generation n, where $n = 0, 1, 2, 3, \ldots$. Assume that those fish lay eggs, or fish larva, and they are cannibalized by the adults A_n within a continuous-time period $0 \le t \le T$ during the nth generation. A fraction of the survivors then turn into adults at time $n+1$. If $L = L(t; n)$ denotes the larval population at generation n, during $0 \le t \le T$, assume that

$$\frac{dL}{dt} = -mA_nL,$$

which states that the per capita mortality rate during the larval period is proportional to the adult population; m is a positive constant. Our assumptions imply that $L(0, n) = A_n$, while $sL(T, n) = A_{n+1}$. Show that

$$A_{n+1} = sA_ne^{-mTA_n},$$

which is the Ricker law. (A discrete model whose value at the next time period is determined by a continuous submodel is called a **metered model.**)

3. Make the same assumptions as in Exercise 2, but now assume that the per capita mortality rate of the larval population is due to intraspecific competition and is proportional to the larval population itself, or

$$\frac{1}{L}\frac{dL}{dt} = -mL,$$

with the same boundary conditions as before. Show that

$$A_{n+1} = \frac{sA_n}{1 + mTA_n},$$

thus deriving a form of the Beverton–Holt law.

2.2 Continuous-Time Models

We first focus attention on continuous-time models for single species, which lead to differential equations. Later in the chapter we take up discrete-time models.

As noted in Section 2.1, the simplest expression of a population law is to posit that the per capita growth rate of a population is constant. This supposition was made by Thomas Malthus in the late 1700s and is called *Malthus's law*. Analytically, if $x = x(t)$ is the number of individuals in a population at time t, then

$$\frac{dx}{dt} = rx, \tag{2.5}$$

where r is the per capita growth rate. Equation (2.5) is a differential equation for the population $x = x(t)$. If we know the population at time $t = 0$, say

$$x(0) = x_0, \tag{2.6}$$

we expect that there is a unique function $x = x(t)$ whose derivative is r times the function itself, and the initial condition (2.6) holds true. The function $x(t)$ that solves this problem is

$$x(t) = x_0 e^{rt}. \tag{2.7}$$

If $r > 0$, the population is growing exponentially, and if $r < 0$, the population is decaying, or dying out, exponentially. Figure 2.1 shows the form of the graph in both cases.

We can visualize the Malthus model graphically by either plotting the growth rate vs. x or the per capita growth rate vs. x, as shown in Fig. 2.2 for $r > 0$. These plots qualitatively tell us a lot, especially the growth rate plot. The growth rate is a linear function of the population x—the larger the population, the larger the growth rate.

When, for any population, the per capita growth rate is constant, we say that the population growth is **density independent**. This means that the per capita growth rate does not depend upon the number of individuals that are already present. The word *density* refers to population density, which is the number of individuals in a fixed region.

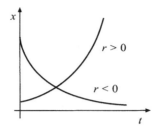

Figure 2.1 Plots of the function $x = x_0 e^{rt}$ for $r > 0$ and $r < 0$. The number x_0 is the value of x at time $t = 0$.

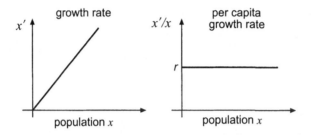

Figure 2.2 Plots of the growth rate (x' vs. x) and the per capita growth rate (x'/x vs. x) for the Malthus model.

The growth–decay model of Malthus applies to quantities other than populations.

Example 2.3

(**Decay of leaves**) Leaves that fall on the ground contain about 50% carbon, measured in dry biomass. When leaves decay, carbon dynamics dominate. Microbial respiration converts the organic carbon to CO_2, or under anaerobic conditions, the carbon is converted to methane CH_4. Methane gas contributes about 22 times the warming potential than that contributed by carbon dioxide when returned to the atmosphere, greatly contributing to greenhouse gases. The simplest decomposition model is

$$C(t) = C(0)e^{-kt},$$

where $C(0)$ is the number of carbon atoms present initially, and k is the decay rate (years^{-1}), which is different for different species. □

Yet another example is the amount of light that passes downward through a canopy in a forest. The intensity of the light is

$$I(d) = I(0)e^{-kd},$$

where d is the distance measured from the top of the canopy. The same type of law holds for light filtering downward into a body of water.

We can certainly argue that populations do not grow exponentially for all times. Eventually, there would be an explosion of individuals, and competition among them for resources would limit growth. Malthus stated that food supply grows only linearly, and thus an exponentially increasing population would experience dire consequences, such as famine, war, and disease. Therefore, we suspect that a better population model might involve a decrease in population growth as the population increases. So the per capita growth rate would not be constant, but would be **density dependent**.

In Section 2.1 we set up the simplest model of density-dependent growth, the **logistic equation**:

$$x' = rx\left(1 - \frac{x}{K}\right), \qquad (2.8)$$

The constants r and K have dimensions of time^{-1} and individuals, respectively, and they are called the **intrinsic growth rate** and the **carrying capacity**. If we plot the growth rate and per capital growth rate vs. x, we obtain the graphs in Fig. 2.3. We see that the carrying capacity K is the population value where both the growth rate and per capita growth rate are zero. Thus, as the population increases, the per capita growth rate decreases and even becomes negative for populations larger than K. On the plot of the growth rate vs. x, we draw arrows on the x axis indicating x values where the population is increasing (positive growth rate) or decreasing (negative growth rate). Note that for $x < K$ the population is increasing, and for $x > K$ the population is decreasing. The x axis with the arrows indicated is referred to as the **phase line**. A phase line is a one-dimensional phase space. The value $x = K$ is special. It is the value where the growth rate is zero, and it represents a constant solution to the differential equation (2.8). Such values are called **equilibrium populations**. The equilibrium populations plot as points on the x axis (on the growth rate plot), where the growth rate is zero. Clearly, $x = 0$ is a second equilibrium population.

Now imagine that the population is exactly at its carrying capacity, $x = K$, and suppose that a few individuals die, perhaps caused by some unexpected environmental event. This would move the population on the phase line just to the left of the value K. But then, as indicated by the arrows, the population would increase back toward K. Similarly, if the population were increased slightly from the carrying capacity, the phase line indicates that the population would decrease back toward K. We say that K is a **stable equilibrium**.

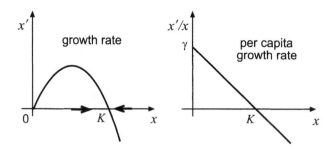

Figure 2.3 Plots of the growth rate (x' vs. x) and the per capita growth rate (x'/x vs. x) for the logistics model. The arrows on the x axis in the growth rate indicate values of increasing or decreasing populations.

In contrast, if the population were at the $x = 0$ equilibrium value, increasing the population slightly would cause it to grow, and it would not return to its equilibrium value. We say that $x = 0$ is an **unstable equilibrium**. We always try to classify equilibrium populations as stable or unstable. We expect that, over time, stable populations are the ones that nature will select.

There is another type of unstable equilibrium population, where one arrow on the phase line points away from the equilibrium and one arrow points toward the equilibrium. This type of unstable equilibrium does not occur with the logistics equation, but it can occur in other models.

We might ask if there is a formula $x = x(t)$ that solves the logistic equation (2.8) with the initial population $x(0) = x_0$ given. The answer is yes, and it is given by (see Exercise 7)

$$x(t) = \frac{K}{1 + (K/x_0 - 1)e^{-rt}}. \tag{2.9}$$

Figure 2.4 shows time series plots of this curve for different values of x_0. In all cases the population approaches the equilibrium $x = K$, the carrying capacity. The plots were obtained using the Euler method with the following m-file. (You can also obtain plots using a graphing calculator.)

```
function logistic
K=20; r=1;
for x0=5:5:25
x=K./(1+(K/x0-1).*exp(-r*t));
t=0:0.01:10;
plot(t,x0)
hold on
end
hold off
```

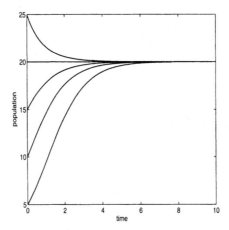

Figure 2.4 Plots of the logistic curve for several initial populations.

xlabel('time')
ylabel('population')

EXERCISES

1. Fix $K = 20$ and $x_0 = 3$, and plot the logistic curve (2.9) for different values of r to observe the effect of the intrinsic growth rate.

2. Write a MATLAB m-file that solves the initial value problem

$$x' = 1.5x(1 - x/30), \quad x(0) = 3,$$

using the Euler algorithm.

3. Solve the problem in Exercise 2 using the MATLAB routine ode23.

4. Reformulate the initial value problem (2.8) with $x(0) = x_0$, in dimensionless form by introducing a dimensionless population and time by the formulas

$$Y = \frac{x}{K}, \quad \tau = rt.$$

In the problem with dimensions there are three parameters. How many are there in the dimensionless formulation?

5. In Chapter 1 we derived the model for the temperature $\theta(t)$ of an animal,

$$\frac{d\theta}{dt} = \frac{q}{mc} - \frac{k}{mc}(\theta - T).$$

Sketch a generic plot of the temperature growth rate $d\theta/dt$ vs. the temperature θ and indicate on your plot the equilibrium temperature. Draw the phase line and determine if the equilibrium is stable or unstable.

6. We showed that the differential equation $dy/dt = ry$ has the solution $y = Ce^{rt}$, where C is any constant. Use this information to find a formula for the solution to the differential equation.

$$\frac{d\theta}{dt} = a\theta + b,$$

where a and b are fixed parameters. Use the following idea: Change the dependent variable from θ to $y = a\theta + b$, and note that $y' = a\theta'$. Then turn the given equation into a differential equation for y. Solve the y equation and then use the y, θ relation to obtain θ.

2.3 Qualitative Analysis of Population Models

The qualitative behavior of solutions of a population law given by a differential equation can be obtained geometrically by finding the equilibria and their stability, just as we did for the logistic equation in Section 2.2.

We consider a population model of the general form

$$\frac{dx}{dt} = f(x), \tag{2.10}$$

where the function $f(x)$ is the growth rate. Using the method of *separation of variables* [e.g., see any calculus text or elementary differential equations text, e.g., Logan (2006, pp 55ff)], it is sometimes possible to find an analytic solution or a formula for $x = x(t)$; this occurs only in cases when $f(x)$ has a very simple form.[1] Almost always we have to resort to a numerical method (e.g., the Euler method) to obtain an approximation. However, the best strategy is to use a geometric, or qualitative, method to determine the essential behavior of the solutions. Usually, the qualitative behavior is all we need to know.

The key to qualitative analysis is a determination of the equilibria of an equation and the dynamics near the equilibria. An **equilibrium solution** of (2.10) is a constant solution, which must clearly be a root of the algebraic

---

[1] The differential equation can be written $[1/f(x)](dx/dt) = 1$. Then, taking the antiderivative of both sides gives

$$\int \frac{1}{f(x)} \frac{dx}{dt} dt = t + C,$$

where C is an arbitrary constant. By a change of variables $x = x(t)$, $dx = x'(t)dt$, the expression becomes

$$\int \frac{1}{f(x)} dx = t + C,$$

which gives, after integration, the solution. See Example 2.6 for additional detail.

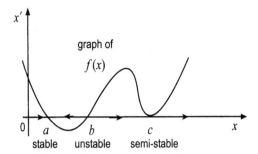

Figure 2.5 Plot of the growth rate $f(x)$ vs. x and the phase line. The plot shows three equilibria, a, b, and c.

equation $f(x) = 0$. Thus, $x(t) = x^*$ is an equilibrium solution if, and only if, $f(x^*) = 0$. The values x^* are the values where the graph of the growth rate $f(x)$ vs. x intersects the x axis. We always assume that the equilibria are *isolated*; that is, if x^* is an equilibrium, there is a small open interval containing x^* that contains no other equilibria. Figure 2.5 shows a generic plot of $f(x)$, where the equilibria are $x^* = a$, b, c. In between the equilibria we can observe the values of x for which x is increasing $[f(x) > 0]$ or decreasing $[f(x) < 0]$. We place arrows on the x axis in between the equilibria to indicate if x is increasing or decreasing as time increases; in this context we call the x axis the **phase line**. If arrows on both sides of an equilibrium point toward that equilibrium point, we say that the equilibrium point is **stable**. If both arrows point away, the equilibrium is **unstable**. If one arrow points toward the equilibrium and one points away, the equilibrium is unstable, but in this case it is often termed **semistable**. If the population is in a stable equilibrium and is given a small *perturbation* (i.e., a change or "bump") to a nearby state, it returns to that state as $t \to +\infty$. It seems plausible that ecological systems will seek out stable states. An unstable equilibrium has the property that a small perturbation can cause the system to go to a different equilibrium or even go off to infinity.

From the phase line plot we can easily sketch the time series plot $x = x(t)$ of the population vs. time. See Fig. 2.6. In between the equilibria, the population is either decreasing or increasing, depending on the direction of the arrow. Notice that solution curves only approach stable equilibria as $t \to \infty$, and they never actually reach it. If a solution curve actually intersected the constant stable equilibrium solution, two solutions would be passing through the same point; this violates a uniqueness condition that dictates that only one solution curve can pass through a fixed point in the tx plane. (This requires proof that is beyond our scope here.)

When we show that an equilibrium x^* is stable, the understanding is that

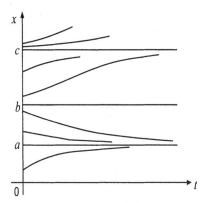

Figure 2.6 Sample time series plots for solutions to equation (2.10) with different initial conditions; the phase line plot is shown in Fig. 2.5.

it is stable with respect to *small* perturbations. To fix the idea, consider a population of fish in a lake that is in a state x^*. A small death event, say caused by a toxic chemical that is dumped into the lake, will cause the population to drop. Stability means that the system will return to the original state x^* over time; we say that the equilibrium is **locally stable**, meaning stable to small changes. If many fish are killed by the pollution event, all bets are off regarding the previous statements; the perturbation is *not* small and there is no guarantee that the fish population will return to the original state x^*; for example, the population may tend to some other equilibrium state or become extinct. If the population returns to the state x^* for all perturbations no matter how large, the state x^* is called **globally stable**.

Graphically, stability can be determined from the slope of the tangent line $f'(x^*)$ to the graph of $f(x)$ vs. x at the equilibrium state. See Fig. 2.5. If $f'(x^*) < 0$, the graph of $f(x)$ through the equilibrium falls from positive to negative values, giving an arrow pattern of a stable equilibrium. Similarly, if $f'(x^*) > 0$, we obtain the arrow pattern of an unstable equilibrium. If $f'(x^*) = 0$, x^* may be an stable, unstable, or semistable.

We can confirm these graphical criteria analytically. Let x^* represent an equilibrium population and $Y = Y(t)$ be a small perturbation from that state; then $x(t) = x^* + Y(t)$, and, approximately,

$$x' = Y' = f(x^* + Y) \approx f(x^*) + f'(x^*)Y.$$

Here, we have approximated $f(x)$ near x^* by a tangent line through $(x^*, f(x^*))$. Thus, we get

$$Y' = f'(x^*)Y,$$

which is called the **linearization** about the equilibrium x^* for the perturbation (the linearization is the first-order Taylor polynomial; see Section 1.2). This is the familiar linear growth–decay equation, such as the Malthus equation, and all of its solutions are given by $Y(t) = Ce^{\lambda t}$, where C is any constant and where $\lambda = f'(x^*)$ is the slope of the tangent line to $f(x)$ at the equilibrium state. If $\lambda < 0$, the perturbation decays, and if $\lambda > 0$, the perturbation grows, implying local stability and instability, respectively.

Example 2.4

(**Logistics model**) The logistics model of population growth is

$$\frac{dx}{dt} = rx\left(1 - \frac{x}{K}\right),$$

where r is the intrinsic growth rate and K is the carrying capacity. The equilibria are $x^* = 0, K$, found by setting the right side $f(x) = rx\left(1 - x/K\right) = 0$. The equilibrium $x^* = K$ is stable because $f'(K) = -r < 0$, and the zero population is unstable because $f'(0) = r > 0$.   ☐

Example 2.5

Suppose that the biomass $p = p(t)$ of a plant species grows according to the logistics law $p' = \gamma p(1 - p/K)$. However, a constant number of herbivores h in the system consume the biomass at a rate proportional to the product of the number of herbivores and the plant biomass p. That is, plant biomass is consumed at the rate ahp. Then we can model the growth rate of the plant biomass by the differential equation

$$\frac{dp}{dt} = \gamma p\left(1 - \frac{p}{K}\right) - ahp.$$

The equilibria are found by setting

$$f(p) = \gamma p\left(1 - \frac{p}{K}\right) - ahp = 0.$$

Factoring gives

$$p\left(\gamma - ah - \frac{\gamma p}{K}\right) = 0,$$

and solving for p gives the equilibria

$$p^* = 0, \quad p^* = \frac{K}{\gamma}(\gamma - ah).$$

The second equilibrium makes sense only when $\gamma > ah$, which gives a positive equilibrium. We could check stability using the analytic criterion: namely,

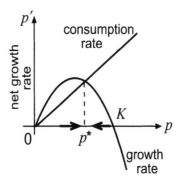

Figure 2.7 Plot of the logistic growth rate and the herbivore consumption rate showing the equilibria 0 and p^* and their stability.

finding the sign of $f'(p^*)$. But let's do it geometrically by plotting, vs. p, the logistic growth rate and the consumption rate ahp on the same set of axes (Fig. 2.7). We do this in lieu of plotting their difference. The intersection of the two curves is the point where the growth rate is zero (growth equals consumption), and this is the nonzero equilibrium p^*. To the left of the equilibrium the growth is larger than consumption, giving $p' > 0$, and to the right of the equilibrium the consumption is larger than the growth, giving $p' < 0$. Thus, we have the associated arrows on the p axis or phase line. Therefore, the equilibrium $p^* = (K/\gamma)(\gamma - ah)$ is stable. Clearly, the equilibrium $p^* = 0$ is unstable. $\square$

A problem of significant interest is to understand how equilibria and their stability change as a parameter in the problem varies. For example, in the present case, how do the plant biomass equilibria change as the level of herbivores h changes? A plot of p^* vs. h is called a *bifurcation diagram*. On such a diagram there may be several branches to the graph, and we usually label them as stable or unstable so that we can read off how the equilibria change as the parameter varies. Here, for $h < \gamma/a$, there are two equilibria, $p^* = 0$ and $p^* = (K/\gamma)(\gamma - ah)$, the first unstable and the second stable. The system will choose the stable equilibrium for $h < \gamma/a$. But as h increases, there is a "bifurcation" at $h = \gamma/a$, and the positive equilibrium disappears, indicating a change in the behavior of the system. Then for $h > \gamma/a$, there is just the zero equilibrium, which becomes stable. Figure 2.8 shows the bifurcation diagram.

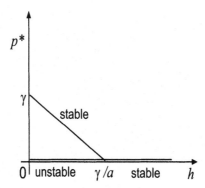

Figure 2.8 Bifurcation diagram showing the branches of the graph of p^* vs. h and their stability.

Example 2.6

(**Separation of variables**) We illustrate the simplest method, separation of variables, for solving a differential equation of the special form

$$\frac{dx}{dt} = g(t)f(x),$$

where g and f are given functions. So the right side of the equation factors into a product of a function of t and a function of x. By "separating the variables," we can write the equation as

$$\frac{dx}{f(x)} = g(t)\,dt.$$

Integrating both sides (taking antiderivatives) gives

$$\int \frac{dx}{f(x)} = \int g(t)\,dt + C,$$

where C is an arbitrary constant. Upon performing the two integrations (if possible) we obtain a formula in terms of x and t, called the *implicit solution*. If we can solve this equation for x to get $x = x(t)$, we obtain an explicit solution. For example, the differential equation

$$\frac{dx}{dt} = (t+1)x$$

can be separated into

$$\frac{dx}{x} = (t+1)\,dt.$$

Integrating yields

$$\int \frac{dx}{x} = \int (t+1)\, dt + C,$$

or

$$\ln |x| = \frac{1}{2}t^2 + t + C.$$

This is the implicit solution. Solving for x gives

$$|x| = e^{(1/2)t^2 + t + C} = e^C e^{(1/2)t^2 + t}.$$

Then

$$x = \pm e^C e^{(1/2)t^2 + t} = K e^{(1/2)t^2 + t},$$

where $K = \pm e^C$ is an arbitrary constant (because C is arbitrary). This is the explicit solution. The value of K can be determined by imposing an initial condition on x at $t = 0$: for example, $x(0) = 2$. Then $2 = K$ and we get $x(t) = 2\exp\left(\frac{1}{2}t^2 + t\right)$. In general, the solution involving an arbitrary constant C is called the *general solution*.

Recall from calculus that the most general antiderivative of a function $f(x)$ can be written as

$$\int_a^x f(s)\, ds + C,$$

an integral with a variable upper limit, where a and C are any constants. This formula can be used when we cannot find an explicit formula for the antiderivative. For example, the antiderivative of e^{-x^2} can be written

$$\int_a^x e^{-s^2}\, ds + C.$$

To solve

$$x'(t) = b(t)x$$

we separate variables to get

$$\frac{dx}{x} = b(t)\, dt,$$

so, integrating both sides gives us

$$\ln|x| = \int_a^t b(s)\, ds + C,$$

from which we can obtain the explicit solution. The value of a in the integral is usually taken to be the value where the initial condition is given. □

EXERCISES

1. Using the method of separation of variables, find the appropriate solution, as indicated, of the following differential equations.

(a) $x' = 2/x$. Find the general solution.

(b) $x' = 2t/x^2$. Find the implicit solution and explicit solutions.

(c) $x' = (x + 3)/(t + 2)$. Find the general solution.

(d) $x' = x(1 + x)$. Find the general solution.

(e) $x' = xe^{-t}/\sqrt{t + 1}$. Find the solution satisfying $x(0) = 1$. Write your answer in terms of an integral.

2. Find the solution for the differential equations in Exercise 1 that satisfy the initial condition $x(0) = 2$.

3. Red blood cells are formed from stem cells in the bone marrow. The red blood cell density x satisfies an equation of the form

$$\frac{dx}{dt} = \frac{bx}{1 + x^n} - cx,$$

where $n > 1$ and $b > c > 0$. Find all the equilibria and determine their stability.

4. A model of tumor growth is the *Gompertz law*,

$$\frac{dR}{dt} = -aR \ln \frac{R}{K},$$

where R is the radius of the tumor and $a, K > 0$. Qualitatively analyze the dynamics of this model and find the general solution $R = R(t)$.

5. A population $N = N(t)$ with $N(0) = N_0$ satisfies the mortality law

$$\frac{dN}{dt} = -m(t)N,$$

where $m(t) > 0$ is a time-dependent mortality rate. Let $S(t)$ be the survivorship function, which is the probability that an individual in the initial population will live to age t [thus, $S(t) = N(t)/N_0$]. The *Weibull model* of mortality is

$$m(t) = \frac{p + 1}{p_0} \left(\frac{t}{t_0} \right)^p,$$

where p_0, t_0, and p are parameters. Find $S(t)$ and plot its shape for the cases $p = 0$, $p = 3$, and $p = 10$. Which one might best model human survivorship? Fish survivorship?

6. Consider the model equation

$$\frac{dx}{dt} = x(a^2 - x^2),$$

where a is a real parameter. Find the equilibria and their stability and draw a bifurcation diagram where a is the varying parameter.

7. Let
$$\frac{dx}{dt} = (\lambda - b)x - ax^3,$$
where a and b are fixed positive parameters and λ is a parameter. If $\lambda < b$, show that there is one stable equilibrium. If $\lambda > b$, find all equilibria and their stability, and draw a bifurcation diagram in this case for varying λ.

8. Draw the bifurcation diagram for the equation $x' = x^3 - x + \lambda$, where λ is a parameter.

9. Draw the bifurcation diagram for the equation $x' = \lambda x + x^3$, where λ is a parameter.

10. The function $f(t) = (\sin t)/t$ has no simple formula for its antiderivative. Find an expression for the most general form of its antiderivative in terms of an integral. Find a function $f(t)$ satisfying $f'(t) = 3[(\sin t)/t]$ and the condition $f(1) = 2$.

11. Find a formula for the solution to the logistics equation
$$\frac{dx}{dt} = rx\left(1 - \frac{x}{K}\right)$$
by changing the dependent variable to $y = 1/x$. [*Hint*: Note that by the chain rule for derivatives, $y' = -(1/x^2)x'$.]

2.4 Dynamics of Predation

Animals or plants do not just grow in an environmental vacuum. There are all types of pressures on them, particularly from other species. There is competition for resources, cooperation, and predation. In this section we set up some of the standard constitutive relations that model consumption, or predation. These formula's, which give the rate of predation R (prey consumed per time per predator) as a function of prey density N (prey per area), are called the predator's **functional responses**. (See also Exercise 8 in Section 1.1.)

The simplest functional response is a linear relation. We think of a predator entering a patch and searching for a certain time with a known search speed. Some prey will come into its visual field, but only some will be detected, pursued and captured, and killed. The assumption is that R_e, the number of prey actually *encountered* by a single predator searching in a region with area A, is equal to the product of the prey density N, the time T_s spent searching the region, and the predator's search speed A_s, measured as area per unit of time. That is,
$$R_e = A_s T_s N.$$

Of those prey encountered, that is, the prey that come into the predator's field of view, only a fraction, d, are actually detected and finally consumed. Thus, the number of prey captured, R_c, is

$$R_c = dA_s T_s N. \tag{2.11}$$

Dividing both sides by time T_s, we therefore obtain

$$R = \frac{R_c}{T} = dA_s N, \tag{2.12}$$

which gives the predation rate. This formula is called a **type I functional response**. As simple and widely used as it is, it is clear that this linear relation cannot hold as the prey density increases indefinitely; the fact is that no animal eats faster and faster the larger the food supply. There is a limit to how much an animal can eat; eventually, it becomes satiated.

The next step is to realize that the time T in the patch involves both searching for prey and handling the prey that are subdued. Once an animal makes a kill, it takes time to consume it and to digest it. So we write

$$T = T_s + T_h,$$

where T is the total time spent in the patch and T_h is the time spent handling the prey captured. The handling time is given by

$$T_h = hR_c,$$

where h is the time required to handle one prey item. Substituting $T_s = T - hR_c$ into (2.11) gives

$$R_c = dA_s(T - hR_c)N.$$

Solving for R_c and dividing by T gives the **type II functional response**

$$R = \frac{aN}{1 + ahN}, \tag{2.13}$$

where $a = dA_s$. The right side plots as the familiar increasing, saturating curve which levels off at the limit $1/h$ as $N \to \infty$ (Fig. 2.9). It is a key fixture in both discrete and continuous nonstructured models (difference equations and differential equations) of predator–prey interactions, and there is probably not a more widely known result in population ecology.

This hyperbolic relationship has had a long history in the life sciences. Michaelis and Menten in 1913 proposed a similar relation between the rate of an enzyme reaction and the concentration of the rate-limiting reactant. Briggs and Haldane, in the mid-1920s, showed a derivation based on chemical kinetics. In the late 1950s, Monod adapted the hyperbolic expression as a model of bacterial

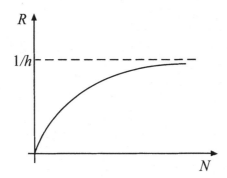

Figure 2.9 Plot of the predation rate (prey per time per predator) vs. the prey density for a Holling type II, or hyperbolic, functional response.

growth. C. Holling in 1959 was the first to apply it to ecology and predator–prey dynamics. In Holling's classical experiment, students (the "predators") were blindfolded and asked to use their fingertips to search for small sandpaper disks (the "prey") glued to a tabletop. Upon finding a disk, student's had to "handle" the prey by removing and discarding it before returning to their search. As a result, (2.13) is also called a Holling type II functional response, or the disk equation. In summary, the basic idea is that the total time breaks up into mutually exclusive times: time to hunt and time to handle. Handling prey itself can be broken into chasing, subduing, biting, chewing, digesting, and so on, but we do not include this detail.

An alternative way to think about search times and handling times is in terms of search times and alternative prey, with instantaneous consumption of the main prey. The view here is from the prey's perspective. If one predator kills n prey per time period, nP is the number of prey killed in the period, where P is the number of predators. Therefore, the risk of an individual prey is

$$\text{risk of a prey} = \frac{nP}{N}.$$

Let F be an alternative food source for the predator, measured in equivalent prey. The presence of this source reduces a single prey's risk to

$$\frac{nP}{F+N}.$$

Therefore, multiplying by N and dividing by P gives the predator's per capita killing rate, given as prey per predator per unit time; that is,

$$R = \frac{N}{P}\frac{nP}{F+N} = \frac{(n/F)N}{1+(1/F)N}.$$

Again, the hyperbolic response with time finding and consuming alternative prey takes the place of the handling time.

Finally, a different satiation response was derived by Ivlev in 1955 based on predator demand. The assumption is that the slope of the response curve $R = f(N)$ is proportional to the relative difference in the predator's demand for food and its eating response. In a sparse population the predator is always hungry and has a high demand rate for food. But when prey are abundant, the predator is always satiated, and increases in prey numbers will have little effect on the eating rates. Quantitatively, if D is the predator's demand rate (in the same units as R), then

$$\frac{dR}{dN} = a\frac{\text{demand} - \text{eating response}}{\text{demand}} = a\frac{D - R}{D},$$

where a is the "apparency," or availability, of the prey. Clearly, $f(0) = 0$, and we can solve the differential equation to obtain

$$R = f(N) = D(1 - e^{aN/D}).$$

This functional response has the same hyperbolic shape as the Holling response, saturating at the value $R = D$ as N gets large. However, the growth is now exponential rather than algebraic.

The type II response (2.13) has been generalized in many directions. For example, a linear, density-dependent encounter rate (i.e., $a = a_0N$,) leads to the type III response,

$$R = \frac{a_0N^2}{1 + a_0hN^2}.$$

This type III response holds for generalist predators that need to ideate an image of the prey before taking an interest. At low densities there is little interest and little consumption, but as the density increases the predator begins to notice the prey's abundance and there is a threshold value at which the predation rate increases substantially. Other modifications and extensions include multiple prey, clumped prey, predator interference, predator–prey ratio dependence, and other spatial effects. In a different direction, functional responses may also depend on environmental variables. For example, daily temperature variations may alter searching times of predators and the activity times of prey, offsetting the interaction opportunities, or changes in temperature may put physiological constraints on the predator. Figure 2.10 compares the the three functional responses, I, II, and II.

Example 2.7

(**Spruce budworm problem**) We now introduce a classical population model for the dynamics of spruce budworms, a pest which, at high densities, is responsible for the defoliation of balsam fir trees in Canada, ultimately killing the

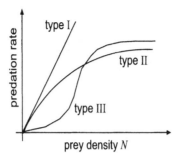

Figure 2.10 Plot of predation rate vs. prey density, comparing a type I, II, and III predator functional response's. The predation rate is measured as prey per unit of time per predator.

trees. The dynamics of the budworm is governed by growth and competition (for the foliage) and predation by birds. The growth–competition effect is taken to be the logistic law, and the predation is taken to be a type III functional response. The governing equation is

$$N' = RN\left(1 - \frac{N}{K}\right) - \frac{bN^2}{a^2 + N^2}. \tag{2.14}$$

The functional response models a generalist predator, such as birds, that switches prey as the prey density changes. As budworms become more abundant, birds begin to form a better image of the prey. The carrying capacity K depends on the foliage on the trees. We treat K as constant, but more detailed models allow for changing foliage.

There are four parameters in the model, and we can significantly reduce this number significantly by rescaling. Letting

$$x = \frac{N}{a}, \quad \tau = \frac{b}{a}t,$$

we obtain

$$\frac{dx}{d\tau} = rx\left(1 - \frac{x}{k}\right) - \frac{x^2}{1 + x^2},$$

where $r = Ra/b$ and $k = K/a$. So the model has been reduced to dimensionless form with two dimensionless parameters.

Clearly, $x = 0$ is an equilibrium, and it easy to see that it is always unstable [either show the derivative condition $f'(0) > 0$, or write out the linearization about $x = 0$]. Positive equilibria therefore occur when

$$r\left(1 - \frac{x}{k}\right) = \frac{x}{1 + x^2}, \tag{2.15}$$

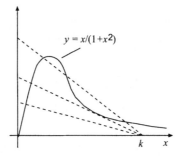

Figure 2.11 Plots showing the possibility of one, two, or three equilibria, depending upon the orientation of the straight line.

which is basically a cubic equation for x. The right side of this equation (per capita predation term) plots as a fixed (solid) curve as shown in Fig. 2.11, and the left side (per capita growth) is a straight line with intercept r and slope $-r/k$, both being parameters. Equilibria occur when the curves intersect. Clearly, from the figure, there is at least the positive equilibrium, and there may be two, or even three, depending on the orientation of the straight line (dashed in the figure).

In case there is a single equilibrium, a phase line diagram (putting direction arrows on the phase line; see Fig. 2.12) shows that it is stable (to the left of equilibrium growth is exceeding predation, so the growth rate is positive). Similarly, the phase line shows that when there are three positive equilibria, the largest and smallest are stable and the intermediate one is unstable. The largest equilibrium is the outbreak equilibrium. When there are two positive equilibria, the straight line must be tangent to the solid curve.

We can find an analytic condition on r and k for tangency, and thus the case of two positive equilibria, by setting the derivatives of both curves equal. We have, by differentiating (2.15),

$$-\frac{r}{k} = \frac{1 - x^2}{(1 + x^2)^2}. \tag{2.16}$$

Substituting r/k into (2.15) and solving for r gives

$$r = \frac{2x^3}{(1 + x^2)^2}. \tag{2.17}$$

Substituting this expression into (2.15) then gives, after some algebra,

$$k = \frac{2x^3}{x^2 - 1}. \tag{2.18}$$

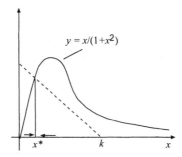

Figure 2.12 Phase line diagram in the case of the positive equilibrium x^*. (Another way, not shown, that this can occur is when the straight line lies above the solid curve.) When the straight line is above the solid curve, x is increasing, and when it is below the solid curve, x is decreasing.

Equations (2.17) and (2.18) are parametric equations for the locus of values (k, r) where there are two equilibria. If we plot these equations on a set of kr axes, we get the result shown in Fig. 2.13. This type of diagram, where the behavior of the system is shown as a function of parameters in the problem, is called a *bifurcation diagram*. When values of the parameters change and induce different behavior in the system, we say that a bifurcation occurs. These results give a complete picture of the dynamics of budworm outbreaks, and some of these are explored in the Exercises. □

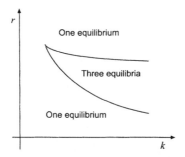

Figure 2.13 Two-dimensional parameter space (k, r) showing values of k and r where one, two, and three positive equilibria occur. The cusp-shaped curves, where two equilibria occur, are described by the parametric equations (2.17)–(2.18). On the upper curve, the parameter x ranges over $1 < x < \sqrt{3}$, and on the lower curve, $\sqrt{3} < x < \infty$. The single equilibrium is stable.

EXERCISES

1. In the spruce budworm example, find the coordinates of the cusp in Fig. 2.13. Verify the properties of the bifurcation diagram regarding the number of equilibria.

2. In the spruce budworm example, hold k fixed and assume that the population is initially in a small, single positive equilibrium $x = a$. Using the bifurcation diagram in Fig. 2.12, describe the dynamics of the budworm population as the parameter r varies slowly from a small value to a large value. Describe the reverse dynamics as r decreases from a large value to a small value, where the population initially is in a large outbreak equilibrium state $x = c$.

3. In the spruce budworm example, hold k fixed and sketch a rough graph of the equilibria vs. the parameter r. On your plot, indicate the stability of each branch of the plot and indicate the values of r where a bifurcation occurs.

4. Consider the plant–herbivore model

$$\frac{dp}{dt} = \gamma p \left(1 - \frac{p}{K}\right) - \frac{ap}{1 + bp}h, \quad a, b, h > 0.$$

Thus, the plant biomass grows logistically and is consumed by herbivores of biomass h via a type II functional response.

(a) Reformulate the problem in dimensionless variables

$$P = \frac{p}{K}, \quad \tau = \gamma t,$$

and show that the model becomes

$$\frac{dP}{d\tau} = P(1 - P) - \frac{\alpha P}{\beta + P},$$

determining the dimensionless parameters α and β. In particular, notice that α is proportional to the herbivore biomass.

(b) In $\alpha\beta$ parameter space (the positive $\alpha\beta$ plane), sketch the regions showing values of α and β where there are two, one, or zero equilibria.

(c) Find the equilibria and their stability, considering all cases. [*Hint*: Plot growth and consumption vs. P in the cases considered in part (b).]

(d) Holding β fixed, sketch a rough graph of the equilibria vs. α, indicating the stability of each branch of the graph. At what values of α do bifurcations occur?

(e) Comment on how the plant biomass crashes (goes to extinction) if there is initially a large plant biomass equilibrium and herbivores are slowly introduced into the system.

5. Consider a large number of identical spatial patches that can be occupied by an animal species, and let $p = p(t)$ be the fraction of those that are occupied at time t. As time progresses, let ϵp be the rate at which occupied patches become unoccupied, and let $cp(1 - p)$ be the rate at which empty patches are recolonized. Hence, the dynamics of patch occupation is

$$\frac{dp}{dt} = cp(1 - p) - \epsilon p, \tag{2.19}$$

which is called a **metapopulation model**. The metapopulation approach is a simple way to include spatial effects in ecological models. In real competitive populations and communities, balance between local extinction and colonization, as well as the ability to disperse, are all positive factors in persistence and survival.

(a) Analyze the model by finding equilibria and stability, and determine when the colonized patches increase or decrease.

(b) If there is an additional mainland population, or source pool, where animals migrate to the patch system, show that an appropriate model is

$$\frac{dp}{dt} = (m + cp)(1 - p) - \epsilon p.$$

Explain the model and investigate realistic steady states and discuss their stability.

(c) What is the effect on model (2.19) if a fraction D of patches is lost because of habitat destruction? Argue that

$$\frac{dp}{dt} = cp(1 - D - p) - \epsilon p$$

and analyze the model.

2.5 Discrete-Time Models

In a differential equation the time t runs continuously. However, many processes are better formulated as **discrete-time models**, where time ticks off in discrete units $t = 0, 1, 2, 3, ...$ (say in hours, days, months, or years). For example, if the money in a savings account is compounded monthly, we need only compute the principal each month. In a fisheries model, the number of fish may be estimated once a year. Or, a wildlife conservationist may census a deer population in the spring and fall to estimate their numbers and make decisions on allowable harvesting rates. Data are usually collected at discrete times.

In a discrete-time model the unknown is a sequence x_t (i.e., x_0, x_1, x_2, $x_3, \ldots$) rather than a continuous function of time. The subscripts denote the time (e.g., in a daily census of mosquitos grown in a laboratory, x_5 would denote the number of mosquitos on the fifth day). We graph the sequence x_t as a set of points (t, x_t) in a tx plane; we often connect the points by straight-line segments to make the population plot easy to visualize.

Example 2.8

(**Interest**) The simplest discrete model is a **growth–decay process**; it is linear and has the form

$$x_{t+1} = Rx_t. \tag{2.20}$$

For example, if x_t is the principal at month t in a savings account that earns 0.3% per month, the principal at the $(t+1)$st month is $x_{t+1} = (1 + 0.003)x_t$, and $R = 1.003$.   □

To solve (2.20) we can perform successive iterations to obtain

$$x_1 = Rx_0,$$
$$x_2 = Rx_1 = R^2 x_0,$$
$$x_3 = Rx_2 = R^3 x_0,$$

and so on. By induction, the solution to (2.20) is

$$x_t = R^t x_0.$$

If $R > 1$, x_t grows geometrically and we have a *growth model*. If $0 < R < 1$, R is a proper fraction and x_t goes to zero geometrically; this is a *decay model*. If $-1 < R < 0$, the factor R is negative and the solution x_t will oscillate between negative and positive values as it converges to zero. Finally, if $R < -1$, the solution oscillates without bound. It is also easily checked by direct substitution that the sequence

$$x_t = CR^t$$

is a solution to the equation for any value of C:

$$x_{t+1} = CR^{t+1} = RCR^t = Rx_t.$$

If x_0 is fixed, then $C = x_0$. Discrete growth–decay models are commonplace in finance, in ecology, and in other areas. For populations in general, the constant

R is given by $R = 1 + b - d + i - \epsilon$, where b, d, i, and ϵ are the birth, death, immigration, and emigration rates, respectively; R is the offspring/parent ratio.

Generally, a one-stage **discrete-time model**, the analog of a first-order differential equation, is an equation of the form

$$x_{t+1} = f(t, x_t), \quad t = 0, 1, 2, 3, ..., \tag{2.21}$$

where f is a given function. Such equations are also called *difference equations* or *recursion relations*. Knowledge of the initial state x_0 allows us to compute the subsequent states recursively, in terms of the states computed previously. Thus, (2.21) is a deterministic update rule that tells us how to compute the next value in terms of the preceding one. If $f(t, x_t) = a_t x_t + b_t$, where a_t and b_t are given fixed sequences, the model is linear; otherwise, (2.21) is nonlinear. In the sequel we mostly examine the autonomous equation

$$x_{t+1} = f(x_t), \quad t = 0, 1, 2, 3, ..., \tag{2.22}$$

where the right side does not depend explicitly on t. Discrete models can also be defined in terms of changes $\Delta x_t = x_{t+1} - x_t$ of the state x_t. Thus,

$$\Delta x_t = g(x_t)$$

defines a discrete model, where g is the given change. Finally, some models are also defined in terms of the relative change, or per capita change, $\Delta x_t / x_t$. Thus,

$$\frac{\Delta x_t}{x_t} = h(x_t),$$

where h is the given per capita change. Any form can be obtained easily from another by simple algebra.

Example 2.9

Example 2.8 showed that the difference equation $x_{t+1} = R x_t$ has general solution $x_t = C R^t$, where C is any constant. Let us modify the equation and consider the linear model

$$x_{t+1} = R x_t + p, \tag{2.23}$$

where p is a constant. We can consider this model to be, say, the monthly growth of principal x_t in a bank account where $R = 1 + r$ (r is the monthly interest rate) and p is a constant monthly addition to the account. In an ecological setting, R is the growth rate of a population and p is a constant recruitment rate. We can solve this equation by recursion to find a formula for x_t. If x_0 is the initial value, iteration yields

$$x_1 = Rx_0 + p,$$
$$x_2 = Rx_1 + p = R^2x_0 + Rp + p,$$
$$x_3 = Rx_2 + b = R^3x_0 + R^2p + Rp + p,$$

$$\vdots$$

$$x_t = R^tx_0 + R^{t-1}p + R^{t-2}p + \cdots + Rp + p.$$

By the formula for the sum of a geometric sum,

$$R^{t-1} + R^{t-2} + \cdots + R + 1 = \frac{1 - R^t}{1 - R},$$

and consequently,

$$x_t = R^tx_0 + p\frac{1 - R^t}{1 - R}, \qquad (2.24)$$

which is the solution to (2.23). $\square$

Example 2.10

(**Logistic model**) In the geometric growth model,

$$x_{t+1} = Rx_t,$$

the population x_t grows unboundedly if the growth rate R, or the offspring/parent ratio, is positive. Such a prediction cannot be accurate over a long period of time. We might expect that for early in the period, when the population is small, there are ample environmental resources to support a high birth rate. But later, as the population grows, there is a higher death rate as individuals compete for space and food (intraspecific competition). Thus, we should argue for a decreasing growth rate as the population increases. As we showed in Example 2.1, the discrete logistic model can be written in the form

$$x_{t+1} = x_t\left(1 + r_0\left(1 - \frac{x_t}{K}\right)\right), \qquad (2.25)$$

which is a nonlinear discrete model. This discrete logistic model is quadratic in the population, and it is one of the simplest nonlinear models that we can develop. Written in terms of the per capita growth rate, it becomes

$$\frac{\Delta x_t}{x_t} = r_0\left(1 - \frac{x_t}{K}\right),$$

which states that the per capita growth rate is decreasing and becomes zero at carrying capacity K. A plot of the per capita growth rate vs. the population

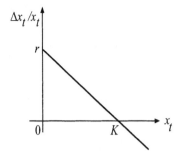

Figure 2.14 Per capita growth rate for the logistics model.

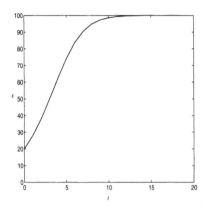

Figure 2.15 Logistics population growth. The graph is continuous because the software package connected the discrete values.

is shown in Fig. 2.14. Solutions can be plotted using MATLAB; see Fig. 2.15, where we have taken $r_0 = 0.5$, $K = 100$, and $x_0 = 20$. It shows a steady increase up to the carrying capacity, where it levels off.

The accompanying MATLAB m-file produces the sequence in Fig. 2.15 for $t = 1, ..., 20$. Many graphing calculators also have a sequence mode that can produce the same graphs.

```
function Logistics
r=0.5; K=100;
x=15; xhistory=x;
for t=1:50;
x=x.*(1+r-r*x/K);
xhistory=[xhistory, x];
end
```

plot(xhistory)   □

In the next section we see that this is not the entire story; different values of the parameters can lead to interesting, unusual, and complex behavior. In fact, we observe behavior in difference equations not seen in differential equations of first order. To some, this fact makes discrete models more complicated that continuous ones.

EXERCISES

1. Solve the following linear difference equations and plot their solutions.

   (a) $x_{t+1} = \frac{1}{2}x_t$, $x_0 = 100$.

   (b) $x_{t+1} = -1.3x_t$, $x_0 = 25$.

   (c) $x_{t+1} = -0.2x_t + 0.7$, $x_0 = 50$.

   (d) $x_{t+1} = 4x_t + 1$, $x_0 = 1$.

2. Write each of the models in Exercise 1 in terms of the per capita rate $\Delta x_t / x_t$.

3. Suppose that you borrow \$12,000 to purchase a new car at a monthly interest rate of 0.8%, and you want to pay off your loan in three years. What is your monthly payment?

4. Use iteration to show that the solution to the difference equation $x_{t+1} = \lambda x_t + p_t$ can be written

$$x_t = x_0 \lambda^t + \sum_{k=0}^{t-1} p_k \lambda^{t-k-1}.$$

   Here p_t is a fixed sequence of values.

5. In this and the next two exercises we investigate second-order discrete models that are the analogs of the differential equation $u'' = pu' - qu$. Consider a second-order linear discrete model of the form

$$x_{t+2} = -px_{t+1} - qx_t$$

   Assume that the solution sequence takes the form $x_t = r^t$ for some r to be determined. Show that r satisfies the quadratic equation (called the **characteristic equation**)

$$r^2 + pr + q = 0.$$

   When there are two distinct real roots r_1 and r_2, we have two distinct solutions, r_1^t and r_2^t. Show that for any two constants A and B, the discrete function $x_t = A r_1^t + B r_2^t$ is a solution to the difference equation. This expression is called the *general solution*, and all solutions have this form. Use this method to find the general solution of

$$x_{t+2} = x_{t+1} + 6x_t.$$

Find a specific solution that satisfies the conditions $x_0 = 1$ and $x_1 = 5$.

6. Referring to Exercise 5, show that if the characteristic equation has two equal real roots ($r = r_1 = r_2$), then $x_t = r^t$ and $x_t = tr^t$ are both solutions to the equation. The general solution is $x_t = A r^t + B t r^t = r^t (A + Bt)$. Find the general solution of the discrete model

$$x_{t+2} = 4x_{t+1} - 4x_t.$$

Find a specific solution that satisfies the conditions $x_0 = 1$ and $x_1 = 0$.

7. Investigate the case when the roots of the characteristic equation in Exercise 5 are complex conjugates; that is, $r = a \pm bi$. Find two real solutions. [*Hint*: Use the exponential Euler representation of the complex number $a + bi$:

$$a + bi = \rho e^{i\theta}.]$$

What is the general solution to the discrete model

$$x_{t+2} = x_{t+1} - x_t?$$

Find and plot the solution if the initial states are $x_0 = 1$ and $x_1 = \frac{1}{2}$.

2.6 Equilibria, Stability, and Chaos

Like continuous-time models, discrete models have equilibria and they may be classified according to their stability. However, in contrast to continuous models, discrete models may exhibit unusual behavior, such as chaos, which is not observed in one-dimensional autonomous differential equations.

A simple model will illustrate the important take-home messages.

Example 2.11

The **Ricker model**, which we derived in Section 2.1, is a nonlinear, discrete time ecological model for the yearly population x_t of a fish stock where adult fish cannibalize the young. The dynamics is

$$x_{t+1} = bx_t e^{-cx_t},$$

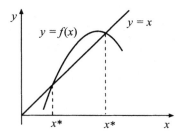

Figure 2.16 Graphical method for finding equilibria at the intersection points of the graphs of $y = x$ and $y = f(x)$.

where $b > 1$ and c is positive. In a heuristic manner, one can think of the model in the following way. In year t there are x_t adult fish, and they would normally give rise to bx_t adult fish the next year, where b is the fecundity, or the number of fish produced per adult. However, if adults eat the younger fish, only a fraction of those will survive to the next year to be adults. We assume that the probability of a fish surviving cannibalism is e^{-cx_t}, which decreases as the number of adults increases. Thus, $bx_t e^{-cx_t}$ is the number of adults the next year. One cannot solve this model to obtain a formula for the fish stock x_t, so we must be satisfied to plot its solution using iteration, or use qualitative analysis to discover its properties. □

An important question for discrete-time models, as it is for continuous-time models, is whether the state of the system approaches an equilibrium as the time elapsed gets large (i.e., as t approaches infinity). Equilibrium solutions are constant solutions, or constant sequences. We say that $x_t = x^*$ is an **equilibrium solution** of $x_{t+1} = f(x_t)$ if

$$x^* = f(x^*). \tag{2.26}$$

Equivalently, x^* is a value that makes the change Δx_t zero. Graphically, we can find equilibria x^* as the intersection points of the graph of $y = f(x)$ and $y = x$ in an xy plane. See Fig. 2.16.

Example 2.12

Setting $\Delta x_t = 0$ in the logistics model gives $r_0 x^* (1 - x^*/K) = 0$, or $x^* = 0$ and $x^* = K$. These two equilibria represent extinction and the population carrying capacity, respectively. Graphically, we can plot $y = x$ and $y = f(x) = x(1 + r_0 - (r/K)x)$; the equilibria are at the intersections points of the two curves. See Fig. 2.17. □

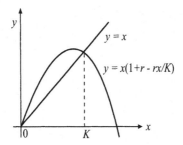

Figure 2.17 Graphical method for finding equilibrium 0 and K for the logistics model.

Example 2.13

An equilibrium state for the Ricker model must satisfy

$$x^* = bx^* e^{-cx^*}$$

or

$$x^*(1 - be^{-cx^*}) = 0.$$

Therefore, one equilibrium state is $x^* = 0$, which corresponds to extinction. Setting the other factor equal to zero gives

$$be^{-cx^*} = 1$$

or

$$x^* = \frac{\ln b}{c}.$$

If $b > 1$ (which we assume), we obtain a positive, viable equilibrium population.
□

If there is an equilibrium solution x^* to a discrete model

$$x_{t+1} = f(x_t),$$

we always ask about its permanence, or stability. For example, suppose that the system is in equilibrium and we perturb it by a small amount (natural perturbations are present in all physical and biological systems). Does the system return to that state, or does it do something else (e.g., go to another equilibrium state, or blow up)? We say that an equilibrium state is **locally asymptotically stable** if small perturbations decay and the system returns to the equilibrium state. If small perturbations of the equilibrium do not cause the system to deviate too far from the equilibrium, we say that the equilibrium is stable. If a small perturbation grows, we say that the equilibrium state is **unstable**. In the next paragraph we use the linearization familiar argument to determine the stability of an equilibrium population.

Let x^* be an equilibrium state. Then (2.26) holds. If y_0 represents a small deviation from x^* at $t = 0$, this perturbation will be propagated in time, having value y_t at time t. Does y_t decay, or does it grow? The dynamics of the state $x^* + y_t$ (the equilibrium state plus the deviation) must still satisfy the dynamical equation. Substituting $x_t = x^* + y_t$ into the dynamical equation gives

$$x^* + y_{t+1} = f(x^* + y_t).$$

We can simplify this equation using the assumption that the deviations y_t are small. We can expand the right side in a Taylor series (Section 1.2) centered about the value x^* to obtain

$$x^* + y_{t+1} = f(x^*) + f'(x^*)y_t + \frac{1}{2!}f''(x^*)y_t^2 + \frac{1}{3!}f''(x^*)y_t^3 + \cdots.$$

Because the deviations are small, we can discard the higher powers of y_t and retain only the linear term. Moreover, $x^* = f(x^*)$, because x^* is an equilibrium. Therefore, small perturbations are governed by the linearized perturbation equation, or **linearization**.

$$y_{t+1} = f'(x^*)y_t.$$

This difference equation is the growth–decay model [note that $f'(x^*)$ is a constant]. For conciseness, let $\lambda = f'(x^*)$. Then the difference equation has the solution

$$y_t = y_0\lambda^t.$$

If $|\lambda| < 1$, the perturbations y_t decay to zero and x^* is locally asymptotically stable; if $|\lambda| > 1$, the perturbations y_t grow and x^* is unstable. If $\lambda = 1$, the linearization gives no information about stability and additional calculations are required. The stability indicator λ is called the **eigenvalue**. Therefore, if the absolute value of the slope of the tangent line to $f(x)$ at the intersection with $y = x$ is less than 1, the equilibrium is asymptotically stable; if the absolute value of the slope is greater than 1, the equilibrium is unstable. In other words, we get stability if f is not too steep at the intersection point.

Example 2.14

Consider the Ricker model

$$x_{t+1} = bx_t e^{-cx_t};$$

where $c > 0$ and $b > 1$. Let us check the equilibrium

$$x^* = \frac{\ln b}{c}$$

for stability. Here $f(x) = bxe^{-cx}$ and we must calculate the eigenvalue, or the derivative of f evaluated at equilibrium. To this end, $f'(x) = (-cx + 1)be^{-cx}$; evaluating at x^* gives

$$\lambda = f'(x^*) = (-cx^* + 1)be^{-cx^*} = 1 - \ln b.$$

We get asymptotic stability when

$$-1 < 1 - \ln b < 1$$

or

$$1 < b < e^2 \approx 7.39.$$

If $b < e^2$, the perturbations decay and the equilibrium is asymptotically stable. Let us follow up on this by performing some simulations with different values of b and observe the effect. Fix $c = 0.001$. Figure 2.18 shows the results for $b = 6.5, 9, 13, 18$. The equilibrium is at $x^* = 1000 \ln b$. For $b = 6.5$, within the stability range, the solution does indeed represent an asymptotically stable, decaying oscillation. For $b = 9$, however, a periodic 2-cycle appears, alternating between high and low population values. When b is increased further, there is some value of b where suddenly the 2-cycle bifurcates into a 4-cycle. The plot shows the periodic 4-cycle when $b = 13$. This period doubling continues as b increases further, giving 8-cycles, 16-cycles, and so on. It can be shown that these cycles are stable limit cycles. But there is a critical value of b where the patterns disappear, and there are seemingly random fluctuations in the population ($b = 18$ shows this case). When b exceeds this critical value, these nonperiodic fluctuations become highly sensitive to initial conditions, a behavior known as **chaos**. The frames in Fig. 2.19 show the solution for conditions $x_0 = 99$ and $x_0 = 101$ with $b = 18$. We observe much different solutions. In the chaotic regime, the solution is deterministic (rather than stochastic) but is highly unstable with respect to initial data. □

Chaotic behavior is common in discrete models, even in one dimension. Such behavior does not appear in differential equations until dimension 3, other than in forced systems.

Finally, we illustrate a graphical procedure that is useful in determining the stability of an equilibrium solution of $x_{t+1} = f(x_t)$. The discussion refers to Fig. 2.20. We first sketch the curves $y = x$ and $y = f(x)$ on a set of axes, as in Fig. 2.20. We then mark the beginning value x_0 on the x axis. To find x_1 we move vertically to the graph of $f(x)$, because $x_1 = f(x_0)$, and mark x_1 on

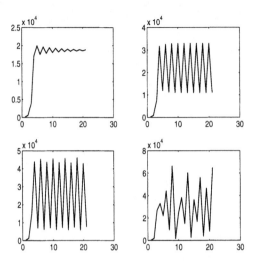

Figure 2.18 Population dynamics for the Ricker model for $b = 6.5, 9, 13, 18$, showing a decaying oscillation, 2-cycle, 4-cycle, and chaos, as b increases.

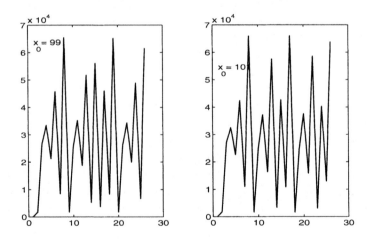

Figure 2.19 Plots of discrete solutions of the Ricker equation in the chaos regime when the initial populations are close.

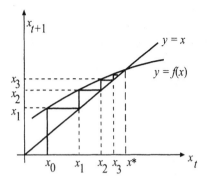

Figure 2.20 Cobweb diagram for a stable equilibrium.

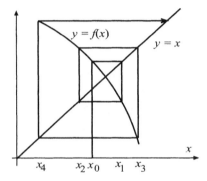

Figure 2.21 Cobweb diagram for an unstable equilibrium.

the y axis. To find x_2 we reflect x_1 back to the x axis through the line $y = x$. Now we have x_1 on the x axis. We repeat the process by moving vertically to the curve $f(x)$, which gives x_2. We reflect it back to the x axis, and so on. This sequence of moves is accomplished by starting at x_0, going vertically to $f(x)$, horizontally to the diagonal, vertically to the curve, horizontally to the diagonal: alternating back and forth. The plot of these vertical and horizontal moves that give the segments between the curve $f(x)$ and the diagonal is called a **cobweb diagram**. If the equilibrium is asymptotically stable, the cobweb will converge to the equilibrium represented by the intersection point. If the equilibrium is unstable, the cobweb will diverge from the intersection point. Figure 2.21 shows a divergent cobweb. By the analytic stability criterion, for stability the curve $f(x)$ cannot cross the 45-degree line $y = x$, which is steeper than the line itself.

EXERCISES

1. The plant biomass P_t at time t is governed by the difference equation

$$P_{t+1} = 2P_t - \frac{3P_t}{1 + 2P_t}.$$

   Find the equilibria and their stability. Draw cobweb diagrams.

2. Consider the discrete population model

$$u_{t+1} = \frac{au_t^2}{b^2 + u_t^2},$$

   where $a > 0$. If $a^2 > 4b^2$, show that the population may be driven to extinction if it becomes less than a critical size.

3. Find the general solution to the **Beverton–Holt model equation**

$$x_{t+1} = \frac{ax_t}{b + x_t}$$

   by making a substitution $w_t = 1/x_t$.

4. Draw a cobweb diagram for the model

$$x_{t+1} = \frac{8x_t}{1 + 2x_t}$$

   when $x_0 = 0.5$.

5. Find the equilibrium for the model

$$x_{t+1} = x_t \ln x_t^2,$$

   and use both the analytic condition and a cobweb diagram to determine its stability.

6. A density-limited population model is

$$x_{t+1} = \frac{ax_t}{(1 + bx_t)^c},$$

   where a, b, and c are positive.

   (a) Use scaling to reduce to two the number of parameters in the model.

   (b) Using the scaled equation, find the equilibria and determine the conditions for stability for each.

   (c) Plot in two-dimensional parameter space the region of stability when there is a positive equilibrium.

7. Let $x = x^* > 0$ and $x = 0$ be the only two equilibria for the two-stage difference equation

$$x_{t+1} = ax_t + F(x_{t-1}),$$

where $0 < a < 1$ and F is a well-behaved positive nonnegative function. Derive a stability condition for the positive equilibrium. (*Hint*: Use small perturbations and linearization.)

8. Use the transformation

$$y_t = \frac{rx_t}{K(1+r)}$$

to transform the discrete logistic equation

$$x_{t+1} = x_t + rx_t\left(1 - \frac{x_t}{K}\right)$$

into the form

$$y_{t+1} = ay_t(1 - y_t),$$

where $a = 1 + r > 0$. Find the equilibria and their stability, and with initial condition $y_0 = 0.25$, plot the solution for $a = 2.75, 3.1, 3.8, 3.828$.

9. Show that the model

$$x_{t+1} = x_t e^{r - x_t - cx_t^2}, \quad r, c > 0,$$

has a unique positive equilibrium and find a condition for which it is stable.

10. For the discrete model

$$x_{t+1} = (\lambda + 1)x_t + x_t^3,$$

draw a bifurcation diagram (expressing the equilibria vs. λ) for values of λ near zero.

11. The *Gilpin–Ayala population model* is given by

$$\Delta P_t = rP_t\left(1 - \left(\frac{P_t}{K}\right)^\theta\right),$$

where r, K, and θ are positive constants.

(a) Find the equilibria.

(b) Find a formula for the per capita population growth.

(c) Write the model in the form $P_{t+1} = f(P_t)$ and determine the stability of the equilibria.

(d) Sketch the region in the $r\theta$ plane that gives values for which the nonzero equilibrium is stable.

12. The growth rate of a population $N(t)$ depends on the population abundance at time $t - T$, where T is a positive fixed time and t is continuous time, and the mortality rate is proportional to the current population $N(t)$. So $N(t)$ solves the differential-delay equation

$$\frac{dN}{dt} = rN(t - T) - mN(t),$$

where m and r are positive. Show that there is a nonzero equilibrium (constant) population. If you try a solution of the form $N(t) = e^{\lambda t}$, what equation must λ satisfy? Under what conditions is there a real value of λ that satisfies this equation?

13. (Differential-delay equation) A model of the form (Hutchinson's equation)

$$\frac{dN}{dt} = rN \left(1 - \frac{1}{K} \int_0^\infty N(t - \tau)k(\tau)\, d\tau \right)$$

is a combined differential-difference equation with a distributed integral term. Here

$$k(\tau) = \frac{1}{T} e^{-\tau/T}$$

denotes the weight given to the population size at a time τ earlier, normalized by $\int_0^\infty k(\tau)\, d\tau = 1$. (Compare to the logistic equation.) The quantity T is the average delay, or $\int_0^\infty \tau k(\tau)\, d\tau = T$.

(a) Show that $N = 0$ and $N = K$ are constant, or equilibrium, solutions.

(b) Show that the linearization about $N = K$ is the equation

$$\frac{dn}{dt} = -r \int_0^\infty n(t - \tau)k(\tau)\, d\tau,$$

where n is a small perturbation from K.

(c) If $n(t) = n_0 e^{\lambda t}$, find an equation for λ, and show that $N = K$ is stable.

2.7 Reference Notes

The mathematical biology and ecology references at the end of Chapter 1 all discuss, at varying levels and detail, population dynamics for single species in both discrete and continuous time. Key classic papers on discrete models are those of R. May and his colleagues: see, for example, see May (1976). The papers by Ricker (1954) and Beverton & Holt (1957) are also very accessible. For more on the budworm problem, see Ludwig et al. (1978). Note that cobweb diagrams can be drawn automatically on, for example, a TI-84 Plus calculator.

References

Beverton, R. J. H. & Holt, S. J. 1957. On the dynamics of exploited fish populations, Series II, *Fish. Investig.* **19**, 1–533.

Ludwig, D., Jones, D. D., & Holling, C. S. 1978. Qualitative analysis of insect outbreak systems: the spruce budworm and forest, *J. Anim. Ecol.* **47**, 315.

May, R. M. 1976. Simple mathematical models with very complicated dynamics, *Nature* **261**, 459–467.

Neuhauser, C. 2004. *Calculus for Biology and Medicine*, 2nd ed., Prentice Hall, Upper Saddle River, NJ.

Ricker, W. E. 1954. Stock and recruitment, *J. Fish. Res. Board Can.* **11**, 559–623.

3

Structure and Interacting Populations

In both the continuous- and discrete-time population models, we tracked a single species and lumped all the individuals into a single quantity without regard to development stage, size, or age. Some animals and plants, however, such as insects, have well-defined life stages, such as egg, larva, pupa, and adult. In this chapter we develop the dynamics of how animals pass through life stages, keeping track of the population of each stage.

This chapter relies heavily on matrix manipulation. For readers who need a review of matrices, or need some basic instruction, we include at the end of the chapter a long section on matrices and linear algebra.

3.1 Structure: Juveniles and Adults

Consider a hypothetical animal that has two life stages, juvenile and adult. Let J_t denote the number of juveniles and A_t the number of adults, where t is given in weeks. Over each time step of one week we assume that juveniles are born, some die, and some graduate to adulthood. Adults are recruited from the graduating juvenile class, and some adults die. For the juvenile population,

$$J_{t+1} = J_t - mJ_t - gJ_t + fA_t. \tag{3.1}$$

That is, the number of juveniles at the next week (J_{t+1}) equals the number at the last week (J_t), minus the fraction that died (mJ_t), minus the fraction that become adults (gJ_t), plus the number of births (fA_t). The constant m

Mathematical Methods in Biology,
By J. David Logan and William R. Wolesensky
Copyright © 2009 John Wiley & Sons, Inc.

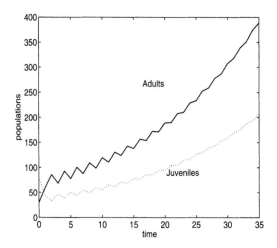

Figure 3.1 Time series plot of population of adults (solid) and juveniles (dashed) vs. time.

is the weekly juvenile mortality rate, g the weekly graduation rate, and f the average weekly fecundity (fertility) of adults. Because adults reproduce, we assume that the number of juveniles produced each week is proportional to the number of adults present. For example, if each adult on the average gives rise to two juveniles each week, then $f = 2$. Then 250 adults would give rise to 500 juveniles per week. In a similar manner, for the adult population,

$$A_{t+1} = A_t - \mu A_t + g J_t, \tag{3.2}$$

where μ is the adult weekly mortality rate. The last term represents those juveniles that become adults. Equations (3.1)–(3.2) represent a system of two coupled difference equations. They provide a way to take information from time t and project it forward to the next time, $t + 1$. Consequently, if we know the initial population in each stage, we can recursively calculate the populations at time $t + 1$. The following MATLAB m-file generated the time series plots shown in Fig. 3.1. For comparison, Fig 3.2 shows a plot (called an *orbit*) of the ordered pairs (J_t, A_t) in a two-dimensional phase plane. To get a phase plot, replace the plot commands in the subsequent program by the single command plot(Jhistory,Ahistory). In either plot we see that the populations of both the juveniles and adults grow over time with an oscillation that decays slowly. (Such plots can also be produced by many graphing calculators.)

```
function stagepopulation
clear
m=0.5; g=0.5; f=2; mu=0.9; timesteps=35;
```

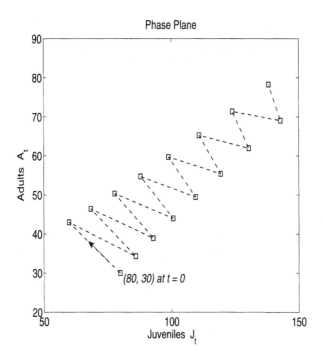

Figure 3.2 Phase plane plot in the (J_t, A_t) plane showing an orbit (J_t, A_t), or sequence of populations, for times $t = 0, 1, 2, \ldots, 15$.

```
J=80; Jhistory=J; A=30; Ahistory=A;
for t=1:timesteps;
x=(1-m-g)*J+f*A;
y=g*J+(1-mu)*A;
J=x; A=y;
Ahistory=[Ahistory J];
Jhistory=[Jhistory A];
end
time=0:timesteps;
plot(time,Jhistory), hold on
plot(time,Ahistory), hold off
```

Let us collect terms and rewrite (3.1)–(3.2) in the form

$$J_{t+1} = (1 - m - g)J_t + fA_t,$$
$$A_{t+1} = gJ_t + (1 - \mu)A_t.$$

Now it becomes clear that the model can be expressed in matrix form as

$$\begin{pmatrix} J_{t+1} \\ A_{t+1} \end{pmatrix} = \begin{pmatrix} 1-m-g & f \\ g & 1-\mu \end{pmatrix} \begin{pmatrix} J_t \\ A_t \end{pmatrix}.$$

Even more concisely, the model can be formulated in vector notation as

$$\mathbf{x}_{t+1} = P\mathbf{x}_t, \tag{3.3}$$

where

$$\mathbf{x}_t = \begin{pmatrix} J_t \\ A_t \end{pmatrix}, \quad P = \begin{pmatrix} 1-m-g & f \\ g & 1-\mu \end{pmatrix}.$$

The matrix P is the **population projection matrix.**. The entries in P have a demographic meaning: The elements on the diagonal are survivorships. That is, $1-m-g$ is the fraction of juveniles that remain in the same class during one time period, and $1-\mu$ is the fraction of adults that remain in the same class; f is the fecundity of adults, or the number of juveniles produced by adults during one time step; and finally, g is the fraction of juveniles that become adults during one time period (the "graduation" rate).

Using iteration we see immediately that the solution to (3.3) is

$$\mathbf{x}_t = P^t\mathbf{x}_0,$$

where $\mathbf{x}_0$ is a vector containing the initial stage populations.

Example 3.1

In matrix notation, if in the example above we had chosen $g = m = 0.5$, $f = 2$, and $\mu = 0.9$, then the model would be

$$\mathbf{x}_{t+1} = \begin{pmatrix} 0 & 2 \\ 0.5 & 0.1 \end{pmatrix} \mathbf{x}_t. \qquad \square$$

3.2 Structured Linear Models

Generalizing the example in Section 3.1, we can consider an expanded population model where the population has n stages or classes. Then we obtain a system of n difference equations that can be expressed in matrix form as

$$\mathbf{x}_{t+1} = A\mathbf{x}_t, \quad t = 0, 1, 2, \dots. \tag{3.4}$$

Here $\mathbf{x}_t$ is an n-vector, or column vector, containing the populations of the n stages at time t, and A is a constant $n \times n$ matrix containing the survivorships, graduation rates, and fecundities. The total population at time t is the sum of the n elements in $\mathbf{x}_t$. Such models arise naturally in stage- and age-structured models in ecology, with A being the population projection matrix. By iteration, if $\mathbf{x}_0$ is an initial vector containing the initial populations in the various stages at $t = 0$, then

$$\mathbf{x}_t = A^t \mathbf{x}_0, \qquad (3.5)$$

which is the solution to 3.4.

We can obtain a more useful representation, which gives additional information about the solution in terms of the eigenvalues and eigenvectors of A. The eigenvalues and eigenvectors of a population projection matrix reveal both the growth rate and the eventual population structure. So they answer questions such as: Does the total population grow or decay? What percentage of the long term total population is in one stage or another?

Suppose that A has n distinct, real eigenvalues λ_k with associated independent eigenvectors $\mathbf{v}_k$, $k = 1, 2, ..., n$. This means that

$$A\mathbf{v}_k = \lambda_k \mathbf{v}_k$$

for all k. Then the eigenvectors form a basis for $\mathbb{R}^n$ and we can express the initial population vector as

$$\mathbf{x}_0 = c_1 \mathbf{v}_1 + \cdots + c_n \mathbf{v}_n,$$

where the c_k are the coordinates of $\mathbf{x}_0$ in the basis. Then, using linearity,

$$\mathbf{x}_t = A^t \mathbf{x}_0 = c_1 A^t \mathbf{v}_1 + \cdots + c_n A^t \mathbf{v}_n = c_1 \lambda_1^t \mathbf{v}_1 + \cdots + c_n \lambda_n^t \mathbf{v}_n, \qquad (3.6)$$

which is the solution (3.5) in terms of eigenvalues and eigenvectors. Here we used the fact that if λ is an eigenvalue of A, λ^t is an eigenvalue of A^t, with the same eigenvector. (Can you show this?)

Parenthetically, we also recall (see Section 3.4) that we can find the power A^t by diagonalizing the matrix A. Let D be the diagonal matrix with the eigenvalues on the diagonal, and let V be the matrix whose columns are the associated eigenvectors. Then, from linear algebra, $V^{-1} A V = D$ (we say that the matrix V diagonalizes A). Thus $A = V D V^{-1}$ and $A^t = (V D V^{-1})^t = V D^t V^{-1}$, where D^t is a diagonal matrix with the entries λ_i^t on the diagonal.

Example 3.2

We return to the example

$$\mathbf{x}_{t+1} = \begin{pmatrix} 0 & 2 \\ 0.5 & 0.1 \end{pmatrix} \mathbf{x}_t.$$

The eigenpairs are

$$\lambda_1 = 1.051, \quad \mathbf{v}_1 = \begin{pmatrix} 0.885 \\ 0.465 \end{pmatrix}; \qquad \lambda_2 = -0.951, \quad \mathbf{v}_2 = \begin{pmatrix} 0.903 \\ -0.429 \end{pmatrix}.$$

From (3.6) the general solution is therefore

$$\mathbf{x}_t = c_1 (1.051)^t \begin{pmatrix} 0.885 \\ 0.465 \end{pmatrix} + c_2 (-0.951)^t \begin{pmatrix} 0.903 \\ -0.429 \end{pmatrix}.$$

The constants c_1 and c_2 are determined by the initial population vector $\mathbf{x}_0$. For example, if $\mathbf{x}_0 = (12, 4)^{\mathrm{T}}$, then

$$c_1 \begin{pmatrix} 0.885 \\ 0.465 \end{pmatrix} + c_2 \begin{pmatrix} 0.903 \\ -0.429 \end{pmatrix} = \begin{pmatrix} 12 \\ 4 \end{pmatrix},$$

which can be solved for c_1 and c_2. $\square$

An important case occurs when there is a real strictly **dominant eigenvalue**, say λ_1. For example, in Example 3.2, $\lambda_1 = 1.051$. This means that $|\lambda_1| > |\lambda_2| \geq |\lambda_3| \geq \cdots \geq |\lambda_n|$. Then we can write the solution as

$$
\begin{aligned}
\mathbf{x}_t &= c_1 \lambda_1^t \mathbf{v}_1 + c_2 \lambda_2^t \mathbf{v}_2 + \cdots + c_n \lambda_n^t \mathbf{v}_n \\
&= \lambda_1^t \left[c_1 \mathbf{v}_1 + c_2 \left(\frac{\lambda_2}{\lambda_1} \right)^t \mathbf{v}_2 + \cdots + c_n \left(\frac{\lambda_n}{\lambda_1} \right)^t \mathbf{v}_n \right].
\end{aligned}
\tag{3.7}
$$

We have $|\lambda_i/\lambda_1| < 1$ for $i = 2, 3, ..., n$, and therefore all the terms $(\lambda_i/\lambda_1)^t$ decay to zero as $t \to \infty$. Consequently, for large t the solution to the linear model behaves approximately as

$$\mathbf{x}_t \approx c_1 \lambda_1^t \mathbf{v}_1.$$

Asymptotically, as $t \to 0$, if $\lambda_1 > 0$, the solution exhibits geometric growth or decay, depending on the magnitude of λ_1. In this case, the dominant eigenvalue λ_1 is called the **growth rate**. The vector $\mathbf{v}_1$ is called the **stable age structure**.

Example 3.3

The dominant eigenvalue in Example 3.3 is $\lambda_1 = 1.051$, so the growth rate is about 5%. The second term in the solution leads to a decaying oscillation caused by the factor $(-0.951)^t$. This is the observation shown in Figs. 3.1 and 3.2. Over a long time period, the juvenile–adult population will approach a stable age structure defined by the eigenvector $(0.885, 0.465)^{\mathrm{T}}$, which[1], when

---

[1] The superscript T denotes the transpose, which makes the row vector into a column vector. Matrix operations are discussed in the appendix on matrices in Section 3.4.

normalized to make the sum of the entries 1, gives $(0.656, 0.344)^{\mathrm{T}}$. This means that in the long run, 65.6% of the population will be juveniles and 34.4% of the population will be adults. That is, there will be about 1.9 juveniles for every adult. □

If some of the eigenvalues of A are complex but the set of eigenvalues is still distinct, the argument above can be repeated. Using Euler's formula, $e^{i\theta} = \cos\theta + i\sin\theta$, every complex term in solution (3.6) can be written in terms of real and imaginary parts, each of which is a real solution. For example, if $\mathbf{v} = \mathbf{u} + i\mathbf{w}$ and $\lambda = \alpha + i\beta = re^{i\theta}$ is a complex eigenpair, where r is the modulus of λ and θ is its argument, then

$$
\begin{aligned}
\mathbf{v}\lambda^t &= (\mathbf{u} + i\mathbf{w})(\alpha + i\beta)^t \\
&= r^t(\mathbf{u} + i\mathbf{w})(\cos\theta t + i\sin\theta t) \\
&= r^t(\mathbf{u}\cos\theta t - \mathbf{w}\sin\theta t) + ir^t(\mathbf{u}\sin\theta t + \mathbf{w}\cos\theta t).
\end{aligned}
$$

The real and imaginary parts of this vector,

$$
r^t(\mathbf{u}\cos\theta t - \mathbf{w}\sin\theta t), \quad r^t(\mathbf{u}\sin\theta t + \mathbf{w}\cos\theta t),
$$

are two linearly independent, real solutions. The complex conjugate eigenpair $\mathbf{v} = \mathbf{u} - i\mathbf{w}$ and $\lambda = \alpha - i\beta$ gives the same two independent solutions. In summary, the two complex terms in (3.6) associated with complex conjugate eigenpairs may be replaced by the two real independent vector solutions. Therefore, we may always write the solution as a real solution. Complex eigenvalues clearly lead to oscillations in the system because of the sine and cosine functions. The analysis of the case of a repeated eigenvalue of A is left to the Exercises.

Notice that (3.7) also implies that the solution to the linear model (3.4) satisfies the condition $\mathbf{x}_t \to \mathbf{0}$ if, and only if, all the eigenvalues have modulus less that 1, or $|\lambda_i| < 1$. In this case we say that $\mathbf{x} = \mathbf{0}$ is an **asymptotically stable equilibrium**. If there is an eigenvalue with $|\lambda| > 1$, the equilibrium $\mathbf{x} = \mathbf{0}$ is **unstable**.

Example 3.4

In the case of four stages, $n = 4$, we have $\mathbf{x}_t = (x_{1t}, x_{2t}, x_{3t}, x_{4t})^{\mathrm{T}}$. Assume that the population projection matrix has the form

$$
A = \begin{pmatrix} s_1 + f_1 & f_2 & f_3 & f_4 \\ g_1 & s_2 & 0 & 0 \\ 0 & g_2 & s_3 & 0 \\ 0 & 0 & g_3 & s_4 \end{pmatrix}, \tag{3.8}
$$

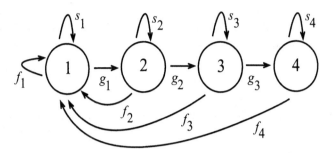

Figure 3.3 Leslie diagram.

where f_i is the fecundity of the ith stage, s_i is the survival rate of the ith stage, and g_i is the fraction of the ith stage that passes to the jth stage during the fixed time step. Note the position of the survival rates (on the diagonal), graduation rates (on the subdiagonal), and the fecundities (on the top row). A **Leslie diagram** may be sketched (see Fig. 3.3) that illustrates how the rates and the stages are connected. In a still more general case, the entries in the projection matrix may depend on t, or they may depend on the populations themselves, making the model nonlinear. □

To review, we have shown that when there is a positive, dominant eigenvalue, say λ_1, of the projection matrix A, we have $\mathbf{x}_t \sim c_1 \mathbf{v}_1 \lambda_1^t$ for large t. Therefore λ_1 is the long-term growth rate and $c_1 \mathbf{v}_1$ represents the stable stage structure. Because c_1 is a constant, we can just use $\mathbf{v}_1$ to determine the proportion, or percentage, of each stage to the entire population. For example, if $\mathbf{v}_1 = (1, 2, 3)^{\mathrm{T}}$ is an eigenvector associated with a dominant eigenvalue, the vector $(\frac{1}{6}, \frac{1}{3}, \frac{1}{2})$ shows the fraction of each stage in the stable age structure.

A **Leslie matrix**, named after P. H. Leslie, a biologist who developed the ideas in the 1940s, was introduced to study age-structured models. Later, in the 1960s, L. P. Lefkovitch generalized the model to stage-structured population models. So a Leslie matrix is the population projection matrix for a structured population model where the stages are ages.

We can ask the obviously important question whether there is a positive dominant eigenvalue for a given model. There are theorems about this question, addressing what types of properties a population projection must have. A simple version of, for example, the **Perron–Frobenius theorem** states that if A is a nonnegative matrix (all of its entries are nonnegative) and if for some positive integer k the matrix A^k has all positive entries, then A has a simple positive dominant eigenvalue with a positive eigenvector. There are other theorems that guarantee the same conclusion under less strenuous hypotheses. See

Caswell (2001) for additional information about the existence of a dominant eigenvalue.

3.2.1 Sensitivity and Elasticity

Another interesting issue in stage- and age-structured population models is to determine how values in the projection matrix, which represent fecundities, graduation rates, and survivorship probabilities, most affect the dominant eigenvalue λ, which is the long-term growth rate. This is particularly important in making management decisions about ecosystems. We can measure the sensitivity of λ to changes in an entry a_{ij} of A by calculating the partial derivative $\partial\lambda/\partial a_{ij}$. (Note that the dominant eigenvalue is a function of the entries in A.) The matrix of these values,

$$S = \left(\frac{\partial\lambda}{\partial a_{ij}}\right),$$

is called the **sensitivity matrix**.

Example 3.5

For a 2×2 projection matrix

$$A = \begin{pmatrix} a & b \\ c & d \end{pmatrix}$$

with nonnegative entries, the dominant eigenvalue is easily found from the roots of the characteristic equation,

$$\lambda^2 - (a+b)\lambda + (ad - bc) = 0$$

or

$$\lambda^2 - \operatorname{tr} A\lambda + \det A = 0.$$

The dominant root is

$$\lambda = \frac{1}{2}\left(\operatorname{tr} A + \sqrt{D}\right), \quad D = (\operatorname{tr} A)^2 - 4\det A.$$

Then, by taking partial derivatives, the sensitivity matrix is

$$S = \begin{pmatrix} \frac{\partial\lambda}{\partial a} & \frac{\partial\lambda}{\partial b} \\ \frac{\partial\lambda}{\partial c} & \frac{\partial\lambda}{\partial d} \end{pmatrix} = \begin{pmatrix} \frac{1}{2}\left(1 + \frac{a-d}{\sqrt{D}}\right) & \frac{c}{\sqrt{D}} \\ \frac{b}{\sqrt{D}} & \frac{1}{2}\left(1 - \frac{a-d}{\sqrt{D}}\right) \end{pmatrix}. \quad \Box$$

Example 3.6

Consider the two-stage (juveniles and adults) model

$$x_{t+1} = 1.7y_t, \quad y_{t+1} = 0.6x_t + 0.1y_t$$

with matrix

$$A = \begin{pmatrix} 0 & 1.7 \\ 0.6 & 0.1 \end{pmatrix}.$$

The value 1.7 is the fecundity of adults and 0.1 is their survival rate; 0.6 is the fraction of juveniles that survive and pass to the adult stage. The dominant eigenvalue is $\lambda = 1.06$, which means that the long-time growth rate of the population is 6% per season. The sensitivity matrix is

$$S = \begin{pmatrix} 0.475 & 0.297 \\ 0.841 & 0.525 \end{pmatrix}.$$

Therefore, we see that the $2, 1$ entry, $c = a_{21}$, the survivorship of juveniles, affects the dominant eigenvalue most. We can estimate the change $\Delta\lambda$ in the long term growth rate when the survivorship is changed by Δc with the linear approximation

$$\Delta\lambda = \frac{\partial\lambda}{\partial c}\Delta c = 0.841 \cdot \Delta c.$$

For example, if the survivorship $c = 0.6$ of juveniles decreases by 0.1 to 0.5 (so $\delta c = -0.1$), the change in the dominant eigenvalue is approximately $\Delta\lambda = 0.841(-0.1) = -0.0841$. Therefore the growth rate becomes $1.06 - 0.0841 = 0.975$, and the population will die out.   □

Now we proceed with the general method to calculate the sensitivity matrix. Consider the n-dimensional model

$$\mathbf{x}_{t+1} = A\mathbf{x}_t$$

with $\lambda, \mathbf{v}$ the dominant eigenpair. Then

$$A\mathbf{v} = \lambda\mathbf{v}.$$

We know that λ is also an eigenvalue of A^T (A-transpose) having some eigenvector $\mathbf{w}$. Therefore, $A^T\mathbf{w} = \lambda\mathbf{w}$, or, taking the transpose,

$$\mathbf{w}^T A = \lambda\mathbf{w}^T.$$

This means that $\mathbf{w}$ is a **left eigenvector** of A.

To calculate the change in the dominant eigenvalue with respect to the change of an entry in the matrix, we change the ijth entry of A by a small amount Δa_{ij}, while holding the remaining entries fixed. Let ΔA be the matrix with Δa_{ij} in the ijth position and zeros elsewhere. Then the perturbed matrix $A + \Delta A$ has dominant eigenvalue $\lambda + \Delta\lambda$ and eigenvector $\mathbf{v} + \Delta\mathbf{v}$, or

$$(A + \Delta A)(\mathbf{v} + \Delta\mathbf{v}) = (\lambda + \Delta\lambda)(\mathbf{v} + \Delta\mathbf{v}).$$

Multiplying out and discarding the higher-order terms (containing products of the small changes) gives, to leading order,

$$A(\Delta\mathbf{v}) + (\Delta A)\mathbf{v} = \lambda(\Delta\mathbf{v}) + (\Delta\lambda)\mathbf{v}.$$

Multiplying by $\mathbf{w}^{\mathrm{T}}$ on the left gives

$$\mathbf{w}^{\mathrm{T}}(\Delta A)\mathbf{v} = (\Delta\lambda)\mathbf{w}^{\mathrm{T}}\mathbf{v}$$

or

$$\Delta\lambda = \frac{\mathbf{w}^{\mathrm{T}}(\Delta A)\mathbf{v}}{\mathbf{w}^{\mathrm{T}}\mathbf{v}} = \frac{w_i v_j \Delta a_{ij}}{\mathbf{w}^{\mathrm{T}}\mathbf{v}}.$$

In the limit we have

$$S = \left(\frac{\partial\lambda}{\partial a_{ij}}\right) = \frac{1}{\mathbf{w}^{\mathrm{T}}\mathbf{v}}(w_i v_j) = \frac{1}{\mathbf{w}^{\mathrm{T}}\mathbf{v}}\mathbf{w}\mathbf{v}^{\mathrm{T}}, \tag{3.9}$$

which is a simple representation of the sensitivity matrix. As a final note, it is always possible to normalize $\mathbf{v}$ and $\mathbf{w}$ so that $\mathbf{w}^{\mathrm{T}}\mathbf{v} = 1$.

Example 3.7

It is clearly difficult to calculate the sensitivity matrix for more than 2×2 matrices. MATLAB is therefore very useful to calculate the sensitivity matrix S. The following command line statements calculate S for the model in Example 3.6.

```
A=[0  1.7; 0.6  0.1]
[V,D]=eig(A)
v=V(:,2)
[W,d]=eig(A')
w=W(:,2)
S=(1/(w'*v))*(w*v')
```
□

Unfortunately, the sensitivity matrix may give information that is difficult to interpret when entries in A have large differences. For example, fecundities are usually high compared to survivorships. Therefore, asking how the dominant eigenvalue changes with respect to a unit change in an entry is not always practical. It is better to compute the *relative* changes, or the elasticity. We define the **elasticity matrix** E by

$$E = \left(\frac{a_{ij}}{\lambda} \frac{\partial \lambda}{\partial a_{ij}} \right) \simeq \left(\frac{\Delta \lambda / \lambda}{\Delta a_{ij} / a_{ij}} \right).$$

In MATLAB this is simply E = A.*S/max(eig(A)). When an entry of E is large, we say that the entry is *elastic*, and when it is small, we say that it is *inelastic*. These terms are familiar terms in economics with regard to the elasticity of demand for a product with respect to its price changes; there,

$$\text{elasticity} = \frac{\% \text{ change in demand}}{\% \text{ change in price}} \simeq \frac{p}{D} \frac{dD}{dp},$$

where the demand D is a function of price p. The definitive treatment by Caswell (2001) has a thorough discussion of matrix population models in the biological sciences.

Remark 3.8

We make one important remark about zero entries in a Leslie-type matrix A. Generally, those entries are, in fact, zero for biological reasons. For example, a survivorship from the first year (stage) to the third year (stage) is zero because animals do not skip a year in aging. Therefore, in computing the sensitivity matrix S, we often do not perturb these zero elements by adding a small amount of survivorship in the 1,3 position to see how the dominant eigenvalue changes. As a result, we just put a zero in that location in the sensitivity matrix. However, the MATLAB algorithm we presented assumes small perturbations in those entries and therefore will often produce nonzero elements in S. The modeler has to decide which strategy to adopt on this issue.   □

EXERCISES

1. Find the solution to the discrete model

$$\begin{aligned} X_{t+1} &= X_t + 4Y_t \\ Y_{t+1} &= \frac{1}{2}X_t \end{aligned}$$

if $X_0 = 6$ and $Y_0 = 0$. Plot times series of the solution, and plot the solution in a phase plane.

2. If λ is an eigenvalue of A, show that λ^t is an eigenvalue of A^t with the same eigenvector for any positive integer t.

3. For each of the following Leslie matrices, find the dominant eigenvalue (if any), the corresponding eigenvector, and the stable age structure. Interpret each entry in the context of a three stage population model.

   (a) $A = \begin{pmatrix} 0 & 1 & 5 \\ 0.3 & 0 & 0 \\ 0 & .5 & 0 \end{pmatrix}$.

   (b) $A = \begin{pmatrix} 0 & 1 & 1 \\ \frac{2}{3} & 0 & 0 \\ 0 & \frac{1}{3} & 0 \end{pmatrix}$.

4. Describe the dynamics of the Leslie model

   $$\mathbf{x}_{t+1} = \begin{pmatrix} 0 & 0 & f \\ a & 0 & 0 \\ 0 & b & 0 \end{pmatrix} \mathbf{x}_t,$$

   where a, b, and f are positive constants. Discuss the two cases $abf > 1$ and $abf < 1$.

5. Consider the linear model

   $$\mathbf{x}_{t+1} = \begin{pmatrix} ac & bc \\ s(1-a) & 0 \end{pmatrix} \mathbf{x}_t,$$

   where s, b, $c > 0$ and $0 < a < 1$.

   (a) Determine conditions (if any) for which there are infinitely many equilibria (constant vector solutions).

   (b) In the case of a single equilibrium, determine conditions on the parameters that ensure that all solutions converge to zero as $t \to \infty$.

6. Prove that the quadratic equation $\lambda^2 - p\lambda + q = 0$ has roots satisfying $|\lambda| < 1$ if, and only if,
   $$|p| < 1 + q < 2.$$

   The condition is called the **Jury condition**.

7. In many bird species the fecundity and survivorship of adults is independent of the age of the adult bird. So we can think of the population as being composed of two classes, juveniles and adults.

   (a) Set up a general population model to describe the dynamics of the female bird population, assuming that juveniles do not reproduce and that the juvenile stage lasts only one year. Adult females may live more than one year.

(b) Find the eigenvalues and eigenvectors of the population matrix for this model.

(c) Identify the growth rate and dominant eigenvalue, and determine the long-term fraction of the total population that are juveniles. Examine all cases.

8. If $\mathbf{w} = (1, w_2, w_3)^{\mathrm{T}}$ is a left eigenvector of the matrix

$$A = \begin{pmatrix} f_1 & f_2 & f_3 \\ s_1 & 0 & 0 \\ 0 & s_2 & 0 \end{pmatrix},$$

find w_2 and w_3.

9. A two-dimensional discrete system is given by

$$\mathbf{x}_{t+1} = \begin{pmatrix} 1 - a & \frac{a}{3} \\ a & 1 - \frac{a}{3} \end{pmatrix} \mathbf{x}_t,$$

where $0 < a < 1$.

(a) Find the eigenvalues and simplify completely.

(b) Find the left and right eigenvectors corresponding to the dominant eigenvalue.

(c) Write down the *form* of the general solution to the system.

(d) Explain the behavior of the system over a long time.

10. Consider the two-dimensional discrete model

$$\mathbf{x}_{t+1} = \begin{pmatrix} 0 & 2 \\ 0.5 & 0.1 \end{pmatrix} \mathbf{x}_t,$$

where $0 < a < 1$.

(a) Find the eigenvalues and eigenvectors.

(b) Find the general solution to the system and describe its long-term behavior.

11. A Leslie matrix is given by

$$\begin{pmatrix} 0 & 0 & 6\alpha^3 \\ \frac{1}{2} & 0 & 0 \\ 0 & \frac{1}{3} & 0 \end{pmatrix},$$

where $\alpha > 0$. Find the eigenvalues and determine if there is a unique dominant eigenvalue.

12. A Leslie matrix is given by

$$
L = \begin{pmatrix} 0 & \frac{3}{2}a^2 & \frac{3}{2}a^3 \\ \frac{1}{2} & 0 & 0 \\ 0 & \frac{1}{3} & 0 \end{pmatrix},
$$

where $a > 0$.

(a) Find the eigenvalues and determine the dominant eigenvalue and eigenvector. [Hint: Guess a root.]

(b) Determine the stable age distribution.

(c) Show that L^k is a positive matrix for some integer k.

13. A population is divided into three age classes, ages 0, 1, and 2. During each time period, 20% of the females of age 0 and 70% of the females of age 1 survive until the end of the following breeding season. Females of age 1 have an average fecundity of 3.2 offspring per female, and the average fecundity of age 2 females is 1.7. No female lives beyond three years.

(a) Set up the Leslie matrix.

(b) If the population consists initially of 2000 females of age 0, 800 females of age 1, and 200 females of age 2, find the age distribution after three years.

(c) Determine the asymptotic behavior of the population.

14. Show that if $\lambda = 0$ is an eigenvalue of a matrix A, then A^{-1} does not exist.

15. Consider a linear model $\mathbf{x}_{t+1} = A\mathbf{x}_t$. Under what condition(s) on A does the system have a single equilibrium?

16. Consider the stage-structured model

$$
\mathbf{x}_{t+1} = \begin{pmatrix} \frac{1}{2} & 1 & \frac{3}{4} \\ \frac{2}{3} & 0 & 0 \\ 0 & \frac{1}{3} & 0 \end{pmatrix} \mathbf{x}_t.
$$

(a) Find the eigenvalues and eigenvectors, and identify the growth rate and stable age structure.

(b) Generate the sensitivity matrix S and determine which entry in A most strongly affects the growth rate.

(c) Compute the elasticity matrix E. Do the entries of E change your conclusion in part (b)?

17. A three-stage model of coyote populations (pups, yearlings, and adults) has a population projection matrix given by

$$\begin{pmatrix} 0.11 & 0.15 & 0.15 \\ 0.3 & 0 & 0 \\ 0 & 0.6 & 0.6 \end{pmatrix}.$$

Repeat parts (a) and (b) of Exercise 16.

18. A two-stage Leslie matrix is

$$\begin{pmatrix} f_1 & f_2 \\ s & 0 \end{pmatrix}.$$

By hand, compute the dominant eigenvalue, the sensitivity matrix S, and the elasticity matrix E.

19. A female population with three stages, juveniles J, youth Y, and adults A, have the following vital statistics. In a time step (one season), no animal remains in the same stage; but on average, during each time step 50% of the juveniles become youth and 60% of the youth survive to become adults. On average, each youth produces 0.8 offspring (juveniles) per time step, and each adult produces 2.2 offspring per time step. Juveniles do not reproduce.

   (a) Draw a Leslie diagram with arrows indicating how the stages interact.

   (b) Write down the discrete model and identify the population projection matrix L.

   (c) If $J_0 = 25$, $Y_0 = 50$, and $A_0 = 25$, draw plots (on the same set of axes) of the populations of each stage for the first 30 seasons.

   (d) What is the the growth rate of the total population? Over a long time, what percentage of the population is in each stage? Explain your observations in part (c) based on the eigenvalues and eigenvectors of the population projection matrix L. What are the absolute values of the eigenvalues?

   (e) Under the same initial conditions, plot the natural logarithms of the populations versus time. Explain the difference observed from part (d).

   (f) Compute the sensitivity matrix S.

   (g) The dominant eigenvalue is most sensitive to a change in what nonzero entry in L?

   (h) Compute the elasticity of the dominant eigenvalue to adult fecundity. Interpret your result in words.

(i) If measures are taken to increase the survivorship of youths by 10%, what would be the approximate population growth rate?

(j) How much would the fecundity of adults have to decrease to drive the population to extinction?

3.3 Nonlinear Interactions

Next we set up and analyze two-dimensional nonlinear discrete systems of the form

$$x_{t+1} = f(x_t, y_t), \tag{3.10}$$

$$y_{t+1} = g(x_t, y_t), \tag{3.11}$$

where f and g are given functions, and $t = 0, 1, 2, \dots$. As in the case of linear models discussed in Section 3.2, we can visualize a solution geometrically as a time series, where the sequences x_t and y_t are plotted against t, or as a sequence of points (x_t, y_t), called an **orbit**, in the two-dimensional xy plane or phase plane. We often connect the points of the sequence by straight-line segments to make the plots continuous and easier to visualize.

Example 3.9

Nonlinear Leslie-type models provide a good class of nonlinear dynamics. These have the form

$$\mathbf{x}_{t+1} = A(x_t, y_t)\mathbf{x}_t,$$

where the population projection matrix depends on the stage populations themselves; for example, the fecundity f of adults may depend on the total population $P_t = x_t + y_t$ when females sense population pressure and reduce their reproductive output. Similarly, survivorships may vary with population. □

Examining equilibria is a key to understanding the behavior of nonlinear systems. An equilibrium solution is a constant solution $x_t = x^*$, $y_t = y^*$. It plots as a single point, called a **critical point**, in the xy phase plane; the behavior of the system is largely determined by the behavior of its orbits near the critical points. Stability, or permanence of an equilibrium, becomes an issue. That is, if the system is in equilibrium, and then given a small perturbation that displaces it a small amount away from equilibrium, does the perturbation decay away and the system return to equilibrium?

An equilibrium solution satisfies the equations

$$x^* = f(x^*, y^*), \quad y^* = g(x^*, y^*).$$

To analyze the orbital behavior near an equilibrium, we consider small perturbations u_t and v_t from x^* and y^*, respectively, or

$$x_t = x^* + u_t, \quad y_t = y^* + v_t.$$

Substituting these expression into the system (3.10)–(3.11), expanding f and g in a Taylor series about the equilibrium, and retaining only linear terms in the perturbations leads to a linear set of equations for the perturbations that can be dealt with using the methods of the Section 3.2. To illustrate, (3.10) becomes

$$
\begin{aligned}
x^* + u_{t+1} &= f(x^* + u_t, y^* + v_t) \\
&= f(x^*, y^*) + f_x(x^*, y^*)u_t + f_y(x^*, y^*)v_t + \cdots,
\end{aligned}
$$

where the three dots denote higher-order terms in u_t and v_t. The subscripts on f denote partial differentiation. Discarding the small terms and using the fact that $x^* = f(x^*, y^*)$, we obtain

$$u_{t+1} = f_x(x^*, y^*)u_t + f_y(x^*, y^*)v_t.$$

In a similar manner,

$$v_{t+1} = g_x(x^*, y^*)u_t + g_y(x^*, y^*)v_t.$$

The last two linear equations for the perturbations u_t and v_t can be written in matrix form as

$$\mathbf{u}_{t+1} = J\mathbf{u}_t, \tag{3.12}$$

where $\mathbf{u}_t = (u_t, v_t)^{\mathrm{T}}$ and $J = J(x^*, y^*)$ is the **Jacobian matrix**

$$J = \begin{pmatrix} f_x(x^*, y^*) & f_y(x^*, y^*) \\ g_x(x^*, y^*) & g_y(x^*, y^*) \end{pmatrix}. \tag{3.13}$$

Therefore, to leading order, the perturbations are governed by a linear model and we can analyze their evolution in time.

As we have seen, the behavior of a linear model is determined by the eigenstructure of its matrix. Let λ_1 and λ_2 denote the eigenvalues of J. From the section on linear systems, we have the following stability result:

(a) If $|\lambda_1| < 1$ and $|\lambda_2| < 1$, the perturbations decay and (x^*, y^*) is locally asymptotically stable.

(b) If either $|\lambda_1| > 1$ or $|\lambda_2| > 1$, some of the perturbations will grow and (x^*, y^*) is unstable.

This result translates into the following general condition for asymptotic stability in terms of the characteristic equation

$$\det(J - \lambda I) = \lambda^2 - (\operatorname{tr} J)\lambda + \det J = 0$$

for the Jacobian matrix. Necessary and sufficient conditions for local asymptotic stability, $|\lambda_1| < 1$ and $|\lambda_2| < 1$, are

$$|\operatorname{tr} J| < 1 + \det J, \quad \det J < 1. \tag{3.14}$$

Conditions (3.14) are called the **Jury conditions**. A derivation is requested in the Exercises.

To plot orbits in the xy phase plane we can use software such as MATLAB, and proceed in a manner similar to that in Section 3.1.

Example 3.10

(**Nicholson–Bailey model**) Many insect predators (e.g., some wasps and flies) are parasitoids. A parasitoid finds prey, called hosts, and lays its eggs inside or on the victim (e.g., a caterpillar). The eggs hatch into larvae, which then consume the host. The host then dies and the larvae metamorphose into a pupal stage and eventually emerge as adults. These creatures are highly abundant on Earth, making up about 10% of all multicelluar species. We can derive a population model based on the probability of the host avoiding detection; it has the general form

$$
\begin{aligned}
x_{t+1} &= r x_t F(y_t), \\
y_{t+1} &= c x_t \left(1 - F(y_t) \right),
\end{aligned}
$$

where x_t is the host population and y_t is the parasitoid, or predator, population at generation t; F is the escapement function, or probability of avoiding detection in one season, r is the geometric growth factor for the host, and c, which is the clutch size, is the number of eggs laid by an adult parasitoid in a single host that will survive to breed in the next generation. Think of F as the fraction of hosts that survive to the next generation, or the fraction *not* parasitized. The dynamics is discrete because both host and parasitoid usually have life cycles sychronized with seasons. The model was developed in the 1930s by A. J. Nicholson and V. A. Bailey, and it is fundamental in population ecology. There are various forms for the escapement function F. The usual model is to base it on a **Poisson random variable** (see Section 5.3). A discrete random variable X is Poisson distributed if

$$\Pr(X = k) = \frac{\lambda^k}{k!} e^{-\lambda}, \quad k = 0, 1, 2, \dots,$$

where λ is a positive constant. A Poisson distribution describes the occurrence of discrete random events, such as encounters between a predator and its prey, and X is the probability of k encounters in a given time period, say over the lifespan of the prey. The constant λ is the average number of encounters during the period. In terms of hosts and parasitoids, the number of encounters n_e is proportional to the product $x_t y_t$ of the population sizes; this is the law of mass action, or a type I functional response. Thus, $n_e = a x_t y_t$, where a is a constant. Therefore, *average* encounter rate for a given host is the number of encounters divided by the total number of hosts, or

$$\lambda = \frac{n_e}{x_t} = a y_t.$$

Using the Poisson distribution, the probability of no encounters is the zeroth-order term $(k = 0)$, or

$$F(y_t) = e^{-a y_t}.$$

Therefore

$$x_{t+1} = r x_t e^{-a y_t}, \tag{3.15}$$

$$y_{t+1} = c x_t \left(1 - e^{-a y_t}\right), \tag{3.16}$$

which is the classic Nicholson–Bailey model.

As it turns out, the predictions of this model are not borne out by experiments. We will see that there is an unstable equilibrium and that a different tack must be taken. When models are not confirmed observations, the deficiencies can be an extraordinary teaching device and they lead to better assumptions and improvements, as discussed in Section 1.1. Some of the inadequacies are these. It is clear at once that there is no density dependence in the host population—in the absence of parasitoids, they grow geometrically, which is probably not true. There is no interference among the parasitoids. There are no host refuges where some can hide. The parasitoids may not search randomly, as assumed, but rather search areas selectively where they have had success.

The equilibria are found from the equations

$$x = r x e^{-a y}, \quad y = c x \left(1 - e^{-a y}\right),$$

which give two equilibria,

$$x^* = y^* = 0; \quad x^* = \frac{r \ln r}{(r - 1) a c}, \quad y^* = \frac{\ln r}{a}.$$

Note that $r > 1$ must be an implicit assumption to the model. To determine stability we must compute the Jacobian matrix (3.13). This exercise in partial differentiation is left to the reader. We obtain

$$J(x^*, y^*) = \begin{pmatrix} r e^{-a y^*} & -a x^* e^{-a y^*} \\ c(1 - e^{-a y^*}) & a c x^* e^{-a y^*} \end{pmatrix}.$$

At extinction,

$$J(0,0) = \begin{pmatrix} r & 0 \\ 0 & 0 \end{pmatrix},$$

which has eigenvalues $r > 1$ and 0. Hence, the extinct population is unstable. The nonzero equilibrium has Jacobian matrix

$$J(x^*, y^*) = \begin{pmatrix} 1 & -\ln r/(c(r-1)) \\ c(r-1)/r & \ln r/(r-1) \end{pmatrix}.$$

To determine stability we apply the Jury conditions (3.14). The trace and determinant are

$$\operatorname{tr} J = 1 + \frac{\ln r}{r-1}, \quad \det J = \frac{r \ln r}{r-1}.$$

We show that $\det J > 1$, which implies that the equilibrium is unstable. Equivalently, we show that $g(r) = r - 1 - r \ln r < 0$ for $r > 1$. But $g(1) = 0$ and $g'(r) = -\ln r < 0$. Therefore, we must have $g(r) < 0$. $\quad \square$

The results of the Nicholson–Bailey model are perplexing. There are no stable coexistent equilibria, yet in nature hosts and parasitoids endure, co-evolving together. Does this mean that it is an invalid model? To get a sense of the dynamics, we simulate the solution with $x_0 = 25$, $y_0 = 10$, taking $a = 0.068$, $c = 1$, and $r = 2$. Figure 3.4 shows the unstable oscillations that result. There have been explanations that attempt to resolve this issue, including Nicholson and Bailey themselves. They conjectured that the host–parasitoid is naturally unstable, but the large population variations in one region cause dispersal of the species to other areas where the model begins all over. So the model requires spatial variations and inhomogeneities to remain valid. But more to the point, the Nicolson–Bailey model has become the basis of many other models equipped with modifications that attempt to stabilize the dynamics. For example, the **Beddington model**

$$\begin{aligned} x_{t+1} &= e^{b(1-x_t/K)} x_t e^{-ay_t}, \\ y_{t+1} &= c x_t \left(1 - e^{-ay_t}\right), \end{aligned}$$

where the growth rate r is replaced by a host-limiting, density-dependent rate $e^{b(1-x_t/K)}$, has a positive stable equilibrium for a large set of parameter values. Other texts in mathematical ecology investigate some of these models in detail. The monograph by Hassell (1978), and many of his papers, offer tremendous insights into these and other systems.

Further, rather than a search among randomly distributed hosts, as described by the Poisson distribution, the parasitoids may search using a more

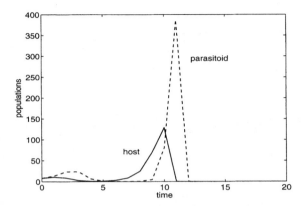

Figure 3.4 Simulations of the Nicholson–Bailey model.

clumped search procedure described by an escapement like the negative binomial distribution,

$$F(x,y) = \left(1 + \frac{ay}{k}\right)^{-k},$$

where k is a "clumping" parameter. (As $k \to \infty$, this distribution approaches the Poisson distribution.)

Example 3.11

(**Epidemics**) Consider a disease in a population of fixed size N, and let S_t, I_t, and R_t be the susceptible, infective, and removed classes at time t, respectively. For definiteness, we assume that t is measured in days. The susceptibles are those who can catch the disease but are currently not infected, the infectives are those who are infected with the disease and are contagious, and the individuals removed have either recovered permanently, are naturally immune, or have died. We have $N = S_t + I_t + R_t$. With homogeneous mixing, the dynamics is given by

$$
\begin{aligned}
S_{t+1} &= S_t - \alpha S_t I_t, \\
I_{t+1} &= I_t + \alpha S_t I_t - \gamma I_t, \\
R_{t+1} &= R_t + \gamma I_t,
\end{aligned}
\tag{3.17}
$$

where the number of individuals in a given class at day $t+1$ equals the number on day t, plus or minus those added or subtracted during that time step. The constant α is the **transmission coefficient**, and γ is the **removal rate**, or the rate at which individuals get over the illness; thus γ^{-1} is the average number of days of duration of the infectious period. This model is the **discrete SIR**

model. To interpret the constants, we can argue as follows. If S_0 and I_0 are the number susceptible and infected initially, $\alpha S_0 I_0$ measures the number that become infected at the outset; thus αS_0 is the number that become infected by contact with a single infected individual during the first day. But this single infected individual has the disease on average γ^{-1} days, and therefore

$$R_0 \equiv \frac{\alpha S_0}{\gamma}$$

is roughly the *number of secondary infections* caused by a single infected individual at time $t = 0$. Epidemiologists refer to the number R_0 as the **basic reproduction number** of the infection. It seems intuitive that the disease will spread if $R_0 > 1$, and it will die out if $R_0 < 1$. To see this analytically, we note from the governing equations that $\Delta S_t \leq 0$, so S_t cannot increase. Hence, $S_0 \geq S_t$ for all t, or $R_0 \geq \alpha S_t / \gamma$ for all t. The infection equation may be written as

$$\Delta I_t = \gamma \left(\frac{\alpha}{\gamma} S_t - 1 \right) I_t.$$

Consequently, if $R_0 > 1$, it follows that $\Delta I_t > 0$ and the number of infectives will grow, causing an epidemic. If $R_0 < 1$, it follows that $\Delta I_t < 0$ and the number of infectives will die out; there will be no epidemic.

Observe that the epidemic peaks when $S_t = \gamma/\alpha$, where ΔI_t changes sign. Simulations of the SIR equations (3.17) are investigated in the Exercises. Both time series plots and orbits in the SI phase plane illustrate the dynamics. Figure 3.5, generated by the accompanying m-file, shows a typical orbit in the phase plane when there is an epidemic. The initial point is on the line $S_0 + I_0 = N$, and the orbit increases up to a maximum, which lies on the vertical line $S_t = \gamma/\alpha$. Then the disease dies out, leaving a number S^* of infectives who never get the disease. □

```
function SIRdiscrete
clear all
N=80; alpha=0.001; gamma=.2; S=500; I=1;
Shist=S; Ihist=2; R=alpha*S/gamma
for j=1:N
u=S-alpha*S.*I; v=I+alpha*S.*I-gamma*I;
S=u; I=v;
Shist=[Shist,S]; Ihist=[Ihist,I];
end
plot(Shist,Ihist)
```

The Exercises show the variety and scope of nonlinear problems in the life sciences.

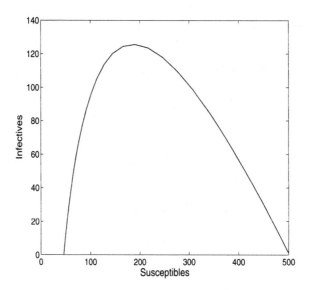

Figure 3.5 An orbit in the SI plane for the discrete SIR model. The data is given in the accompanying m-file SIRdiscrete. Here, one infective is introduced into a susceptible population of 500. We have an epidemic with $R_0 = 2.5$, and about 50 individuals do not get the disease.

EXERCISES

1. Consider the linear model

$$\Delta P_t = P_t - 2Q_t, \quad \Delta Q_t = -P_t.$$

   (a) Find the equilibria.

   (b) Draw the nullclines (where $\Delta P_t = 0$, and where $\Delta Q_t = 0$) and arrows indicating the direction of the changes on the nullclines and in the regions between the nullclines.

   (c) On the line $Q_t = P_t$ in the phase plane, draw arrows indicating how P_t and Q_t are changing.

   (d) Does the equilibrium look stable or unstable, or can you tell?

   (e) Construct a rough phase plane diagram that indicates the orbits.

   (f) Find the equations of the orbits by finding the solution P_t and Q_t in terms of t.

2. In 1998 van der Meijden described the following model connecting the cinnabar moth and ragweed. The life cycle of the moth is as follows. It lives for one year, lays it eggs on the plant, and dies. The eggs hatch the next spring. The number of eggs is proportional to the plant biomass the preceding year, or

$$E_{t+1} = aB_t.$$

The plant biomass the next year depends on the biomass the current year and the number of eggs according to

$$B_{t+1} = ke^{-cE_t/B_t}.$$

(a) Explain why the last equation is a reasonable model (plot biomass vs. eggs and biomass vs. biomass).

(b) What are the dimensions of k, c, and a?

(c) Rewrite the model equations in terms of the "rescaled" variables $X_t = B_t/k$ and $Y_t = E_t/ka$. Call $b = ac$. What are the dimensions of X_t, Y_t, and b?

(d) Find the equilibrium and determine a condition on b for which the equilibrium is stable.

(e) If the system is perturbed from equilibrium, describe how it returns to equilibrium in the stable case.

3. Analyze the following discrete model, which is similar to the continuous time Lotka–Volterra predator–prey model:

$$x_{t+1} = rx_t - x_t y_t, \quad y_{t+1} = -sx_t + x_t y_t,$$

where r and s are positive constants.

4. Examine the nonlinear model

$$\begin{aligned} \Delta x_t &= rx_t \left(1 - \frac{x_t}{K}\right) - sx_t y_t, \\ \Delta y_t &= -dy_t + \epsilon x_t y_t \end{aligned}$$

with respect to equilibria and their stability. Here all the parameters are positive.

5. Consider the nonlinear fisheries model

$$\begin{pmatrix} x_{t+1} \\ y_{t+1} \end{pmatrix} = \begin{pmatrix} ame^{-\beta x_t} & \frac{3}{2}mb \\ ae^{-\beta x_t} & 0 \end{pmatrix} \begin{pmatrix} x_t \\ y_t \end{pmatrix},$$

where a, m, b, $\beta > 0$.

(a) Find the equilibria [use the notation $R = am\,(1 + 3b/2)$]. Find a condition on R which guarantees that the nontrivial equilibrium lies in the first quadrant.

(b) For a positive equilibrium, determine conditions on R, a, and b under which the equilibrium is asymptotically stable.

(c) For a fixed a, plot the stability region in the b, $\ln R$ parameter plane and identify the stability boundary.

6. Suppose that one infected individual in a population of 501 is infected. The likelihood that an individual becomes infected is $\alpha = 0.1\%$, and an infective is contagious for an average of 10 days.

(a) Compute the basic reproduction number R_0 of the disease and determine if an epidemic occurs.

(b) Plot the time series S_t, I_t, and R_t over a 60-day period. At what time does the disease reach a maximum? How many do not get the disease? Over a long time, how many do not get the disease?

(c) Plot the orbit in an SI phase plane.

3.4 Appendix: Matrices

In mathematical biology we use matrices and linear algebra to describe and analyze many of the models that arise. Matrix theory provides a convenient language and notation to express many of the ideas concisely. Complicated formulas are simplified considerably in this framework, and matrix notation is more or less independent of dimension. In this section we present a brief introduction to square matrices. Some definitions and properties are given for general $n \times n$ matrices, but our focus is on two- and three-dimensional cases. Thus, this section is not a thorough treatment of matrix theory, but rather a limited discussion focusing on tools necessary to discuss many topics in mathematical biology.

3.4.1 Matrix Notation and Algebra

A square array A of numbers having n rows and n columns is called a **square matrix** of size n, or an $n \times n$ matrix (we say, "n by n matrix"). The number, or entry, in the ith row and jth column is denoted by a_{ij}. General 2×2 and

3×3 matrices have the form, respectively,

$$A = \begin{pmatrix} a_{11} & a_{12} \\ a_{21} & a_{22} \end{pmatrix}, \quad A = \begin{pmatrix} a_{11} & a_{12} & a_{13} \\ a_{21} & a_{22} & a_{23} \\ a_{31} & a_{32} & a_{33} \end{pmatrix}.$$

The **main diagonal** of a square matrix A is the set of elements $a_{11}, a_{22}, ..., a_{nn}$. We often write matrices using the notation $A = (a_{ij})$. An n-**vector x** is a list of n numbers $x_1, x_2, ..., x_n$, written as a *column*; so "vector" means "column list." The numbers $x_1, x_2, ..., x_n$ in the list are called its **components**. For example,

$$\mathbf{x} = \begin{pmatrix} x_1 \\ x_2 \end{pmatrix}$$

is a 2-vector. Vectors are denoted by lowercase boldface letters (i.e., **x**, **y**, etc.), and matrices are denoted by capital letters l(i.e., A, B, etc.). To minimize space in typesetting, we often write, for example, a 2-vector **x** as $(x_1, x_2)^{\mathrm{T}}$, where the T denotes *transpose*, meaning turn the row into a column.

Two square matrices having the same size can be added entrywise. That is, if $A = (a_{ij})$ and $B = (b_{ij})$ are both $n \times n$ matrices, then the **sum** $A + B$ is an $n \times n$ matrix defined by $A + B = (a_{ij} + b_{ij})$. A square matrix $A = (a_{ij})$ of any size can be multiplied by a constant c by multiplying all the elements of A by the constant; in symbols, this **scalar multiplication** is defined by $cA = (ca_{ij})$. Thus, $-A = (-a_{ij})$, and it is clear that $A + (-A) = 0$, where 0 is the **zero matrix** having all zero entries. If A and B have the same size, then **subtraction** is defined by $A - B = A + (-B)$. Also, $A + 0 = A$ if 0 has the same size as A. Addition, when defined, is both commutative and associative. Therefore the arithmetic rules of addition for $n \times n$ matrices are the same as the usual rules for addition of numbers.

Similar rules hold for addition of column vectors of the same length and multiplication of column vectors by scalars; these are the definitions encountered in multivariable calculus, where n-vectors are regarded as elements of $\mathbf{R}^n$. Vectors add componentwise, and multiplication of a vector by a scalar multiplies each component of that vector by that scalar.

Example 3.12

Let

$$A = \begin{pmatrix} 1 & 2 \\ 3 & -4 \end{pmatrix}, \quad B = \begin{pmatrix} 0 & -2 \\ 7 & -4 \end{pmatrix}, \quad \mathbf{x} = \begin{pmatrix} -4 \\ 6 \end{pmatrix} \quad \mathbf{y} = \begin{pmatrix} 5 \\ 1 \end{pmatrix}.$$

Then

$$A + B = \begin{pmatrix} 1 & 0 \\ 10 & -8 \end{pmatrix}, \quad -3B = \begin{pmatrix} 0 & 6 \\ -21 & 12 \end{pmatrix},$$

$$5\mathbf{x} = \begin{pmatrix} -20 \\ 30 \end{pmatrix}, \quad \mathbf{x} + 2\mathbf{y} = \begin{pmatrix} 6 \\ 8 \end{pmatrix}.$$

The following MATLAB commands can be used to produce the same results. The percent sign precedes a comment, and what follows is not executed.   □

```
A = [1 2; 3 -4];     % create the A matrix
B = [0 -2; 7 -4];    % create the B matrix
x = [-4 6]';         % create the x column vector
y = [5 1]';          % create the y column vector
A+B                  % add the matrices A and B
-3*B                 % multiply the B matrix by the scalar −3
5*x                  % multiply the x vector by 5
x + 2*y              % add x plus 2 times y
```

The product of two square matrices of the same size is *not* found by multiplying entrywise. Rather, **matrix multiplication** is defined as follows. Let $A = (a_{ij})$ and $B = (b_{ij})$ be $n \times n$ matrices. Then the matrix AB is defined as the $n \times n$ matrix $C = (c_{ij})$, where the ij entry of the product C is found by taking the product (dot product, as with vectors) of the ith row of A and the jth column of B. In symbols, $AB = C$, where

$$c_{ij} = \mathbf{a}_i \cdot \mathbf{b}_j = a_{i1}b_{1j} + a_{i2}b_{2j} + \cdots + a_{in}b_{nj},$$

where $\mathbf{a}_i$ denotes the ith row of A, and $\mathbf{b}_j$ denotes the jth column of B. Generally, matrix multiplication is *not* commutative (i.e., $AB \neq AB$); so the order in which matrices are multiplied is important. However, the associative law $AB(C) = (AB)C$ does hold, so products of matrices may be regrouped at will. The distributive law connecting addition and multiplication, $A(B + C) = AB + AC$, also holds. The powers of a square matrix are defined by $A^2 = AA$, $A^3 = AA^2$, and so on.

Example 3.13

Let

$$A = \begin{pmatrix} 2 & 3 \\ -1 & 0 \end{pmatrix}, \quad B = \begin{pmatrix} 1 & 4 \\ 5 & 2 \end{pmatrix}.$$

Then

$$AB = \begin{pmatrix} 2 \cdot 1 + 3 \cdot 5 & 2 \cdot 4 + 3 \cdot 2 \\ -1 \cdot 1 + 0 \cdot 5 & -1 \cdot 4 + 0 \cdot 2 \end{pmatrix} = \begin{pmatrix} 17 & 14 \\ -1 & -4 \end{pmatrix}.$$

Also

$$A^2 = \begin{pmatrix} 2 & 3 \\ -1 & 0 \end{pmatrix} \begin{pmatrix} 2 & 3 \\ -1 & 0 \end{pmatrix}$$

$$= \begin{pmatrix} 2 \cdot 2 + 3 \cdot (-1) & 2 \cdot 3 + 3 \cdot 0 \\ -1 \cdot 2 + 0 \cdot (-1) & -1 \cdot 3 + 0 \cdot 0 \end{pmatrix} = \begin{pmatrix} -1 & 6 \\ -2 & -3 \end{pmatrix}. \quad \square$$

The following MATLAB commands can be used to produce the preceding results.

```
A = [2 3; -1 0];   % create the A matrix
B = [1 4; 5 2];    % create the B matrix
A*B                % multiply the two matrices
A^2                % square the A matrix
```

Next we define multiplication of an $n \times n$ matrix A times an n-vector $\mathbf{x}$. The product $A\mathbf{x}$, with the matrix on the left, is defined to be the n-vector whose ith component is $\mathbf{a}_i \cdot \mathbf{x}$. In other words, the ith element in the list $A\mathbf{x}$ is found by taking the "dot" product of the ith row of A with the vector $\mathbf{x}$. The product $\mathbf{x}A$ is not defined.

Example 3.14

When $n = 2$ we have

$$A\mathbf{x} = \begin{pmatrix} a & b \\ c & d \end{pmatrix} \begin{pmatrix} x \\ y \end{pmatrix} = \begin{pmatrix} ax + by \\ cx + dy \end{pmatrix}.$$

For a numerical example, take

$$A = \begin{pmatrix} 2 & 3 \\ -1 & 0 \end{pmatrix}, \quad \mathbf{x} = \begin{pmatrix} 5 \\ 7 \end{pmatrix}.$$

Then

$$A\mathbf{x} = \begin{pmatrix} 2 \cdot 5 + 3 \cdot 7 \\ -1 \cdot 5 + 0 \cdot 7 \end{pmatrix} = \begin{pmatrix} 31 \\ -5 \end{pmatrix}. \quad \square$$

To carry out these calculations in MATLAB we use the following commands:

```
A = [2 3; -1 0];   % create the A matrix
x = [5 7]';        % create the x column vector
A*x                % compute the product Ax
```

The special square matrix that has ones on the main diagonal and zeros elsewhere else is called the **identity matrix** and is denoted by I. For example, the 2×2 and 3×3 identities are

$$I = \begin{pmatrix} 1 & 0 \\ 0 & 1 \end{pmatrix} \quad \text{and} \quad I = \begin{pmatrix} 1 & 0 & 0 \\ 0 & 1 & 0 \\ 0 & 0 & 1 \end{pmatrix}.$$

It is easy to show that if A is any square matrix and I is the identity matrix of the same size, then $AI = IA = A$. Therefore multiplication by the identity matrix does not change the result, a situation similar to multiplying real numbers by the unit number 1. If A is an $n \times n$ matrix and there exists a matrix B for which $AB = BA = I$, then B is called the **inverse** of A and we denote it by $B = A^{-1}$. If A^{-1} exists, we say that A is a **nonsingular** matrix; otherwise, it is called **singular**. One can show that the inverse of a matrix, if it exists, is unique. We never write $1/A$ for the inverse of A.

A useful number associated with a square matrix A is its determinant. The **determinant** of a square matrix A, denoted by $\det A$ (also by $|A|$) is a number found by combining the elements of the matrix in a special way. The determinant of a 1×1 matrix is just the single number in the matrix. For a 2×2 matrix we define

$$\det A = \det \begin{pmatrix} a & b \\ c & d \end{pmatrix} = ad - cb,$$

and for a 3×3 matrix we define

$$\det \begin{pmatrix} a & b & c \\ d & e & f \\ g & h & i \end{pmatrix} = aei + bfg + cdh - cef - bdi - ahf. \tag{3.18}$$

Example 3.15

We have

$$\det \begin{pmatrix} 2 & 6 \\ -2 & 0 \end{pmatrix} = 2 \cdot 0 - (-2) \cdot 6 = 12.$$

The MATLAB command det(A) produces the determinant. □

Although there is a general inductive formula that defines the determinant of an $n \times n$ matrix as a sum of $(n-1) \times (n-1)$ matrices, we strongly recommend using MATLAB or a scientific calculator for this task. A more thorough discussion of computing the determinant for any $n \times n$ matrix can be found in any elementary linear algebra text (see the Reference Notes).

Using the determinant, we can give a simple formula for the *inverse* of a 2×2 matrix A. Let

$$A = \begin{pmatrix} a & b \\ c & d \end{pmatrix}$$

and suppose that $\det A \neq 0$. Then

$$A^{-1} = \frac{1}{\det A} \begin{pmatrix} d & -b \\ -c & a \end{pmatrix}. \tag{3.19}$$

So the inverse of a 2×2 matrix is found by interchanging the main diagonal elements, putting a minus sign on the off-diagonal elements, and dividing by the determinant. There is a similar formula for the inverse of larger matrices, but prefer to use MATLAB or a scientific calculator for this task.

Example 3.16

If

$$A = \begin{pmatrix} 1 & 2 \\ 4 & 3 \end{pmatrix},$$

then

$$A^{-1} = \frac{1}{\det A} \begin{pmatrix} 3 & -2 \\ -4 & 1 \end{pmatrix} = \frac{1}{-5} \begin{pmatrix} 3 & -2 \\ -4 & 1 \end{pmatrix} = \begin{pmatrix} -\frac{3}{5} & \frac{2}{5} \\ \frac{4}{5} & -\frac{1}{5} \end{pmatrix}.$$

The reader can easily check that $AA^{-1} = I$. In MATLAB, we compute the inverse of the matrix A by

```
A = [1 2; 4 3];   % create the matrix A
B = inv(A)        % B contains is the inverse of the matrix A   □
```

Equation (3.19) is revealing because it seems to indicate that the inverse matrix exists only when the determinant is nonzero (you can't divide by zero). In fact, these two statements are equivalent for any square matrix, regardless of its size: A^{-1} exists if, and only if, $\det A \neq 0$. This is a major theoretical result in matrix theory, and it is a convenient test for invertibility of small matrices. Again, for larger matrices it is more efficient to use mathematical software or a scientific calculator to calculate determinants and inverses. The reader should remember the equivalences in the following important theorem.

Theorem 3.17

Let A be an $n \times n$ matrix. Then

$$A^{-1} \text{exists} \Leftrightarrow A \text{ is nonsingular} \Leftrightarrow \det A \neq 0. \quad \square$$

3.4.2 Linear Algebraic Equations

Matrices were developed to represent and study linear algebraic systems (n linear algebraic equations in n unknowns) in a concise, economical way. For example, consider two equations in two unknowns x_1, x_2, given in standard form by

$$a_{11}x_1 + a_{12}x_2 = b_1$$
$$a_{21}x_1 + a_{22}x_2 = b_2$$

Using matrix notation, we can write this as

$$\begin{pmatrix} a_{11} & a_{12} \\ a_{21} & a_{22} \end{pmatrix} \begin{pmatrix} x_1 \\ x_2 \end{pmatrix} = \begin{pmatrix} b_1 \\ b_2 \end{pmatrix},$$

or simply as

$$A\mathbf{x} = \mathbf{b}, \tag{3.20}$$

where

$$A = \begin{pmatrix} a_{11} & a_{12} \\ a_{21} & a_{22} \end{pmatrix}, \quad \mathbf{x} = \begin{pmatrix} x_1 \\ x_2 \end{pmatrix}, \quad \mathbf{b} = \begin{pmatrix} b_1 \\ b_2 \end{pmatrix}.$$

A is the **coefficient matrix**, $\mathbf{x}$ is a column vector of unknowns, and $\mathbf{b}$ is a column vector representing the right side. If $\mathbf{b} = \mathbf{0}$, the zero vector, then the system (3.20) is called **homogeneous**. In a two-dimensional system each equation represents a line in the plane. When $\mathbf{b} = \mathbf{0}$ the two lines pass through the origin. A solution vector $\mathbf{x}$ is represented by a point that lies on both lines. There is a unique solution when both lines intersect at a single point; there are infinitely many solutions when both lines coincide; there is no solution if the lines are parallel and different. In the case of three equations in three unknowns, each equation in the system has the form $\alpha x_1 + \beta x_2 + \gamma x_3 = d$ and represents a plane in three-dimensional space. If $d = 0$, the plane passes through the origin. The three planes represented by the three equations can intersect in many ways, giving no solution (no common intersection points), a unique solution (when they intersect at a single point), a line of solutions (when they intersect in a common line), and a plane of solutions (when all the equations represent the same plane).

The following key theorem tells us when a linear system $A\mathbf{x} = \mathbf{b}$ of n equations in n unknowns is solvable.

Theorem 3.18

Let A be a $n \times n$ matrix. If A is nonsingular, the system $A\mathbf{x} = \mathbf{b}$ has a unique solution given by $\mathbf{x} = A^{-1}\mathbf{b}$; in particular, the homogeneous system $A\mathbf{x} = \mathbf{0}$

has only the trivial solution $\mathbf{x} = \mathbf{0}$. If A is singular, then the homogeneous system $A\mathbf{x} = \mathbf{0}$ has infinitely many solutions, and the nonhomogeneous system $A\mathbf{x} = \mathbf{b}$ may have no solution or infinitely many solutions. $\square$

Example 3.19

Consider the homogeneous linear system

$$\begin{pmatrix} 4 & 1 \\ 8 & 2 \end{pmatrix} \begin{pmatrix} x_1 \\ x_2 \end{pmatrix} = \begin{pmatrix} 0 \\ 0 \end{pmatrix}.$$

The coefficient matrix has determinant zero, so there will be infinitely many solutions. The two equations represented by the system are

$$4x_1 + x_2 = 0, \quad 8x_1 + 2x_2 = 0,$$

which are clearly not independent; one is a multiple of the other. Therefore we need only consider one of the equations, say $4x_1 + x_2 = 0$. With one equation in two unknowns we are free to pick a value for one of the variables and solve for the other. Let $x_1 = 1$; then $x_2 = -4$ and we get a single solution $\mathbf{x} = (1, -4)^{\mathrm{T}}$. More generally, if we choose $x_1 = \alpha$, where α is any real parameter, $x_2 = -4\alpha$. Therefore, all solutions are given by

$$\mathbf{x} = \begin{pmatrix} x_1 \\ x_2 \end{pmatrix} = \begin{pmatrix} \alpha \\ -4\alpha \end{pmatrix} = \alpha \begin{pmatrix} 1 \\ -4 \end{pmatrix}, \quad a \in \mathbf{R}.$$

Thus, all solutions are multiples of $(1, -4)^{\mathrm{T}}$, and the solution set lies along the straight line through the origin defined by this vector. Geometrically, the two equations represent two lines in the plane that coincide. $\square$

The **row reduction method** for solving linear systems is a method where we combine the equations in a systematic way to produce a simpler, equivalent system. We illustrate the method when $n = 3$. Consider the algebraic system $A\mathbf{x} = \mathbf{b}$, or

$$\begin{aligned} a_{11}x_1 + a_{12}x_2 + a_{13}x_3 &= b_1, \\ a_{21}x_1 + a_{22}x_2 + a_{23}x_3 &= b_2, \\ a_{31}x_1 + a_{32}x_2 + a_{33}x_3 &= b_3. \end{aligned} \tag{3.21}$$

At first we assume that the coefficient matrix $A = (a_{ij})$ is nonsingular, so that the system has a unique solution. The basic idea is to transform the system into the simpler *triangular form*

$$\begin{aligned} \tilde{a}_{11}x_1 + \tilde{a}_{12}x_2 + \tilde{a}_{13}x_3 &= \tilde{b}_1, \\ \tilde{a}_{22}x_2 + \tilde{a}_{23}x_3 &= \tilde{b}_2, \\ \tilde{a}_{33}x_3 &= \tilde{b}_3. \end{aligned}$$

This triangular system is easily solved by back substitution. That is, the third equation involves only one unknown and we can instantly find x_3. That value is substituted back into the second equation where we can then find x_2, and those two values are substituted back into the first equation and we can find x_1. The process of transforming (3.21) into triangular form is carried out by three admissible operations which do not affect the solution structure:

1. Any equation may be multiplied by a nonzero constant.

2. Any two equations may be interchanged.

3. Any equation may be replaced by that equation plus (or minus) a multiple of any other equation.

We observe that any equation in the system (3.21) is represented by its coefficients and the right side, so we only need work with the numbers, which saves writing. We organize the numbers in an **augmented array**,

$$\begin{pmatrix} a_{11} & a_{12} & a_{13} & b_1 \\ a_{21} & a_{22} & a_{23} & b_2 \\ a_{31} & a_{32} & a_{33} & b_3 \end{pmatrix}.$$

The admissible operations listed above translate into row operations on the augmented array: Any row may be multiplied by a nonzero constant, any two rows may be interchanged, and any row may be replaced by itself plus (or minus) any other row. By performing these row operations, we transform the augmented array into a triangular array with zeros in the lower left corner below the main diagonal. The process is carried out one column at a time, beginning from the left.

Example 3.20

Consider the system

$$\begin{aligned} x_1 + x_2 + x_3 &= 0, \\ 2x_1 - 2x_3 &= 2, \\ x_1 - x_2 + x_3 &= 6. \end{aligned} \qquad (3.22)$$

The augmented array is

$$\begin{pmatrix} 1 & 1 & 1 & 0 \\ 2 & 0 & -2 & 2 \\ 1 & -1 & 1 & 6 \end{pmatrix}.$$

Begin working on the first column to get zeros in the 2,1 and 3,1 positions by replacing the second and third rows by themselves plus multiples of the first row. So we replace the second row by the second row minus twice the first row and replace the third row by third row minus the first row. This gives

$$\begin{pmatrix} 1 & 1 & 1 & 0 \\ 0 & -2 & -4 & 2 \\ 0 & -2 & 0 & 6 \end{pmatrix}.$$

Next, work on the second column to get a zero in the 3,2 position, below the diagonal entry. Specifically, replace the third row by the third row minus the second row:

$$\begin{pmatrix} 1 & 1 & 1 & 0 \\ 0 & -2 & -4 & 2 \\ 0 & 0 & 4 & 4 \end{pmatrix}.$$

This is triangular, as desired. To make the arithmetic easier, multiply the third row by $1/4$ and the second row by $-1/2$, to get

$$\begin{pmatrix} 1 & 1 & 1 & 0 \\ 0 & 1 & 2 & -1 \\ 0 & 0 & 1 & 1 \end{pmatrix},$$

with 1's on the diagonal. This triangular, augmented array represents the system

$$\begin{aligned} x_1 + x_2 + x_3 &= 0, \\ x_2 + 2x_3 &= -1, \\ x_3 &= 1. \end{aligned}$$

Therefore, using back substitution, $x_3 = 1$, $x_2 = -3$, and $x_1 = 2$, which is the unique solution, representing a point $(2, -3, 1)$ in $\mathbf{R}^3$. When A is nonsingular, the following MATLAB commands can be used to solve the system. □

```
A = [1 1 1;2 0 -2;1 -1 1];   % create the coefficient matrix A
b = [0 2 6]';                % create the column vector b
x = A \ b                    % solve the system Ax = b
```

If the coefficient matrix A is singular, we can end up with different types of triangular forms: for example,

$$\begin{pmatrix} 1 & * & * & * \\ 0 & 1 & * & * \\ 0 & 0 & 0 & * \end{pmatrix}, \quad \begin{pmatrix} 1 & * & * & * \\ 0 & 0 & * & * \\ 0 & 0 & 0 & * \end{pmatrix}, \quad \text{or} \quad \begin{pmatrix} 1 & * & * & * \\ 0 & 0 & 0 & * \\ 0 & 0 & 0 & * \end{pmatrix},$$

where the $*$ denotes an entry. These augmented arrays can be translated back into equations. Depending on the values of those entries, we will get no solution (the equations are inconsistent) or infinitely many solutions. For example, suppose that there are three systems with triangular forms at the end of the process given by

$$
\begin{pmatrix} 1 & 1 & 3 & 0 \\ 0 & 1 & 2 & 5 \\ 0 & 0 & 0 & 7 \end{pmatrix}, \quad \begin{pmatrix} 1 & 0 & 3 & 3 \\ 0 & 0 & 1 & 1 \\ 0 & 0 & 0 & 0 \end{pmatrix}, \quad \text{or} \quad \begin{pmatrix} 1 & 2 & 0 & 1 \\ 0 & 0 & 0 & 0 \\ 0 & 0 & 0 & 0 \end{pmatrix}.
$$

There would be no solution for the first system (the last row states that $0 = 7$), and infinitely many solutions for the second and third systems. Specifically, the second system would have solution $x_3 = 1$ and $x_1 = 0$, with $x_2 = a$, which is arbitrary. Therefore the solution to the second system can be written

$$
\begin{pmatrix} x_1 \\ x_2 \\ x_3 \end{pmatrix} = \begin{pmatrix} 0 \\ a \\ 1 \end{pmatrix} = a \begin{pmatrix} 0 \\ 1 \\ 0 \end{pmatrix} + \begin{pmatrix} 0 \\ 0 \\ 1 \end{pmatrix},
$$

with a an arbitrary constant. This represents a line in $\mathbf{R}^3$. A line is a one-dimensional geometrical object described in terms of one parameter. The third system above reduced to $x_1 + 2x_2 = 1$. So we may pick x_3 and x_2 arbitrarily, say $x_2 = a$ and $x_3 = b$, and then $x_1 = 1 - 2a$. The solution to the third system can then be written

$$
\begin{pmatrix} x_1 \\ x_2 \\ x_3 \end{pmatrix} = \begin{pmatrix} 1 - 2a \\ a \\ b \end{pmatrix} = a \begin{pmatrix} -2 \\ 1 \\ 0 \end{pmatrix} + b \begin{pmatrix} 0 \\ 0 \\ 1 \end{pmatrix} + \begin{pmatrix} 1 \\ 0 \\ 0 \end{pmatrix},
$$

which is a plane in $\mathbf{R}^3$. A plane is a two-dimensional object in $\mathbf{R}^3$ requiring two parameters for its description.

MATLAB can also be used to row-reduce the augmented matrix. This can be particularly useful when the coefficient matrix is singular.

Example 3.21

Consider the system

$$
\begin{aligned}
x_1 + 3x_2 + 7x_3 &= 5, \\
-x_1 + 4x_2 + 4x_3 &= 2, \\
x_1 + 10x_2 + 18x_3 &= 12.
\end{aligned}
$$

The augmented matrix is

$$
\begin{pmatrix} 1 & 3 & 7 & 5 \\ -1 & 4 & 4 & 2 \\ 1 & 10 & 18 & 12 \end{pmatrix}.
$$

The MATLAB commands below row-reduce the augmented matrix to give

$$\begin{pmatrix} 1.000 & 0 & 2.2857 & 2.000 \\ 0 & 1.000 & 1.5714 & 1.000 \\ 0 & 0 & 0 & 0 \end{pmatrix}. \ \square$$

```
A = [1 3 7;-1 4 4;1 10 18];   % coefficient matrix A
b = [5 2 12]'                 % column vector b of the right side
C = [A b];                    % augmented matrix C
rref(C)                       % row-reduced augmented matrix
```

The set of all solutions to a homogeneous system $A\mathbf{x} = \mathbf{0}$ is called the **nullspace** of A. The nullspace may consist of a single point $\mathbf{x} = \mathbf{0}$ when A is nonsingular, or it may be a line or plane passing through the origin when A is singular.

Finally, we introduce the notion of independence of column vectors. A set of vectors is said to be a linearly independent set if any one of them cannot be written as a combination of the others. We can express this statement mathematically as follows. A set (p of them) of n-vectors $\mathbf{v}_1, \mathbf{v}_2, ..., \mathbf{v}_p$ is a **linearly independent set** if the equation[2]

$$c_1\mathbf{v}_1 + c_2\mathbf{v}_2 + \cdots + c_p\mathbf{v}_p = \mathbf{0}$$

forces all the constants to be zero (i.e., $c_1 = c_2 = \cdots = c_p = 0$). If all the constants are not forced to be zero, we say that the set of vectors is **linearly dependent**. In this case there would be at least one of the constants, say c_r, which is not zero, at which point we could solve for $\mathbf{v}_r$ in terms of the remaining vectors.

3.4.3 The Eigenvalue Problem

Of particular interest in linear models is the **eigenvalue problem**. This problem consists of determining those values of the constant λ for which nontrivial solutions $\mathbf{x}$ exist to

$$A\mathbf{x} = \lambda\mathbf{x}. \tag{3.23}$$

The values of λ for which nontrivial solutions exist are called the **eigenvalues** of A, and the corresponding vector solutions are known as the **eigenvectors**. The pair λ, $\mathbf{x}$ is called an **eigenpair**. Geometrically, we think of the eigenvalue problem in $\mathbb{R}^2$ like this: A represents a transformation that maps vectors in

---

[2] A sum of constant multiples of a set of vectors is called a **linear combination** of those vectors.

the plane to vectors in the plane; a vector $\mathbf{x}$ gets transformed to a vector $A\mathbf{x}$. An eigenvector of A is a special vector that is mapped to a multiple (λ) of itself (i.e., $A\mathbf{x} = \lambda\mathbf{x}$). The eigenvectors define subspaces (e.g., a line) that are invariant under the transformation A; that is, a vector in the invariant space remains in that space after the action of A.

To solve the eigenvalue problem, we rewrite (3.23) as a homogeneous linear system

$$(A - \lambda I)\mathbf{x} = \mathbf{0}. \tag{3.24}$$

This system will have the desired nontrivial solutions if the coefficient matrix is singular: that is, if the determinant of the coefficient matrix is zero, or

$$\det(A - \lambda I) = 0. \tag{3.25}$$

When A is the 2×2 matrix

$$\begin{pmatrix} a & b \\ c & d \end{pmatrix},$$

this system (3.24) has the form

$$\begin{pmatrix} a - \lambda & b \\ c & d - \lambda \end{pmatrix} \begin{pmatrix} x_1 \\ x_2 \end{pmatrix} = \begin{pmatrix} 0 \\ 0 \end{pmatrix},$$

where the coefficient matrix $A - \lambda I$ is the matrix A with λ subtracted from the diagonal elements. Equation (3.25) is, explicitly,

$$\det \begin{pmatrix} a - \lambda & b \\ c & d - \lambda \end{pmatrix} = (a - \lambda)(d - \lambda) - cb = 0,$$

or equivalently,

$$\lambda^2 - (a + b)\lambda + (ad - bc) = 0.$$

This last equation can be memorized easily if it is written

$$\lambda^2 - (\operatorname{tr} A)\lambda + \det A = 0, \tag{3.26}$$

where $\operatorname{tr} A = a + d$ is called the **trace** of A, defined to be the sum of the diagonal elements of A. Equation (3.26) is called the **characteristic equation** associated with A, and it is a quadratic equation in λ. Its roots, found by factoring or from the quadratic formula, are the two eigenvalues. The eigenvalues may be real and unequal, real and equal, or complex conjugates; if the eigenvalues are complex, the two corresponding eigenvectors are also complex conjugates. A single real eigenvalue for a 2×2 matrix may have two associated independent eigenvectors.

Once the eigenvalues are computed, we can substitute them in turn into the system (3.24) to determine corresponding eigenvectors $\mathbf{x}$. Note that any multiple of an eigenvector is again an eigenvector for that same eigenvalue; this follows from the calculation

$$A(c\mathbf{x}) = cA\mathbf{x} = c(\lambda\mathbf{x}) = \lambda(c\mathbf{x}).$$

Thus, an eigenvector corresponding to a given eigenvalue is not unique; we may multiply them by constants.

Example 3.22

Consider the 2×2 matrix

$$A = \begin{pmatrix} 13 & -4 \\ -4 & 7 \end{pmatrix}.$$

The characteristic equation is

$$\lambda^2 - 20\lambda + 75 = 0,$$

which has roots (the eigenvalues) $\lambda_1 = 5$, $\lambda_2 = 15$. The eigenvalues are found by solving $A\mathbf{x} - \lambda I = 0$. For $\lambda_1 = 5$ we find that $\mathbf{x} = c(1, 2)$, where c is an arbitrary scalar. For $\lambda_2 = 15$ we find that $\mathbf{x} = c(-2, 1)$; again, c an arbitrary scalar. □

Example 3.23

Consider the 3×3 matrix

$$A = \begin{pmatrix} 1 & 3 & -3 \\ -3 & 7 & -3 \\ -6 & 6 & -2 \end{pmatrix}. \tag{3.27}$$

The MATLAB commands

```
A = [1 3 -3;-3 7 -3;-6 6 -2];   % Create matrix A
d = eig(A)                       % find the eigenvalues of A
```

yields $\mathbf{d} = (4, -2, 4)$, which is a vector containing the eigenvalues of A. In this case the eigenvalue $\lambda_1 = 4$ is a repeated eigenvalue of multiplicity 2. To find the eigenvectors, we use the additional MATLAB command [V D] = eig(A) , which returns two matrices, V and D. In the example below, the eigenvalue 4 has two independent associated eigenvectors; but an eigenvalue with multiplicity 2 can have only one associated eigenvalue,. in which case we say that the matrix is deficient. The diagonal elements of D are the eigenvalues of A, and the columns of the matrix V are the corresponding eigenvectors of A. In this case, MATLAB returns

$$V = \begin{pmatrix} 0.4082 & -0.4082 & -0.2774 \\ -0.4082 & -0.4082 & -0.8037 \\ -0.8165 & -0.8165 & -0.5264 \end{pmatrix},$$

and

$$D = \begin{pmatrix} 4 & 0 & 0 \\ 0 & -2 & 0 \\ 0 & 0 & 4 \end{pmatrix}. \quad \square$$

An $n \times n$ matrix of the form

$$D = \begin{pmatrix} a_{11} & 0 & 0 & 0 & 0 \\ 0 & a_{22} & 0 & 0 & 0 \\ 0 & 0 & . & 0 & 0 \\ 0 & 0 & 0 & . & 0 \\ 0 & 0 & 0 & 0 & a_{nn} \end{pmatrix}$$

is called a **diagonal matrix**. If a matrix A has n linearly independent eigenvectors, we can write A in the special form

$$A = VDV^{-1},$$

where D is a diagonal matrix containing the eigenvalues of A on the diagonal, and V is a matrix whose columns are eigenvectors of A. The process of finding the matrices V and D is called the **diagonalization** of a matrix.

Example 3.24

Referring back to Example 3.23, we see that $A = VDV^{-1}$. $\quad \square$

In many instances we need to compute powers of a matrix. The diagonalization process coupled with the following theorem allow us easily to find powers of a matrix.

Theorem 3.25

Suppose that A is diagonalizable, say $A = VDV^{-1}$, where D is a diagonal matrix. Then $A^k = VD^kV^{-1}$. $\quad \square$

If D is a diagonal matrix with entries λ_i on the diagonal, its power D^k is a diagonal matrix whose diagonal elements are the powers λ_i^k of the diagonal elements of D.

Example 3.26

(**Differential equations**) A system of two linear differential equations

$$x' = ax + by, \quad y' = cx + dy$$

can be written in matrix form as

$$\mathbf{x}' = A\mathbf{x},$$

where $\mathbf{x} = (x, y)^{\mathrm{T}}$, $\mathbf{x}' = (x', y')^{\mathrm{T}}$, and the coefficient matrix is

$$A = \begin{pmatrix} a & b \\ c & d \end{pmatrix}.$$

If we assume a solution of the form $\mathbf{x} = \mathbf{v}e^{\lambda t}$, where $\mathbf{v}$ is an unknown vector and λ is an unknown scalar, then after canceling the nonzero factor $\exp(\lambda t)$ from both sides, $\mathbf{x}' = A\mathbf{x}$ becomes $\lambda \mathbf{v} = A\mathbf{v}$. This is the eigenvalue problem for the matrix A. If we assume that the eigenpairs are $\lambda_1, \mathbf{v}_1$ and $\lambda_2, \mathbf{v}_2$, and that the eigenvalues are real and unequal, then we obtain two independent solutions to the system of differential equations:

$$\mathbf{v}_1 e^{\lambda_1 t}, \quad \mathbf{v}_2 e^{\lambda_2 t}.$$

It can be shown that the general solution, that is, all solutions of the system is a linear combination of those two. Hence,

$$\mathbf{x}(t) = c_1 \mathbf{v}_1 e^{\lambda_1 t} + c_2 \mathbf{v}_2 e^{\lambda_2 t},$$

where c_1 and c_2 are arbitrary constants. In Chapter 4 we show how the eigenstructure of A determines the general solution of a linear system in all cases. □

EXERCISES

1. With

$$A = \begin{pmatrix} 1 & 3 \\ 2 & 4 \end{pmatrix}$$

and $\mathbf{b} = (2, 1)^{\mathrm{T}}$, solve the system $A\mathbf{x} = \mathbf{b}$ using A^{-1}. Then solve the system by row reduction.

2. Let

$$A = \begin{pmatrix} 0 & 2 & -1 \\ 1 & 6 & -2 \\ 2 & 0 & 3 \end{pmatrix}, \quad B = \begin{pmatrix} 1 & -1 & 0 \\ 2 & 1 & 4 \\ -1 & -1 & 1 \end{pmatrix}, \quad \mathbf{x} = \begin{pmatrix} 2 \\ 0 \\ -1 \end{pmatrix}.$$

Find $A + B$, $B - 4A$, BA, A^2, $B\mathbf{x}$, $\det A$, AI, $A - 3I$, and $\det(B - I)$.

3. Find all values of the parameter λ that satisfy the equation $\det(A-\lambda I) = 0$, where A is given as in Exercise 1.

4. Let
$$A = \begin{pmatrix} 2 & -1 \\ -4 & 2 \end{pmatrix}.$$

Compute $\det A$. Does A^{-1} exist? Find all solutions to $A\mathbf{x} = \mathbf{0}$ and plot the solution set in the plane.

5. Use the row reduction method to determine all values m for which the algebraic system
$$2x + 3y = m, \quad -6x - 9y = 5,$$

has no solution, a unique solution, or infinitely many solutions.

6. Use row reduction to determine the value(s) of m for which the following system has infinitely many solutions:

$$\begin{aligned} x + y &= 0, \\ 2x + y &= 0, \\ 3x + 2y + mz &= 0. \end{aligned}$$

7. Construct simple homogeneous systems $A\mathbf{x} = \mathbf{0}$ of three equations in three unknowns that have (a) a unique solution; (b) an infinitude of solutions lying on a line in $\mathbb{R}^3$; (c) an infinitude of solutions lying on a plane in $\mathbb{R}^3$. Is there a case in which there is no solution?

8. Find the inverse of A and use it to solve $A\mathbf{x} = \mathbf{b}$, where $\mathbf{b} = (1, 0, 4)^{\mathrm{T}}$ and

$$A = \begin{pmatrix} 1 & 0 & 3 \\ -2 & 1 & 4 \\ 0 & -2 & -2 \end{pmatrix}.$$

9. Solve $A\mathbf{x} = \mathbf{b}$ in part (b) using row reduction.

10. Find all solutions to the homogeneous system $A\mathbf{x} = \mathbf{0}$ if

$$A = \begin{pmatrix} -2 & 0 & 2 \\ 2 & -4 & 0 \\ 0 & 4 & -2 \end{pmatrix}.$$

11. Use the definition of linear independence to show that the 2-vectors $(2, -3)^{\mathrm{T}}$ and $(-4, 8)^{\mathrm{T}}$ are linearly independent.

12. Use the definition to show that the 3-vectors $(0, 1, 0)^{\mathrm{T}}$, $(1, 2, 0)^{\mathrm{T}}$, and $(0, 1, 4)^{\mathrm{T}}$ are linearly independent.

13. Use the definition to show that the 3-vectors $(1, 0, 1)^{\mathrm{T}}$, $(5, -1, 0)^{\mathrm{T}}$, and $(-7, 1, -2)^{\mathrm{T}}$ are linearly dependent.

14. Find the eigenvalues and eigenvectors of the following matrices:

$$A = \begin{pmatrix} -1 & 4 \\ -2 & 5 \end{pmatrix}; \quad B = \begin{pmatrix} 2 & 3 \\ 4 & 6 \end{pmatrix}; \quad C = \begin{pmatrix} 2 & -8 \\ 1 & -2 \end{pmatrix}.$$

15. For the matrices in Exercise 14, perform matrix diagonalization where appropriate. That is, find V such that $A = VDV^{-1}$, where D is a diagonal matrix.

16. Write the general solution of the linear system $\mathbf{x}' = A\mathbf{x}$ if A has eigenpairs 2, $(1, 5)^{\mathrm{T}}$ and -3, $(2, -4)^{\mathrm{T}}$.

3.5 Reference Notes

Because matrix theory is one of the key courses in the undergraduate curriculum in many areas, there is an industry of texts on linear algebra and matrix theory. Our list consists of a few elementary books we have found useful and easy to read. Moreover, many texts on mathematical biology have short sections or appendices on matrices, and even some of the inexpensive, brief, outline series on linear algebra present enough information to proceed. We have also cited Caswell (2001), which is considered the definitive treatment of the role of structure and matrices in biology; this is an advanced book. Cushing (1998) has a thorough discussion of the theory of structured models. To get a real sense of the underlying theoretical ideas in linear algebra, we strongly recommend the classic book by Halmos (1987) or the treatment by Axler (2004). Trefethen & Bau (1997) may be consulted for numerical methods.

References

Axler, S. 1997. *Linear Algebra Done Right*, 2nd ed., Springer-Verlag, New York.

Caswell, H. 2001. *Matrix Population Models*, 2nd ed., Sinauer Associates, Sunderland, MA.

Cushing, J. M. 1998. *An Introduction to Structured Population Dynamics*, SIAM, Philadelphia.

Halmos, P. R. 1987. *Finite-Dimensional Vector Spaces*, Springer-Verlag, New York (reprint of the 1958 version published by Litton Educational Publishing, Inc.).

Hassell, M. P. 1978. *The Dynamics of Arthropod Predator–Prey Systems*, Princeton University Press, Princeton, NJ.

Kwak, J. H. & Hong, S. 2004. *Linear Algebra*, 2nd ed., Birkhauser, Boston.

Shores, T. S. 2007. *Linear Algebra and Matrix Theory*, Springer-Verlag, New York.

Trefethen, L. N. & Bau, D. 1997. *Numerical Linear Algebra*, SIAM, Philadelphia.

4

Interactions in Continuous Time

So far we have examined discrete-time and continuous-time dynamical problems in which there was a single unknown state, which could be the population of an animal or plant species, a chemical concentration, temperature, or any other quantity that changes with time. Next, we extended the analysis to discrete time systems, both linear and nonlinear, where it takes two or more states to describe the configuration of the system. Now we move to continuous-time systems governed by sets of coupled differential equations with several unknown states. The focus is primarily on two-dimensional systems with two unknowns: for example, predator–prey models or any consumer–resource model. The governing differential equations describe the interactions between the constituents. For example, the equations tell us the dynamics of births and deaths of each species and define in detail the nature and dynamics of the interactions.

4.1 Interacting Populations

One of the classical problems in mathematical ecology is the Lotka–Volterra predator–prey system:

$$N' = rN - cNP, \tag{4.1}$$

$$P' = bNP - mP, \tag{4.2}$$

where $N = N(t)$ and $P = P(t)$ are the number (or density) of prey and predators, respectively, and r, c, b, and m are positive constants. Historically, this

Mathematical Methods in Biology,
By J. David Logan and William R. Wolesensky
Copyright © 2009 John Wiley & Sons, Inc.

problem often defines the beginning of the modern era of mathematical analysis in ecology. In the mid-1920s A. Lotka, a physical chemist, and V. Volterra, a mathematical physicist, wrote this model to describe simple dynamics between a predator and a prey. Originally, it came from Volterra's analysis of a question posed to him regarding the cause of the increase in selachians (predatory sharks) caught in fish nets in the northern Mediterranean Sea during World War I. Desspite its historical interest, it turns out that the Lotka–Volterra system is structurally unstable, meaning that a small change in the model can produce drastically different results; for this reason, it is considered limited as an accurate mathematical model. In the first equation, the first term on the right side implies that in the absence of predators, the prey population will grow exponentially with growth rate r. The second term, $-cNP$, is the rate at which prey are lost due to predation. The first term in the second equation, bNP, is the rate at which predation produces more predators, and the $-mP$ term implies that predators decrease exponentially in the absence of prey; m is the per capita mortality rate. The interaction term, or predation term, $-cNP$, is proportional to the product of the number of prey and the number of predators. This assumption makes sense because if there are N prey and P predators, the total number of possible interactions is NP; the constant c gives the fraction of those interactions that actually result in a kill. If there are actually cNP prey harvested, the number of predators produced is $\varepsilon(cNP)$, where ε is the conversion efficiency of prey biomass into predators. In our equations, we have defined $b = \varepsilon c$. Many will recognize this interaction term as the **law of mass action** in chemistry, which states that the *rate* of the chemical reaction $A + B \to$ products is proportional to the product $[A][B]$ of the concentrations of the reactants entering the reaction. The reader should realize that the assumptions underlying the Lotka–Volterra model are actually unrealistic; the only factor that governed the dynamics is the interaction between the predator and the prey. For example, there is no interference among the predators, there is no time required to handle a kill, and there are no alternative prey when the predator is exposed to low prey densities. Also there is no structure in the system—all-sized predators consume all-sized prey, again unrealistic. There are many other deficiencies.

Nevertheless, the model of Lotka and Volterra is considered to have ushered in the modern age of mathematical modeling and theoretical ecology. Although there were earlier mathematical models (e.g., the logistic model of Verhulst or the laws of genetics of Mendel), the ideas following the work of Lotka and Volterra became attempts to understand the general qualitative features of ecological systems rather than focusing on making only quantitative predictions.

We note that there are four constants in (4.1)–(4.2). It is a good exercise for the reader to determine the dimensions of these constants. From Chapter 1, we recall that the number of independent constants in a model can often be reduced by nondimensionalizizing the problem. Therefore, let us replace N and P by the dimensionless populations x and y defined by

$$x = \frac{N}{m/b}, \quad y = \frac{P}{r/c}.$$

Then (4.1)–(4.2) become

$$x' = rx(1 - y), \tag{4.3}$$
$$y' = my(x - 1), \tag{4.4}$$

which contains only two parameters. We have not selected a dimensionless time variable (although we could, and it would reduce the system to just one constant).

Now, how do we analyze (4.3)–(4.4)? Like most systems of differential equations, the equations cannot be solved in closed form to obtain the populations $x = x(t)$ and $y = y(t)$. If m and r are known numerical constants, and we impose the initial populations $x(0)$ and $y(0)$, we could resort immediately to computation and obtain a numerical solution by Euler's method, or use a software package. (At the end of the chapter, before the Reference Notes, we present a MATLAB m-file that uses the Euler method to draw time series or orbits of a system of two differential equations.) We would get time series plots of the populations $x = x(t)$ and $y = y(t)$. But this is unsatisfying in some sense because time series plots of solutions depending on specific values of parameters do not always give a good qualitative picture of the dynamics. Therefore, we tack in a different direction and use geometrical methods to discover the qualitative behavior.

Recall, for example, our strategy in dealing with a single differential equation, say $u' = f(u)$. By setting $f(u) = 0$, we identified the possible equilibrium solutions (constant solutions), and then we linearized the equation about those equilibria to determine their stability. A phase line diagram showing the u axis, the equilibria, and arrows indicating how solutions $u = u(t)$ evolve in time between the equilibria is easily constructed, giving a complete picture of the dynamics of the unknown state u. We can do a similar thing in two dimensions where we construct a phase plane instead of a phase line. It can be extended to three and even higher dimensions, although we lose our geometric visualization in more than three dimensions.

To find the equilibrium solutions of (4.3)–(4.4), we can set the right sides equal to zero to get the simultaneous equations

$$rx(1 - y) = 0, \quad my(x - 1) = 0.$$

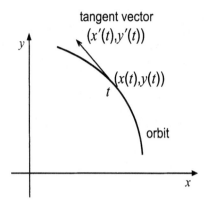

Figure 4.1 Orbit $(x(t), y(t))$ in phase space. At each point along the orbit the tangent vector represents its direction.

The only two solutions are $x = y = 0$ and $x = y = 1$. These are the constant solutions representing $x(t) = 0$, $y(t) = 0$, and $x(t) = 1$, $y(t) = 1$. The origin $(0,0)$ in the xy plane represents extinction of both species, and the nonzero equilibrium $(1, 1)$ represents a possible coexistent state.

Generally, a solution $x = x(t)$, $y = y(t)$ of a system of two differential equations such as (4.3)–(4.4) can be regarded as the parametric equations of a curve in the xy phase plane. The solution curve itself is called an **orbit** (also, a path or a trajectory) of the system, and the vector of derivatives, $(x'(t), y'(t))$, which is defined by the right sides of differential equations themselves, is the tangent vector to the curve $x = x(t)$, $y = y(t)$ at time t. See Fig. 4.1. An orbit is traced out in increasing time, indicated by the direction of its tangent vectors. Every point in the plane has some orbit passing through it. We usually plot a set, or field, of tangent vectors at various points (x, y) in the plane to indicate the directions of the orbits. The orbits, or solutions, are the curves that fit into the vector field in such a way that they have these vectors as their tangents. A collection of orbits in the phase plane is called a **phase portrait**.

To understand the shape of the orbits for the Lotka–Volterra system we can plot some tangent vectors (x', y') and find where they point. Systematically, we can proceed as follows. The x **nullclines** for (4.3)–(4.4) are defined as the set of points in the plane where $x' = 0$ (x is not changing in time, which means that the tangent vectors are vertical); in this case the x nullclines are $x = 0$ and $y = 1$. Thus, orbits must cross these two lines vertically. The y nullclines, where $y' = 0$, are $y = 0$ and $x = 1$; the orbits cross the y nullclines horizontally. The equilibria are the intersections of the x and y nullclines because at an equilibrium both $x' = 0$ and $y' = 0$. See Fig. 4.2. Next, on each nullcline we

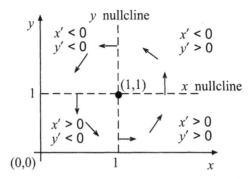

Figure 4.2 Nullclines and vector field for the Lotka–Volterra system.

can find the direction of the orbit. For example, on the x nullcline to the right of
the equilibrium, where $x > 1$, $y = 1$, we have $y' = my(x-1) > 0$. Therefore, the
tangent vectors point up and the orbits cross the nullcline vertically upward.
Similarly, we can determine the directions of the orbits as they cross the other
three rays out of the equilibrium $(1, 1)$. These are shown in Fig. 4.2 with arrows
denoting the directions of the orbits. We can next determine the direction of the
orbits in the regions between the nullclines. We can do this either by selecting
an arbitrary point in that region and calculating x' and y' from the differential
equations, or just by noting the sign of x' and y' in that region from information
obtained from the system. For example, in the northeast quadrant, above and
to the right of the nonzero equilibrium, it is easy to see, because $x > 1$ and
$y > 1$, that $x' < 0$ and $y' > 0$; so the orbit travels upward and to the left,
as shown in Fig. 4.2. This task can be performed for each quadrant and we
obtain the directions shown by arrows in Fig. 4.2. Having the direction of the
orbits along the nullclines and in the regions bounded by the nullclines tells
us the directions of the orbits. Near $(0,0)$ the orbits appear to veer away from
the extinction equilibrium, implying that $(0,0)$ is not stable; that is, if you are
located at $(0,0)$ (no predators or prey) and then the system is given a small
perturbation (adding a small number of predators and prey), that puts the
population on an orbit that goes away from the equilibrium. It appears that
orbits wind around the nonzero equilibrium $(1,1)$ in some counterclockwise
fashion, but precisely how must be verified. As it turns out, for this system
we can obtain the equation of the orbits by the following device (this does not
work for all systems). We divide the two equations (4.3)–(4.4) to get

$$\frac{y'}{x'} = \frac{dy/dt}{dx/dt} = \frac{dy}{dx} = \frac{my(x-1)}{rx(1-y)}.$$

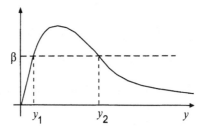

Figure 4.3 Plot of $y^r e^{-ry} = \beta$, showing the existence of at most two roots y_1 and y_2.

We can therefore separate variables to get

$$r\frac{1-y}{y}\,dy = m\frac{x-1}{x}\,dx.$$

Integrating gives

$$r\ln y - ry = mx - m\ln x + C,$$

where C is an arbitrary constant. This is an algebraic equation stating how x and y vary along the orbits; we have lost time dependence. If an initial point is prescribed, where the orbit starts at $t = 0$, a unique value of C is determined, and thus a unique orbit. It is obscure geometrically what these orbits are, because it is not possible to solve for either of the variables. But we can proceed as follows. Fix C (an orbit). If we exponentiate the last equation, we can write the orbit as

$$y^r e^{-ry} = e^C e^{mx} x^{-m}.$$

Now consider the y nullcline where x is fixed at a value 1. The right side of the last equation is a positive number β, so $y^r e^{-ry} = \beta$. Using elementary calculus techniques for graphing, we can plot the left side of this equation (a power function times a decaying exponential); see Fig. 4.3. We observe that there can be at most two solutions for y. Hence, along the y nullcline $x = 1$, the orbit can cross at most twice, preventing it from spiraling into or out from the equilibrium point because that would mean that many values of y would be possible. We conclude that the orbits are periodic closed curves that surround the equilibrium.

A phase diagram is shown in Fig. 4.4. When the prey population is high, the predators have a high food source and their numbers start to increase, thereby eventually driving down the prey population. Then the prey population drops low, ultimately reducing the number of predators because of lack of food. Then the process repeats, giving cycles. The Lotka–Volterra model is the simplest

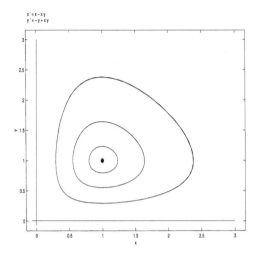

Figure 4.4 Counterclockwise periodic orbits of the Lotka–Volterra system.

model in ecology, showing how populations can cycle, and it was one of the first strategic models to explain qualitative observations in some natural systems. Note that the nonzero equilibrium $(1,1)$ is stable in that a small perturbation from equilibrium puts the populations on one of the periodic orbits; so, the populations remain close to the equilibrium $(1,1)$. But the system does not return to equilibrium.

As one can imagine, the Lotka–Volterra model is just one of a myriad of dynamic population models or consumer–resource interactions. A simple variation of the Lotka–Volterra model can be considered by replacing the exponential growth of the prey by a self-limiting law such as logistic growth. Then we have

$$
\begin{aligned}
N' &= rN\left(1 - \frac{N}{K}\right) - cNP, \\
P' &= bNP - mP.
\end{aligned}
$$

In this case one can show that populations do not cycle, but rather, there can exist, depending on the parameter values, a stable equilibrium that is approached by nearby paths. This is discussed in detail in a later example.

A further complication is to replace mass action kinetics with a Holling type II functional response. We obtain the Rosenzweig–MacArthur model,

$$
\begin{aligned}
N' &= rN\left(1 - \frac{N}{K}\right) - \frac{cNP}{1 + hN}, \\
P' &= \frac{bNP}{1 + hN} - mP,
\end{aligned}
$$

which displays interesting bifurcations as the parameter values change.

When there is competition between two species, say x and y, the presence of each inhibits the growth of the other. For example,

$$x' = rx\left(1 - \frac{x + \alpha y}{K}\right),$$

$$y' = sy\left(1 - \frac{y + \beta x}{L}\right)$$

is a competition model where r and s represent growth rates for x and y, respectively, and K and L represent their carrying capacities. The coefficients α and β define the magnitude of the competition term, which is negative for each species. This model allows us to prove the **competitive exclusion principle**, which states that two very similar species trying to occupy the same niche cannot coexist.

Animals, or an animal and a plant, can exist in a mutualistic, or cooperative, arrangement. That is, each benefits from the presence of the other. A model of mutualism is

$$x' = rx\left(1 - \frac{x - \alpha y}{K}\right),$$

$$y' = sy\left(1 - \frac{y - \beta x}{L}\right),$$

where the interaction terms are now positive.

In addition to consumer–resource interactions, there are many important models involving epidemiology and disease dynamics. In this case, for example, the unknowns may represent the number of individuals S in a population that are susceptible to a disease, those numbers I who are infected with the disease, and the number R that have been removed from the dynamics (either gotten over the disease and become immune, or died). The SIR model is

$$S' = -bSI, \tag{4.5}$$

$$I' = bSI - rI, \tag{4.6}$$

$$R' = rI, \tag{4.7}$$

where b is the transmission rate and r is the rate at which infectives get over the disease; it is assumed that the population is closed, or $S + I + R = N$, where N is the constant total population. The last equation for R' is decoupled from the first two, so the system reduces to just two equations and two unknowns. Observe that the interaction term between susceptibles and infectives is again described by mass action kinetics. One can extend this model to include demographics (birth and death), vaccination, latent periods where there is a group that carries the disease but is not yet infected, and so on. This is a rich area for study, and models for HIV and AIDS, West Nile virus, and most other diseases have been analyzed in detail, often giving essential information about the prevention of epidemics and other dynamics. We examine some specific questions about SIR models in the Exercises in Section 4.5.

4.2 Phase Plane Analysis

We now outline in more detail a general context for carrying out a geometrical analysis of a two-dimensional system. Consider the model

$$x' = f(x, y), \quad y' = g(x, y), \qquad (4.8)$$

in two unknowns $x = x(t)$ and $y = y(t)$. As a mathematical condition, the functions f and g are assumed to possess continuous partial derivatives of all orders in a domain D of the xy plane. A system of the type (4.8), in which the independent variable t does not appear in f and g is said to be **autonomous**. Under these assumptions on f and g, it can be shown that there is a unique solution $x = x(t)$, $y = y(t)$ to the **initial value problem**

$$x' = f(x, y), \quad y' = g(x, y), \qquad (4.9)$$
$$x(t_0) = x_0, \quad y(t_0) = y_0,$$

where t_0 is an instant of time (usually, $t_0 = 0$) and $(x_0, y_0) \in D$. The solution is defined in some interval $\alpha < t < \beta$ containing t_0. Most often, $-\infty < t < \infty$. There are two ways to plot the solution. It can be plotted as time series in state space (graphing x and y vs. t). Or, we can plot the curve defined by parametric equations $x = x(t)$, $y = y(t)$ in the xy phase plane. This curve is called an **orbit** (path, trajectory) of the system (4.8). The orbit is directed in the sense that it is traced out in a certain direction as t increases; this direction is often indicated by an arrowhead on the curve. Every initial point (x_0, y_0) leads to an oribit. Because solutions to the initial value problem are unique, it follows that at most one orbit passes through each point of the phase plane, and all of the orbits cover the entire phase plane without intersecting each other. A constant solution $x(t) = x_0$, $y(t) = y_0$ of (4.8) is called an **equilibrium solution**. Such a solution does not define an orbit in the phase plane, but rather a point, often called a **critical point**. Clearly, critical points occur where both f and g vanish; that is,

$$f(x_0, y_0) = g(x_0, y_0) = 0.$$

It is evident that no orbit can pass through a critical point; otherwise, uniqueness would be violated. The totality of all the orbits, their directions, and

critical points graphed in the xy phase plane is called the **phase diagram** of
the system (4.8). Indeed, the qualitative behavior of all the orbits in the phase
plane is determined to a large extent by the location of the critical points and
the local behavior of the orbits near those points. One can prove the following
results, which essentially form the basis of the Poincaré[1]–Bendixson theorem
(Section 4.3). This important theorem limits the types of orbits that can occur
in two-dimensional autonomous systems.

1. An orbit cannot approach a critical point in finite time; that is, if an orbit
   approaches a critical point, then necessarily, $t \to \pm\infty$.

2. The following alternatives hold: Either an orbit is a critical point or it is a
   simple closed curve; or as $t \to \pm\infty$, an orbit approaches a critical point or
   approaches a closed orbit; or an orbit leaves every bounded set.

A closed orbit corresponds to a **periodic solution**, or **cycle**, of (4.8). These
ideas are now illuminated by some examples.

Example 4.1

Consider the system

$$x' = y, \quad y' = -x. \tag{4.10}$$

Here $f(x,y) = y$ and $g(x,y) = -x$, so $(0,0)$ is the only critical point; it corresponds to the equilibrium solution $x(t) = 0$, $y(t) = 0$. For this simple equation we can find two independent solutions $x(t) = \cos t$, $y(t) = -\sin t$, and
$x(t) = \sin t$, $y(t) = \cos t$. The linear combinations of these two solutions,

$$
\begin{aligned}
x &= x(t) = c_1 \cos t + c_2 \sin t, \\
y &= y(t) = -c_1 \sin t + c_2 \cos t,
\end{aligned}
$$

form the general solution of (4.10), where c_1 and c_2 are constants. These constants can be found from initial conditions, if they are prescribed; then a single
solution is selected. We can plot the resulting time series, which are 2π-periodic,
oscillatory functions, versus t. What are the shape of the orbits in the phase
plane? Here we can eliminate the parameter t by squaring and adding the last
two equations to obtain

$$x^2 + y^2 = c_1^2 + c_2^2 = C^2.$$

---

[1] Henri Poincaré (1860–1934) was called the *last universalist*. His contributions to
nearly all areas of mathematics made him one of the most influential mathematicians of all time. His work in dynamical systems paved the way to modern topo-
logical methods.

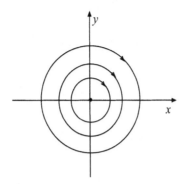

Figure 4.5 Periodic orbits surrounding a center.

Therefore, the orbits plot as circles centered at the origin, and the phase diagram is shown in Fig. 4.5. The arrows on the orbits show the direction of the curves in increasing time; they were determined by noting that $x' > 0$ (x is increasing) for $y > 0$ (in the upper half-plane), and $x' < 0$ (x is decreasing) for $y < 0$ (in the lower half-plane). The critical point, which is the origin, is called a **center** when it is surrounded by closed orbits. Alternatively, we may divide the two equations in the system (4.10) to obtain

$$\frac{dy}{dx} = -\frac{x}{y}.$$

This can be integrated directly using separation of variables to obtain the orbits

$$x^2 + y^2 = C.$$

In summary, all the orbits are closed curves and each corresponds to a 2π periodic solution because

$$x(t + 2\pi) = x(t), \quad y(t + 2\pi) = y(t). \qquad \square$$

Example 4.2

The autonomous system

$$x' = 2x, \quad y' = 3y,$$

is *decoupled*, and each equation can be solved separately to obtain the solution

$$x = c_1 e^{2t}, \quad y = c_2 e^{3t}.$$

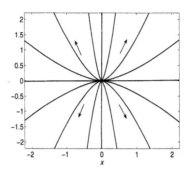

Figure 4.6   Phase diagram showing an unstable node where all the orbits are emitted from the origin.

The origin $(0,0)$ is the only critical point. The orbits can be found by eliminating t, or by integrating

$$\frac{dy}{dx} = \frac{3y}{2x}.$$

Taking the latter approach, we obtain $y^2 = Cx^3$, where C is a constant. The orbits are plotted in Fig. 4.6. A critical point having this behavior is called a **node**. Because the orbits are emanating out from the origin, the origin is called an **unstable critical point**, or **repeller**.   □

Geometrically, the functions f and g on the right side of (4.8) define a vector field $\mathbf{v}(x,y) = (f(x,y), g(x,y)) = f(x,y)\mathbf{i} + g(x,y)\mathbf{j}$ in the plane; that is, to each point (x,y) there is attached a specific vector $\mathbf{v}(x,y)$ from that field. Because it satisfies (4.8), an orbit, $x = x(t)$, $y = y(t)$, must fit into the vector field in such a way that a member of the field is tangent to the curve at each of its points. The critical points are points where the vector field vanishes, or $\mathbf{v} = \mathbf{0}$. In plotting the orbital structure in the phase plane, we usually plot, separately, the locus $f(x,y) = 0$ (where $x' = 0$, or the vector field is vertical) and the locus $g(x,y) = 0$ (where $y' = 0$, or the vector field is horizontal). These loci are called the x and y **nullclines**, respectively. Notice that x and y nullclines intersect at a critical point. Sketching nullclines and critical points and plotting the directions of the vector field in between all the nullclines facilitate drawing the phase plane diagram. Of course, software can plot the orbits and the vector field for us if the parameters have specific values.

As stated, the critical points of (4.8) are found by solving the simultaneous equations $f(x,y) = 0$ and $g(x,y) = 0$. The orbits in the phase plane can

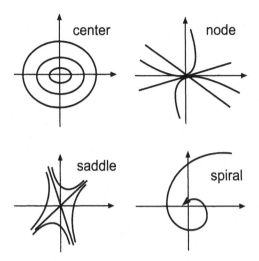

Figure 4.7 The types of critical points for linear systems.

sometimes be found by integrating the differential relationship

$$\frac{dy}{dx} = \frac{g(x, y)}{f(x, y)}.$$

A critical point of (4.8) is said to be **isolated** if there is a small circle around the critical point whose interior contains no other critical points. There are four types of isolated critical points that occur in *linear* problems, where both f and g are linear functions of x and y; these are called *centers, nodes, saddles,* and *spirals*. In Fig. 4.7 these critical points are drawn schematically, showing the local orbital structure. In nonlinear systems, critical points of unusual type can occur where, for example, a nodal structure is on one side and a saddle structure is on the other, as shown in Fig. 4.8.

Critical points represent equilibrium solutions of (4.8), and another important issue is that of stability. That is, does the equilibrium solution have some degree of permanence to it when it is subjected to small perturbations? Roughly speaking, a critical point is stable if all orbits that start sufficiently close to the point remain close to the point. To formulate this in mathematical terms, let us suppose that the origin $(0, 0)$ is an isolated critical point of (4.8). We say that $(0, 0)$ is **stable,** if for each $\varepsilon > 0$ there exists a positive number δ_ε such that every orbit that is inside the circle of radius δ_ε at some time t_0 remains inside the circle of radius ε for all $t > t_0$ (see Fig. 4.9). A critical point is **locally asymptotically stable** if it is stable and there exists a circle of radius δ_ε such that every orbit that is inside this circle at some time $t = t_0$ approaches $(0, 0)$ as $t \to \infty$. If $(0, 0)$ is not stable, it is **unstable**. We note that

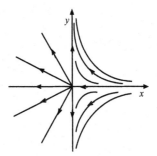

Figure 4.8 One possibility for the orbital structure near a critical point in a nonlinear system.

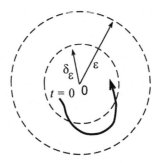

Figure 4.9 Definition of stability.

a center is stable, but not asymptotically stable; a saddle is unstable. A spiral is either asymptotically stable or unstable, depending on the direction of the orbits, and similarly for nodes.

To determine analytic criteria for stability we can proceed as we did for a single differential equation: namely, to linearize near the critical point. Suppose that (x_0, y_0) is an isolated critical point of (4.8). This means that $x_0(t) = x_0$, $y_0(t) = y_0$, is an equilibrium solution. Let

$$x(t) = x_0 + u(t),$$
$$y(t) = y_0 + v(t),$$

where $u(t)$ and $v(t)$ represent small perturbations, or changes, from the equilibrium state. Substituting into (4.8) gives the perturbation equations

$$\frac{du}{dt} = f(x_0 + u, y_0 + v), \qquad \frac{dv}{dt} = g(x_0 + u, y_0 + v).$$

We can expand the right sides by Taylor's theorem:

$$f(x_0 + u, y_0 + v) = f(x_0, y_0) + f_x(x_0, y_0)u + f_y(x_0, y_0)v + O(u^2 + v^2)$$
$$= au + bv + O(u^2 + v^2),$$

where $a = f_x(x_0, y_0)$, $b = f_y(x_0, y_0)$, and $O(u^2 + v^2)$ represents terms that are at least quadratic in u and v. Similarly,

$$g(x_0 + u, y_0 + v) = cu + dv + O(u^2 + v^2),$$

where $c = g_x(x_0, y_0)$ and $d = g_y(x_0, y_0)$. Neglecting the higher-order terms, we obtain the linearized perturbation equations, or the **linearization** near the critical point:

$$\frac{du}{dt} = au + bv, \quad \frac{dv}{dt} = cu + dv. \tag{4.11}$$

These equations tell us approximately how the small perturbations, or changes, evolve in time. The matrix associated with this linearization,

$$J(x_0, y_0) = \begin{pmatrix} a & b \\ c & d \end{pmatrix} = \begin{pmatrix} f_x(x_0, y_0) & f_y(x_0, y_0) \\ g_x(x_0, y_0) & g_x(x_0, y_0) \end{pmatrix},$$

is called the **Jacobian matrix** of the system (4.8) at the equilibrium (x_0, y_0). It seems reasonable that the stability of the equilibrium (x_0, y_0) of the nonlinear system (4.8) ought to be indicated by the stability of $(0, 0)$ of the associated linearized system (4.11). For example, if u and v go to zero, then if you move away from the critical point (x_0, y_0) a small amount, you will return to the critical point. Under suitable conditions, discussed in subsequent sections, this is true.

EXERCISES

1. Find equations of the orbits in the xy plane for the system $x' = y^2$, $y' = -\frac{2}{3}x$, and sketch the phase diagram. Is the origin stable or unstable?

2. Find the linearization of the system

$$x' = y - x,$$
$$y' = -y + \frac{5x^2}{4 + x^2},$$

   about the equilibrium point $(1, 1)$.

3. The Lotka–Volterra model holds in the absence of fishing. With fishing, the model becomes

$$N' = rN - cNP - \rho EN,$$
$$P' = bNP - mP - \sigma EP,$$

where E is the fishing effort and ρ and σ are the catchability coefficients for the prey and predator, respectively. Find the biologically realistic equilibria and compare to the nonfishing model. Show that intervention in this predator–prey system which removes both prey and predators at a constant per capita rate has the effect of increasing the average predator population. (You may want to nondimensionalize the model.)

4. Discuss the wisdom of spraying a pesticide in a field to control aphids where there is already some biological control of the aphids by lady beetles. (Assume pesticides remove insects at a constant rate.)

5. In the Lotka–Volterra model suppose that the prey have a refuge where some of them can hide. Show that the model becomes

$$
\begin{aligned}
N' &= rN - c(N - R)P, \\
P' &= b(N - R)P - mP,
\end{aligned}
$$

where R is a constant. What effect does a prey refuge have on the equilibria?

4.3 Linear Systems

We showed in Section 4.2 how the behavior of orbits near a critical point might be dictated by its linearization, which is a linear system for the small perturbations from the critical point. In this section we show how to analyze linear systems. The reader may want to review the on matrices in Chapter 2.

Consider a two-dimensional linear system

$$
x' = ax + by, \quad y' = cx + dy, \tag{4.12}
$$

where a, b, c, and d are constants. We assume that the coefficient matrix

$$
A = \begin{pmatrix} a & b \\ c & d \end{pmatrix}
$$

is nonsingular, or

$$
\det A = ad - bc \neq 0. \tag{4.13}
$$

This condition guarantees that the algebraic system

$$
\begin{aligned}
ax + by &= 0, \\
cx + dy &= 0
\end{aligned}
$$

has an isolated critical point at the origin. Otherwise, there is an entire line of nonisolated critical points. It is easier to examine (4.12) using matrix notation.

This is advantageous because the notation generalizes immediately to higher dimensions. Letting

$$\mathbf{x} = \begin{pmatrix} x \\ y \end{pmatrix}, \quad A = \begin{pmatrix} a & b \\ c & d \end{pmatrix},$$

equations (4.12) can be written

$$\mathbf{x}' = A\mathbf{x}. \tag{4.14}$$

Following our experience with linear equations with constant coefficients, we try a solution of (4.14) of the form

$$\mathbf{x} = \mathbf{v}e^{\lambda t}, \tag{4.15}$$

where the vector $\mathbf{v}$ and the number λ are to be determined. Substituting (4.15) into (4.14) gives, after simplification,

$$A\mathbf{v} = \lambda\mathbf{v}, \tag{4.16}$$

which is the **algebraic eigenvalue problem**. Consequently, λ is an eigenvalue of A and $\mathbf{v}$ is an associated eigenvector. We are looking for nontrivial solutions of (4.16), or, equivalently,

$$(A - \lambda I)\mathbf{v} = 0. \tag{4.17}$$

This equation will have a nontrivial solution $\mathbf{v}$ if the coefficient determinant vanishes, or

$$\det(A - \lambda I) = 0. \tag{4.18}$$

This equation determines the eigenvalues λ. The corresponding eigenvector(s) are then determined by solving the homogeneous system

$$(A - \lambda I)\mathbf{v} = 0. \tag{4.19}$$

Equation (4.18) is a quadratic equation in λ, called the characteristic equation. We leave it to an exercise to show that the **characteristic equation** has the form

$$\lambda^2 - (\operatorname{tr} A)\lambda + \det A = 0,$$

where $\operatorname{tr} A = a + d$ is the trace of A. It will have two roots. Hence, there are several cases to consider, depending on whether the roots λ_1 and λ_2 are real, equal or unequal, of the same or different signs, are complex, or are purely imaginary. We work out one of the cases in detail and indicate the solution for the others, leaving those details to the reader.

1. $0 < \lambda_1 < \lambda_2$ (λ_1 and λ_2 are real, distinct, and positive). Let $\mathbf{v}_1$ and $\mathbf{v}_2$ be independent eigenvectors corresponding to λ_1 and λ_2, respectively. Then $\mathbf{v}_1 e^{\lambda_1 t}$ and $\mathbf{v}_2 e^{\lambda_2 t}$ are independent solutions of (4.14), and the general solution is given by

$$\mathbf{x}(t) = c_1 \mathbf{v}_1 e^{\lambda_1 t} + c_2 \mathbf{v}_2 e^{\lambda_2 t}, \tag{4.20}$$

where c_1 and c_2 are arbitrary constants. If we set $c_2 = 0$, then $\mathbf{x} = c_1 \mathbf{v}_1 e^{\lambda_1 t}$ gives two **linear orbits**, consisting of two rays emanating out of the origin in the directions defined by $\mathbf{v}_1$ and $-\mathbf{v}_1$. Similarly, if $c_1 = 0$, then $\mathbf{x} = c_2 \mathbf{v}_2 e^{\lambda_2 t}$ represents two linear orbits emanating out of the origin in the directions $\mathbf{v}_2$ and $-\mathbf{v}_2$. If $c_1 \neq 0$ and $c_2 \neq 0$, (4.20) represents curved orbits that enter $(0,0)$ as t goes backward in time (i.e., $t \to -\infty$), and go to infinity as $t \to +\infty$. To see how orbits enter the origin, we note that for large negative t the term $e^{\lambda_1 t}$ dominates $e^{\lambda_2 t}$ because $\lambda_2 > \lambda_1$. Thus,

$$\mathbf{x}(t) \approx c_1 \mathbf{v}_1 e^{\lambda_1 t}, \quad \text{large negative } t,$$

so the orbits enter the origin tangent to $\mathbf{v}_1$. For large positive t all the orbits (4.20) go to infinity with slope asymptotic to $\mathbf{v}_2$, because in that case $e^{\lambda_2 t}$ dominates $e^{\lambda_1 t}$. In this case, the origin $(0,0)$ is an **unstable node**.

2. $\lambda_2 < \lambda_1 < 0$ (real, distinct, negative). This case is similar to the preceding case, but $(0,0)$ is an **asymptotically stable node**. The direction of time along the orbits is reversed.

3. $\lambda_1 < 0 < \lambda_2$ (real, distinct, opposite signs). The general solution is

$$\mathbf{x}(t) = c_1 \mathbf{v}_1 e^{\lambda_1 t} + c_2 \mathbf{v}_2 e^{\lambda_2 t}.$$

As in the first two cases, the eigenvectors $\mathbf{v}_1$ and $\mathbf{v}_2$ and their negatives define linear orbits. One set, corresponding to the eigenpair $\lambda_1, \mathbf{v}_1$, enter the origin as $t \to \infty$. The other set, corresponding to the eigenpair $\lambda_2, \mathbf{v}_2$, leave the origin as $t \to \infty$. No orbit other than $c_1 \mathbf{v}_1 e^{\lambda_1 t}$ can enter the origin at $t \to \infty$, and all orbits are asymptotic to the linear orbits as $t \to \pm\infty$. The origin is a **saddle point** in this case, which is unstable.

4. $\lambda_1 = \lambda_2$ (real and equal). When there is a single eigenvalue $\lambda \equiv \lambda_1 = \lambda_2$ of multiplicity 2, the solution depends on whether there are one or two linearly independent eigenvectors. (a) If $\mathbf{v}_1$ and $\mathbf{v}_2$ are linearly independent eigenvectors corresponding to λ, the general solution of (4.14) is

$$\mathbf{x}(t) = c_1 \mathbf{v}_1 e^{\lambda t} + c_2 \mathbf{v}_2 e^{\lambda t}$$
$$= (c_1 \mathbf{v}_1 + c_2 \mathbf{v}_2) e^{\lambda t}.$$

Because $\mathbf{v}_1$ and $\mathbf{v}_2$ are independent, every direction is covered and every orbit is therefore a linear ray entering the origin ($\lambda < 0$) or leaving the origin ($\lambda > 0$). Depending on the sign of λ, the origin is an unstable node or an asymptotically stable node. (b) If $\mathbf{v}$ is the only eigenvector corresponding to λ, then $\mathbf{v}e^{\lambda t}$ is a solution of (4.14). A second linearly independent solution has the form $(\mathbf{w} + \mathbf{v}t)e^{\lambda t}$ for a vector $\mathbf{w}$ satisfying $(A - \lambda I)\mathbf{w} = \mathbf{v}$, and so the general solution to (4.14) is

$$\mathbf{x}(t) = (c_1\mathbf{v} + c_2\mathbf{w} + c_2\mathbf{v}t)e^{\lambda t}.$$

For large t we have $\mathbf{x}(t) \approx c_2 t\mathbf{v}e^{\lambda t}$, so the orbits enter the origin with direction $\mathbf{v}$ when $\lambda < 0$, or they leave the origin when $\lambda > 0$; the origin is an asymptotically stable **node** or unstable node, respectively.

5. $\lambda_1 = \alpha + i\beta, \lambda_2 = \alpha - i\beta$ (complex conjugate roots). Let $\mathbf{w} + i\mathbf{v}$ be an eigenvector corresponding to λ_1. Then a complex solution of (4.14) is

$$(\mathbf{w} + i\mathbf{v})e^{(\alpha+i\beta)t},$$

or after expanding with Euler's formula,

$$e^{\alpha t}(\mathbf{w}\cos\beta t - \mathbf{v}\sin\beta t) + ie^{\alpha t}(\mathbf{w}\sin\beta t + \mathbf{v}\cos\beta t).$$

The real and imaginary parts of a complex solution are independent real solutions, and therefore the general solution is

$$\mathbf{x}(t) = c_1 e^{\alpha t}(\mathbf{w}\cos\beta t - \mathbf{v}\sin\beta t) + c_2 e^{\alpha t}(\mathbf{w}\sin\beta t + \mathbf{v}\cos\beta t).$$

The other eigenvalue gives the same two independent solutions. Notice that the two functions $\mathbf{w}\cos\beta t - \mathbf{v}\sin\beta t$ and $\mathbf{w}\sin\beta t + \mathbf{v}\cos\beta t$ are periodic functions with period $2\pi/\beta$, and the factor $e^{\alpha t}$ is like an amplitude factor. If $\alpha = 0$, the two components $u(t)$ and $v(t)$ of $\mathbf{x}$ are strictly oscillatory functions with period $2\pi/\beta$. Therefore, the orbits are closed curves and $(0,0)$ is a **center**. If $\alpha < 0$, the amplitude of $\mathbf{x}$ decreases and the orbits form spirals that wind into the origin; these are decaying oscillations and the origin is called an asymptotically stable **spiral point**. If $\alpha > 0$, the amplitude of $\mathbf{x}$ increases and the orbits are unstable spirals, representing growing oscillations, and the origin is an unstable spiral point. We often refer to a center as a **neutrally stable** equilibrium. Note that, graphically, a spiral orbit usually enters or leaves the equilibrium very quickly and does not spiral about the equilibrium several times; it goes to infinity very quickly because of the exponential growth or decay amplitude of its solution formula. So the spiral is not tight.

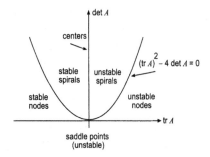

Figure 4.10 Character of a linear system with matrix A in terms of its trace and determinant. The parabolic curve is the discriminant representing degenerate nodes, and the positive vertical axis, tr $= 0$, det > 0, represents centers, and det < 0 represents saddles.

The preceding discussion completely characterizes the solution, orbital structure, and the stability properties of an isolated critical point $(0,0)$ of the linear system (4.14). We can summarize all this information in a diagram (Fig. 4.10). The eigenvalues of (4.14) are determined from the characteristic equation (4.3) in terms of the determinant and trace of the coefficient matrix. Figure 4.10 shows a plot of det A vs. tr A, indicating the regions where the different cases occur. The reader is invited to confirm this diagram by examining the eigenvalues, which from the quadratic formula are:

$$\lambda = \frac{1}{2}(\text{tr } A \pm \sqrt{(\text{tr } A)^2 - 4\det A}).$$

A key observation is summarized in the following theorem.

Theorem 4.3

The critical point $(0,0)$ of the linear system

$$\frac{d\mathbf{x}}{dt} = A\mathbf{x}, \quad \det A \neq 0,$$

is stable if, and only if, the eigenvalues of $\mathbf{A}$ have nonpositive real parts; it is asymptotically stable if, and only if, the eigenvalues have negative real parts. □

Example 4.4

Consider the linear system

$$x' = 3x - 2y,$$
$$y' = 2x - 2y.$$

The coefficient matrix

$$A = \begin{pmatrix} 3 & -2 \\ 2 & -2 \end{pmatrix}$$

has the characteristic equation

$$\det \begin{pmatrix} 3 - \lambda & -2 \\ 2 & -2 - \lambda \end{pmatrix} = \lambda^2 - \lambda - 2 = 0,$$

and therefore the eigenvalues are $\lambda = -1, 2$. Therefore, the origin is an unstable saddle point. The eigenvectors are found by solving the linear homogeneous system

$$\begin{pmatrix} 3 - \lambda & -2 \\ 2 & -2 - \lambda \end{pmatrix} \begin{pmatrix} v_1 \\ v_2 \end{pmatrix} = \begin{pmatrix} 0 \\ 0 \end{pmatrix}.$$

When $\lambda = -1$,

$$4v_1 - 2v_2 = 0, \quad 2v_1 - v_2 = 0,$$

and an eigenvector corresponding to $\lambda = -1$ is $[v_1, v_2]^T = [1, 2]^T$. When $\lambda = 2$,

$$v_1 - 2v_2 = 0, \quad 2v_1 - 4v_2 = 0,$$

which gives $[v_1, v_2]^T = [2, 1]^T$. The general solution to the system is

$$\mathbf{x}(t) = c_1 \begin{pmatrix} 1 \\ 2 \end{pmatrix} e^{-t} + c_2 \begin{pmatrix} 2 \\ 1 \end{pmatrix} e^{2t}.$$

The eigenvectors define the directions of the separatrices, or the straightline solutions. To get a more accurate plot, it is helpful to plot the nullclines, or loci of points where $x' = 0$ and where $y' = 0$. A phase diagram is shown in Fig. 4.11. □

EXERCISES

1. Find the general solution and sketch phase diagrams for the following systems; characterize the equilibria as to type (node, saddle, etc.) and stability.

   a) $x' = x - 3y, \ y' = -3x + y$.

   b) $x' = -x + y, \ y' = y$.

   c) $x' = 4y, \ y' = -9x$.

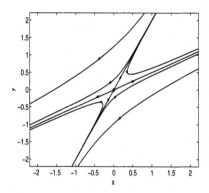

Figure 4.11 Saddle point.

d) $x' = x + y, \ y' = 4x - 2y.$

e) $x' = 3x - 4y, \ y' = x - y.$

f) $x' = -2x - 3y, \ y' = 3x - 2y.$

2. Determine the behavior of solutions near the origin for the system

$$\mathbf{x}' = \begin{pmatrix} 3 & b \\ 1 & 1 \end{pmatrix} \mathbf{x}$$

for different values of b.

3. Prove that the critical point $(0,0)$ of the linear system (4.12) is asymptotically stable if the only if $\operatorname{tr} A < 0$ and $\det A > 0$, where $\operatorname{tr} A = a + d$ is the trace of A.

4. For the linear system (4.14), let $p = \operatorname{tr} A$ and $q = \det A$. Shade the region in the pq plane representing values of p and q where the origin is a saddle point. For which values is the origin a stable spiral? An unstable node? A center?

5. Consider the system

$$\mathbf{x}' = \begin{pmatrix} -a & -1 \\ 1 & -a \end{pmatrix} \mathbf{x}.$$

What are the possible behaviors of solutions to this system, depending on the parameter a?

6. Write down a simple two-dimensional linear system whose matrix has zero determinant. What are the eigenvalues? What are the equilibrium solutions? Draw a phase plane diagram for your system.

4.4 Nonlinear Systems

We now revisit the question of whether the linearization

$$\mathbf{x}' = J\mathbf{x}, \quad J = \begin{pmatrix} f_x(x_0, y_0) & f_y(x_0, y_0) \\ g_x(x_0, y_0) & g_y(x_0, y_0) \end{pmatrix} \tag{4.21}$$

of a nonlinear system

$$x' = f(x, y), \quad y' = g(x, y) \tag{4.22}$$

at a critical point (x_0, y_0) predicts its type (node, saddle, center, spiral point) and stability. The following result of Poincaré gives a partial answer to this question. The proof may be found in the references.

Theorem 4.5

Let (x_0, y_0) be an isolated critical point for the nonlinear system (4.22) and let $J = J(x_0, y_0)$ be the Jacobian matrix for linearization (4.21), with $\det J \neq 0$. Then (x_0, y_0) is a critical point of the same type as the origin $(0, 0)$ for the linearization in the following cases.

(i) The eigenvalues of J are real, either equal or distinct, and have the same sign (node).

(ii) The eigenvalues of J are real and have opposite signs (saddle).

(iii) The eigenvalues of J are complex, but not purely imaginary (spiral). □

Therefore, the exceptional case is when the linearization has a center. The orbital structure for the nonlinear system near the critical points mirrors that of the linearization in the nonexceptional cases, with a slight distortion of the orbits due to the nonlinearity.

Example 4.6

The nonlinear system

$$x' = -y - x^3, \quad y' = x,$$

has a stable spiral at the origin. The linearization has Jacobian

$$J = \begin{pmatrix} 0 & -1 \\ 1 & 0 \end{pmatrix},$$

with purely imaginary eigenvalues $\pm i$. Hence, $(0, 0)$ is a center for the linearization. This is the exceptional case in Theorem 4.5. □

The question of stability is answered by the following theorem, which is a corollary of the Poincaré theorem.

Theorem 4.7

If $(0,0)$ is asymptotically stable for (4.21), it is asymptotically stable for (4.22).
□

Example 4.8

Consider the nonlinear system

$$\dot{x} = 2x + 3y + xy,$$
$$\dot{y} = -x + y - 2xy^3,$$

which has an isolated critical point at $(0,0)$. The Jacobian matrix J at the $(0,0)$ is

$$J = \begin{pmatrix} -2 & 3 \\ -1 & 1 \end{pmatrix},$$

which has eigenvalues $-\frac{1}{2} \pm (\sqrt{3}/2)i$. Thus, the linearization has an asymptotically stable spiral at $(0,0)$, and therefore the nonlinear system has an asymptotically stable spiral at $(0,0)$.   □

Example 4.9

Consider the model

$$x' = y,$$
$$y' = 3x^2 - 1.$$

The orbits can be found immediately by dividing the equations and separating variables to get

$$ydy = (3x^2 - 1)dx.$$

Integrating gives

$$\frac{1}{2}y^2 = x^3 - x + C$$

or

$$y = \pm\sqrt{2}\sqrt{x^3 - x + C}.$$

They are plotted in Fig. 4.12 for different values of C. There are two equilibria, $x = \sqrt{\frac{1}{3}}$, $y = 0$ and $x = -\sqrt{\frac{1}{3}}$, $y = 0$. The equilibrium solution $x = -\sqrt{\frac{1}{3}}$, $y = 0$ has the structure of a center, and for small initial values the system

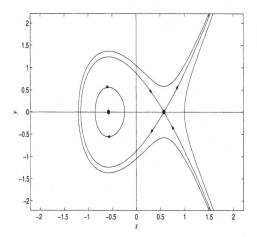

Figure 4.12 Orbits.

will oscillate. The other equilibrium, $x = \sqrt{\frac{1}{3}}$, $y = 0$, has the structure of an unstable saddle point. Because $x' = y$, for $y > 0$ we have $x' > 0$, and the orbits are directed to the right in the upper half-plane. For $y < 0$ we have $x' < 0$, and the orbits are directed to the left in the lower half-plane. The x nullcline is the y axis, where the vector field is vertical; the y nullclines are the lines $x = -\sqrt{\frac{1}{3}}$ and $x = \sqrt{\frac{1}{3}}$, where the vector field is horizontal. For large initial values the system does not oscillate but rather goes to $x = +\infty$, $y = +\infty$. The Jacobian matrix is

$$J(x, y) = \begin{pmatrix} 0 & 1 \\ 6x & 0 \end{pmatrix}.$$

Then

$$J\left(-\sqrt{\frac{1}{3}}, 0\right) = \begin{pmatrix} 0 & 1 \\ -2\sqrt{3} & 0 \end{pmatrix}, \quad J\left(\sqrt{\frac{1}{3}}, 0\right) = \begin{pmatrix} 0 & 1 \\ 2\sqrt{3} & 0 \end{pmatrix}.$$

The eigenvalues of the first matrix are purely imaginary, so the linearization has a center at $\left(-\sqrt{\frac{1}{3}}, 0\right)$. This is the exceptional case and the theorem gives no information about the nonlinear system. However, for this problem we were able to find equations for the orbits and we found that the nonlinear system does have a center at $\left(-\sqrt{\frac{1}{3}}, 0\right)$. The second matrix has real eigenvalues of opposite sign, and therefore $\left(\sqrt{\frac{1}{3}}, 0\right)$ is a saddle point, confirming the analysis.
□

Example 4.10

Consider the scaled Lotka–Volterra model

$$x' = rx(1 - y),$$
$$y' = my(x - 1),$$

with two equilibria, $(0,0)$ and $(1,1)$. The Jacobian matrix is

$$J(x, y) = \begin{pmatrix} r(1 - y) & -rx \\ my & m(x - 1) \end{pmatrix}.$$

Clearly,

$$J(0, 0) = \begin{pmatrix} r & 0 \\ 0 & -m \end{pmatrix}$$

has eigenvalues r and $-m$, and so $(0,0)$ is indeed a saddle point. At $(1,1)$,

$$J(1, 1) = \begin{pmatrix} 0 & -r \\ m & 0 \end{pmatrix},$$

which has purely imaginary eigenvalues. Thus, the linearization is inconclusive and additional work is required to determine the nature of the equilibrium $(1,1)$. Earlier we gave an argument which showed that $(1,1)$ is a center.   □

A central problem in the theory of nonlinear systems is to deal with the exceptional case, or to determine whether a nonlinear system admits any cycles, or closed orbits. These orbits represent periodic solutions of the system, or oscillations. In general, existence of closed orbits may be difficult to decide. However, the following negative result due to Bendixson and Dulac, is easy to prove.

Theorem 4.11

For system (4.22), if $f_x + g_y$ is of one sign in a region of the phase plane, system (4.22) cannot have a closed orbit in that region.   □

By way of contradiction, assume that the region contains a closed orbit Γ given by

$$x = x(t), \ y = y(t), \quad 0 \le t \le T,$$

and denote the interior of Γ by R. By Green's theorem from multivariable calculus,

$$\int_{\Gamma} f \, dy - g \, dx = \int \int_{R} (f_x + g_y) \, dx \, dy \ne 0.$$

On the other hand,

$$\int_{\Gamma} f \, dy - g \, dx = \int_{0}^{T} (fy' - gx') \, dt = \int_{0}^{T} (fg - gf) \, dt = 0,$$

which is a contradiction.   □

This theorem is generalized in the Exercises. Another negative criterion is due to Poincaré.

Theorem 4.12

A closed orbit of the system (4.22) surrounds at least one critical point of the system.   □

Positive criteria that ensure the existence of closed orbits for general systems are not abundant, particularly criteria that are practical and easy to apply. For certain special types of equations, such criteria do exist. The following fundamental theorem, referred to earlier as the **Poincaré–Bendixson theorem**, is a general theoretical result.

Theorem 4.13

Let R be a closed bounded region in the plane containing no critical points of (4.22). If Γ is an orbit of (4.22) that lies in R for some t_0 and remains in R for all $t > t_0$, then Γ is either a closed orbit or it spirals toward a closed orbit as $t \to \infty$.   □

The Poincaré–Bendixson theorem guarantees in the plane that either an orbit leaves every bounded set as $t \to \pm\infty$, is a closed curve, approaches a critical point or a closed curve as $t \to \pm\infty$, or is a critical point. Thus, in the plane, the only *attractors* are closed orbits or critical points. Interestingly enough, the Poincaré–Bendixson theorem does not generalize directly to higher dimensions. In three dimensions, for example, there exist *strange attractors* that have neither the character of a point, a curve, or a surface, which act as attractors to all orbits.

Example 4.14

(**Predator–prey with self-limiting prey**) We consider the predator–prey model introduced in Section 4.1, where in the Lotka–Volterra model we replaced the exponential prey growth rate by a self-limiting logistic growth rate:

$$N' = rN\left(1 - \frac{N}{K}\right) - cNP,$$
$$P' = bNP - mP.$$

There are five parameters, so scaling is advisable. We scale the prey by the carrying capacity K and the predator by the constant r/c, the latter of which has predator dimensions. We scale time by the prey growth rate r. Thus, we introduce new dimensionless variables via

$$x = \frac{N}{K}, \quad y = \frac{P}{r/c}, \quad \tau = rt.$$

Then, in these new variables the problem becomes (Exercise!)

$$\frac{dx}{d\tau} = x(1-x) - xy, \tag{4.23}$$

$$\frac{dy}{d\tau} = Lxy - ay, \tag{4.24}$$

where L and a are dimensionless constants defined by

$$L = \frac{bK}{r}, \quad a = \frac{m}{r}.$$

Notice that a is the ratio of two time scales, the per capita mortality rate of the predator to the per capita growth rate of the prey. The constant L is a mixture of dimensionless constants. As expected, in dimensionless variables there is a considerable economy in the number of parameters with which we must deal.

We now go through a thorough analysis of (4.23)–(4.24). The method and calculation can be used as a template for addressing all two-dimensional phase plane problems. The plan is:

– Determine the nullclines.

– Determine the critical points (equilibria).

– Determine the direction of the orbits along the nullclines and in the regions in between.

– Analyze the Jacobian at each critical point to determine its type and stability.

– Draw sample orbits to illustrate the phase diagram.

The x nullclines, where $dx/d\tau = 0$, are the straight lines

$$x = 0, \quad y = 1 - x,$$

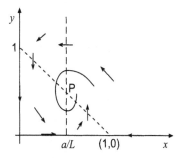

Figure 4.13 Phase diagram for $a/L < 1$. The critical point P turns out to be either a stable spiral (shown) or a stable node.

and the y nullclines, where $dy/d\tau = 0$, are the straight lines

$$y = 0, \quad x = \frac{a}{L}.$$

When you attempt to plot these lines, you see that there is an obvious decision regarding the size of the parameter a/L. Is it less than, equal to, or greater than 1? We study the case

$$a/L < 1$$

and leave the other two cases as exercises. When $a < L$ we have the phase diagram in Fig. 4.13, which shows the nullclines. The critical points, where the nullclines intersect (where both $dx/d\tau = 0$ and $dy/d\tau = 0$) are

$$\text{Critical points: } (0,0), \ (1,0), \ \left(\frac{a}{L}, \frac{L-a}{L}\right).$$

In the sequel, for conciseness, we use *prime* to denote the τ derivative $d/d\tau$. Next, we see that along the nullcline $x = 0$ (the y axis) we clearly have, from (4.24), $y' < 0$; so the orbits go directly downward. Along the nullcline $y = 0$ (the x axis) we have, from (4.23), $x' > 0$ when $x < 1$ and $x' < 0$ when $x > 1$. The appropriate arrows are shown on the x axis in Fig. 4.13. Now, in the large unbounded region in the northeast corner we clearly have both x and y large; then, from (4.23) and (4.24), the quadratic terms dominate and in that region $x' < 0$, $y' > 0$. This means that orbits in that region go up and to the left, as shown in Fig. 4.13. The orbital structure determined so far fixes the directions of the orbits on the remaining two nullclines, $x = a/L$ and $y = 1 - x$, and in the remaining regions between the nullclines. We follow on Fig. 4.13 a typical orbit. Beginning at a point Q in the northeast region, an orbit must go up and to the left, crossing the vertical line $x = a/L$ from right to left; then it must go down and to the left, consistent with the downward flow along the y axis. This means that it crosses the nullcline $y = 1 - x$ from top to bottom. It then veers away from the origin, going down and to the right, crossing $x = a/L$ again, from left to right. It then veers upward, still to the right, crossing $y = 1 - x$ vertically upward. This evolution continues and it appears that the orbit winds around the critical point $(a/L, (L - a)/L)$ in a counterclockwise manner. If these intuitive deductions are unclear to the reader, one can always go directly back to the differential equations and use inequalities to check the signs of the derivatives in each region separated by the nullclines, and on the nullclines. In our heuristic way of reasoning we used the fact that orbits cannot collide and that the vector field defining the flow must be continuous, which precludes nearby orbits going in opposite directions.

Now we check the Jacobian and the types of critical points. These results will give another check on the consistency of the reasoning above. The Jacobian at (x, y) is

$$J(x, y) = \begin{pmatrix} 1 - 2x - y & -x \\ Ly & Lx - a \end{pmatrix}.$$

At the origin,

$$J(0, 0) = \begin{pmatrix} 1 & 0 \\ 0 & -a \end{pmatrix},$$

which has real eigenvalues (1 and $-a$) of opposite sign; so $(0, 0)$ is a saddle point. At $(1, 0)$,

$$J(1, 0) = \begin{pmatrix} -1 & -1 \\ 0 & L - a \end{pmatrix}.$$

The eigenvalues are -1 and $L - a > 0$, so $(1, 1)$ is a saddle point. Finally,

$$J\left(\frac{a}{L}, \frac{L - a}{L}\right) = \begin{pmatrix} -a/L & -a/L \\ L - a & 0 \end{pmatrix}.$$

Here, $\operatorname{tr} J < 0$ and $\det J > 0$. By the theorem we know that $(a/L, (L - a)/L)$ is therefore asymptotically stable. Therefore, it is a stable node or a stable spiral, depending on whether the eigenvalues are real or complex. For this critical point the characteristic equation is

$$\lambda^2 - (\operatorname{tr} J)\lambda + \det J = \lambda^2 + \frac{a}{L}\lambda + \frac{a}{L}(L - a) = 0.$$

The discriminant is

$$(\operatorname{tr} J)^2 - 4 \det J = \left(\frac{a}{L}\right)^2 - 4\frac{a}{L}(L - a).$$

Therefore, we get a stable node when

$$\frac{a}{L} > 4(L - a),$$

and a stable spiral when

$$\frac{a}{L} < 4(L - a).$$

Figure 4.13 shows a typical orbit in the spiral case.

It is interesting to note that there are no cycles, as in the Lotka–Volterra model; including a small amount of prey density dependence has removed the periodic orbits. This result confirms our earlier remark that the Lotka–Volterra model is structurally unstable; the appearance of even the smallest change in the Lotka–Volterra system destroyed the structure. □

4.5 Bifurcation

As observed from most all the models formulated so far, there are always parameters present. We have also seen that altering the values of parameters can cause significant changes in the solution structure. Therefore, let us examine systems of equations of the form

$$x' = f(x, y, \mu), \quad y' = g(x, y, \mu), \tag{4.25}$$

depending on a single, real parameter μ. As μ varies, the nature of the solution often changes at special values of μ, giving fundamentally different critical point structures. For example, what was once a stable equilibrium can bifurcate into an unstable one. A simple example illustrates the idea.

Example 4.15

Consider the linear system

$$x' = x + \mu y, \quad y' = x - y,$$

where μ is a parameter. The eigenvalues λ are determined from

$$\det \begin{pmatrix} 1 - \lambda & \mu \\ 1 & -1 - \lambda \end{pmatrix} = 0$$

or

$$\lambda^2 - (1 + \mu) = 0.$$

Hence

$$\lambda = \pm\sqrt{1 + \mu}.$$

If $\mu > -1$, then the eigenvalues are real and have opposite signs, and therefore the origin is a saddle point. If $\mu = -1$, then the system becomes

$$x' = x - y, \quad y' = x - y$$

and there is a line of equilibrium solutions $y = x$. If $\mu < -1$, the eigenvalues are purely imaginary and the origin is a center. Consequently, as μ decreases, the nature of the solution changes at $\mu = -1$; the equilibrium state $(0,0)$ evolves from an unstable saddle to a stable center as μ passes through the critical value -1. We say that a bifurcation occurs at $\mu = -1$. $\square$

These types of phenomena occur frequently in biological systems. For example, in a predator–prey interaction in ecology we may ask how coexistent states, or equilibrium populations, depend on the carrying capacity K of the prey species. The parameter K then acts as a bifurcation parameter. For a given system, we may find a coexistent, asymptotically stable state for small carrying capacities. But as the carrying capacity increases, the coexistent state can suddenly bifurcate to an unstable equilibrium, and the dynamics could change completely. This type of behavior, if present, is an important characteristic of differential equations in the natural sciences.

Example 4.16

Consider the system

$$x' = -y - x(x^2 + y^2 - \mu), \quad y' = x - y(x^2 + y^2 - \mu). \tag{4.26}$$

The origin is an isolated equilibrium, and the linearization is

$$x' = \mu x - y, \quad y' = x + \mu y.$$

It is easy to see that the eigenvalues λ of the Jacobian matrix are given by

$$\lambda = \mu \pm i.$$

If $\mu < 0$, then $\mathrm{Re}\,\lambda < 0$, and $(0,0)$ is a stable spiral; if $\mu = 0$, then $\lambda = \pm i$, and $(0,0)$ is a center; if $\mu > 0$, then $\mathrm{Re}\,\lambda > 0$, and $(0,0)$ is an unstable spiral. From results in the preceding sections regarding the relationship between a nonlinear system and its linearization, we know that $(0,0)$ is a stable spiral for the nonlinear system when $\mu < 0$ and an unstable spiral when $\mu > 0$. Therefore, there is a bifurcation at $\mu = 0$ where the equilibrium exchanges stability. $\square$

The system in the last Example 4.16 deserves a more careful look. It is possible to solve the nonlinear problem (4.26) directly if we transform to polar coordinates $x = r\cos\theta$, $y = r\sin\theta$. First, it is straightforward to show that

$$xx' + yy' = rr', \quad xy' - yx' = r^2\theta'.$$

If we multiply the first equation by x and the second by y and add, we obtain

$$\dot{r} = r(\mu - r^2). \tag{4.27}$$

If we multiply the first equation by $-y$ and the second by x and then add, we obtain

$$\dot{\theta} = 1. \tag{4.28}$$

The equivalent system (4.27)–(4.28) in polar coordinates may be integrated directly. Clearly,

$$\theta = t + t_0, \tag{4.29}$$

where t_0 is a constant. Therefore, the polar angle θ winds around counterclockwise as time increases.

Next we examine the r-equation when $\mu > 0$. By separating variables and using the partial fraction decomposition

$$\frac{1}{r(r^2 - \mu)} = \frac{-1/\mu}{r} + \frac{1/2\mu}{r - \sqrt{\mu}} + \frac{1/2\mu}{r + \sqrt{\mu}},$$

we obtain

$$r = \frac{\sqrt{\mu}}{\sqrt{1 + c\ \exp(-2\mu t)}}, \tag{4.30}$$

where c is a constant of integration. When $c = 0$ we obtain the solution

$$r = \sqrt{\mu},$$
$$\theta = t + t_0,$$

which represents a periodic solution whose orbit is the circle $r = \sqrt{\mu}$ in the phase plane. If $c < 0$, then (4.30) represents an orbit that spirals counterclockwise toward the circle $r = \sqrt{\mu}$ from the outside, and if $c > 0$, the solution represents an orbit that spirals counterclockwise toward the circle from the inside. This means that the origin is an unstable spiral, as indicated by the linear analysis, and the periodic solution $r = \sqrt{\mu}$ is approached by all other solutions. When a periodic orbit, or cycle, is approached by orbits from both the inside and outside, we say that the orbit is a stable **limit cycle**. See Fig. 4.14.

In the case $\mu = 0$ the r-equation becomes

$$\frac{dr}{r^3} = -dt,$$

and direct integration gives the solution

$$r = \frac{1}{\sqrt{2t + c}}, \quad \theta = t + t_0.$$

This implies that the origin is a stable spiral.

Finally, in the case $\mu < 0$, let $k^2 = -\mu > 0$. Then the radial r-equation is

$$\frac{dr}{r(r^2 + k^2)} = -dt. \tag{4.31}$$

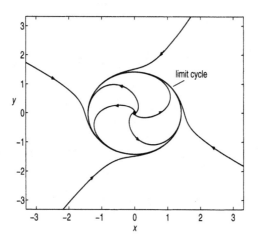

Figure 4.14 Stable limit cycle.

The partial fraction decomposition is

$$\frac{1}{r(r^2 + k^2)} = \frac{1/k^2}{r} - \frac{(1/k^2)r}{r^2 + k^2},$$

and (4.31) integrates to

$$r^2 = \frac{ck^2 e^{-2k^2 t}}{1 - ce^{-2k^2 t}}.$$

It is easy to verify that this solution represents spirals that approach the origin at $t \to +\infty$. See Fig. 4.14.

Let us review the results. As μ passes through the critical value $\mu = 0$ from negative to positive, the origin changes from a stable spiral to an unstable spiral and there appears a new time periodic solution. This type of bifurcation is common, and it is an example of a **Hopf bifurcation**. A graph in the complex plane of the eigenvalues $\lambda = \mu + i$ and $\lambda = \mu - i$ of the linearized system, as functions of the parameter μ, shows that the complex conjugate pair crosses the imaginary axis at $\mu = 0$. Generally, this signals bifurcation to a time-periodic solution.

Another important consequence of the Poincaré–Bendixson theorem predicting the existence of limit cycles is the following. Suppose that there is a closed region R in the plane where at each point (x, y) on the boundary of R the vector field $(f(x, y), g(x, y))$ for the system points inward into R. We call R a **basin of attraction**. If R contains no critical points, there must be a periodic orbit in R.

4.5.1 Rosenzweig–MacArthur Model

Now we consider the Rosenzweig–MacArthur predator–prey model,

$$N' = rN\left(1 - \frac{N}{K}\right) - \frac{cNP}{h+N},$$

$$P' = \frac{aNP}{h+N} - dP,$$

where the prey grow logistically and the predation interaction term is a type II response. Because of the large number of parameters, it is wise to rescale. We choose dimensionless quantities via

$$x = \frac{N}{K}, \ y = \frac{P}{rK/c}, \ \tau = at.$$

Then the system becomes

$$\varepsilon \frac{dx}{d\tau} = x\left(1 - x - \frac{y}{b+x}\right),$$

$$\frac{dy}{d\tau} = y\left(\frac{x}{b+x} - m\right),$$

where

$$b = \frac{h}{K}, \ m = \frac{d}{a}, \ \varepsilon = \frac{a}{r}.$$

We assume that $m < 1$ (the mortality rate of the prey is smaller than its yield from predation), $b < 1$ (the half saturation of predation is smaller than the carrying capacity of prey), and $\varepsilon < 1$ (usually, in fact, $\varepsilon \ll 1$). The first step is to note the nullclines:

$$x \text{ nullclines:} \quad x = 0, \quad y = (1 - x)(b + x).$$

$$y \text{ nullclines:} \quad y = 0, \quad x = x^* = \frac{bm}{1 - m}.$$

When we sketch these nullclines in the xy phase plane we immediately observe that there are three possible cases, depending on where the vertical line $x = x^*$ is placed [note that $(1 - b)/2$ is the x value where the parabolic x nullcline has a maximum]:

(i) $x^* > 1$.

(ii) $(1 - b)/2 < x^* < 1$.

(iii) $x^* < (1 - b)/2$.

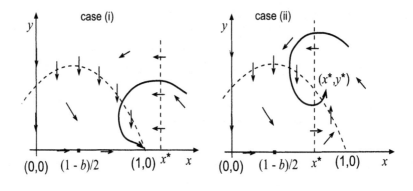

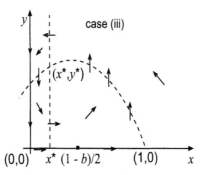

Figure 4.15 Three cases for the Rosenzweig–MacArthur model: when the vertical nullcline is to the right of $x = 1$, when it lies between the maximum of the parabolic arc and $x = 1$, and when it lies to the left of the maximum of the parabolic arc.

These three cases are shown in Fig. 4.15. Nullclines are dashed, and the direction fields are indicated. The reader should take time to verify these diagrams. If $x^* > 1$, there are two critical points: $(0,0)$ and $(1,0)$. If $x^* < 1$, there are three critical points: $(0,0)$, $(1,0)$, and (x^*, y^*), where $y^* = (1 - x^*)(b + x^*)$. To calculate the type and stability of these critical points, we rewrite (with hindsight) the system in a more tractable form: namely,

$$\varepsilon \frac{dx}{d\tau} = \frac{x}{b + x}\left((1 - x)(b + x) - y\right),$$
$$\frac{dy}{d\tau} = y\left(\frac{x}{b + x} - m\right).$$

Then

$$\varepsilon \frac{dx}{d\tau} = F(x)\left(G(x) - y\right),$$

$$\frac{dy}{d\tau} = y\left(F(x) - m\right),$$

where

$$F(x) = \frac{x}{b + x}, \quad G(x) = (1 - x)(b + x).$$

Then the Jacobian is

$$J(x, y) = \begin{pmatrix} \varepsilon^{-1}[F(x)G'(x) + F'(x)(G(x) - y)] & -\varepsilon^{-1}F(x) \\ yF'(x) & F(x) - m \end{pmatrix}.$$

First, at $(0, 0)$ we have

$$J(0, 0) = \begin{pmatrix} \varepsilon^{-1} & 0 \\ 0 & -m \end{pmatrix},$$

which has eigenvalues ε^{-1} and $-m$ that are real and of opposite sign. Therefore, $(0, 0)$ is always a saddle point. At $(1, 0)$ we have

$$J(1, 0) = \begin{pmatrix} -\varepsilon^{-1} & -1/(\varepsilon(1 + b)) \\ 0 & (1 - m)(1 - x^*)/(1+b) \end{pmatrix}.$$

The eigenvalues are $-\varepsilon^{-1}$ and $(1 - m)(1 - x^*)/(1 + b)$. Therefore, if $x^* > 1$, both eigenvalues are negative and $(1, 0)$ is a stable node; if $x^* < 1$, the eigenvalues have opposite signs and $(1, 0)$ is a saddle point.

The third critical point is $(x^*, y^*) = (x^*, G(x^*))$, which appears when $x^* < 1$. The Jacobian becomes

$$J(x^*, G(x^*)) = \begin{pmatrix} \varepsilon^{-1}F(x^*)G'(x^*) & -\varepsilon^{-1}F(x^*) \\ G(x^*)F'(x^*) & F(x^*) - m \end{pmatrix}.$$

But one easily checks that $F(x^*) = m$, so

$$J(x^*, G(x^*)) = \begin{pmatrix} m\varepsilon^{-1}G'(x^*) & -m\varepsilon^{-1} \\ G(x^*)F'(x^*) & 0 \end{pmatrix}.$$

To check stability we calculate the determinant and trace. We have

$$\det J(x^*, G(x^*)) = m\varepsilon^{-1}G(x^*)F'(x^*) > 0$$

and

$$\operatorname{tr} J(x^*, G(x^*)) = m\varepsilon^{-1}G'(x^*) = m\varepsilon^{-1}(1 - b - 2x^*).$$

Therefore, $\operatorname{tr} J(x^*, G(x^*)) < 0$ when $x^* > \frac{1}{2}(1 - b)$, and $\operatorname{tr} J(x^*, G(x^*)) < 0$ when $x^* < \frac{1}{2}(1 - b)$. This means that (x^*, y^*) is asymptotically stable whenever $x^* > \frac{1}{2}(1 - b)$, and (x^*, y^*) is unstable whenever $x^* < \frac{1}{2}(1 - b)$.

Let us summarize what we have shown for the three cases listed above:

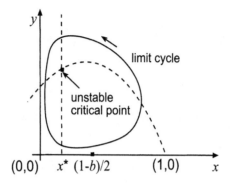

Figure 4.16 Limit cycle for the Rosenzweig–MacAuthur model when $x^* < (1 - b)/2$.

(i) $x^* > 1$. In this case, $(0,0)$ is a saddle and $(1,0)$ is a stable node.

(ii) $(1 - b)/2 < x^* < 1$. In this case, $(0,0)$ and $(1,0)$ are both saddle points, and (x^*, y^*) is asymptotically stable.

(iii) $x^* < (1 - b)/2$. In this case, $(0,0)$ and $(1,0)$ are both saddle points, and (x^*, y^*) is unstable.

Therefore, we have the following interesting behavior as the parameter x^* decreases from 1: At first the critical point (x^*, y^*) is asymptotically stable; then, when x^* hits the value $\frac{1}{2}(1-b)$, there is a bifurcation and (x^*, y^*) becomes unstable, so there is a change in stability. By definition of x^*, it decreases whenever b decrease. But b is inversely proportional to the carrying capacity K. Thus, the bifurcation occurs as the carrying capacity increases. This means that enriching the environment for the prey causes a destabilization in coexistence between the predator and the prey.

But the situation is even more strange. When the bifurcation occurs, there is a Hopf bifurcation, as in (4.26). That is, there suddenly appears a limit cycle that attracts the orbits, or populations. This limit cycle (see Fig. 4.16) comes very close to the x and y axes, and therefore the populations of both predator and prey come dangerously close to extinction. This seems to be a paradox: Increasing the prey's carrying capacity may drive the populations to extinction. This is, in fact, the **Rosenzweig enrichment paradox**, and it suggests that enriching an environment for the prey can have highly unexpected consequences.

It is beyond our scope to prove the existence of the limit cycle. However, we do state a general, well-known theorem of A. Kolmogorov from 1936 that guarantees its existence. Its proof is based on the Poincaré–Bendixson theorem.

Theorem 4.17

(*Kolmogorov*) In the region x, $y > 0$ consider the system

$$x' = xf(x,y), \quad y' = yg(x,y),$$

where the following four conditions hold:

- $f_y < 0$ and $xf_x + yf_y < 0$.

- $g_y \leq 0$ and $xg_x + yg_y > 0$.

- $f(0,0) > 0$.

- There are positive constants a, b, and c such that $f(0,a) = 0$, $f(b,0) = 0$, $g(c,0) = 0$ with $b > c$.

Then there exists either a stable critical point or a stable limit cycle.   □

We ask the reader to supply the argument that the Rosenzweig–MacArthur system satisfies these conditions (Exercise 16).

EXERCISES

1. Determine the nature and stability properties of the critical points of the systems, and sketch the phase diagram.

   (a) $x' = x^2 + y^2 - 4$, $y' = y - 2x$.

   (b) $x' = y^2$, $y' = -\frac{2}{3}x$.

   (c) $x' = 8x - y^2$, $y' = -y + x^2$.

2. Consider the system

$$x' = 4x + 4y - x(x^2 + y^2),$$
$$x' = -4x + 4y - y(x^2 + y^2).$$

   a) Show there is a closed orbit in the region $1 \leq r \leq 3$, where $r^2 = x^2 + y^2$.

   b) Find the general solution.

3. Determine if the following systems admit periodic solutions.

   (a) $x' = y$, $y' = (x^2 + 1)y - x^5$.

   (b) $x' = x + x^3 - 2y$, $y' = -3x + y^5$.

   (c) $x' = y$, $y' = y^2 + x^2 + 1$.

   (d) $x' = y$, $y' = 3x^2 - y - y^5$.

(e) $x' = 1 + x^2 + y^2$, $y' = (x-1)^2 + 4$.

4. Find the values of μ where solutions bifurcate and examine the stability of the origin in each case.

(a) $x' = x + \mu y$, $y' = \mu x + y$.

(b) $x' = y$, $y' = -2x + \mu y$.

(c) $x' = x + y$, $y' = \mu x + y$.

(d) $x' = 2y$, $y' = 2x - \mu y$.

(e) $x' = y$, $y' = \mu x + x^2$.

(f) $x' = y$, $y' = x^2 - x + \mu y$.

(g) $x' = \mu y + xy$, $y' = -\mu x + \mu y + x^2 + y^2$.

(h) $x' = \mu x - x^2 - 2xy$, $y' = (\mu - \frac{5}{3})y + xy + y^2$.

5. Show that the orbits of $x' = y$, $y' = -f(x)$ in the xy phase plane are periodic if $f(x)$ is an increasing function and $f(0) = 0$. [Hint: Show that $\frac{1}{2}y^2 + V(x) = C$, where $V(x) = \int_0^x f(u)\, du$.]

6. Prove a generalization of the Bendixson–Dulac theorem : For the system (4.22), if there is a smooth function $h(x, y)$ for which $(hf)_x + (hg)_y$ is of one sign in a region R of the phase plane, then system (4.22) cannot have a closed orbit in R. Use this result to show that the system

$$x' = x(A - ax + by), \quad y' = y(B - cy + dx), \quad a,\, c > 0,$$

has no periodic orbits in the first quadrant. [Hint: Take $h(x, y) = 1/xy$.]

7. If $p(0,0)$ and $h(0)$ are positive, prove that the origin is an asymptotically stable critical point for the dynamical equation

$$
\begin{aligned}
x' &= y, \\
y' &= -p(x, y)y - xh(x).
\end{aligned}
$$

8. Examine the stability of the equilibrium solutions of the system

$$\frac{dx}{dt} = \gamma - y^2 - vx, \quad \frac{dy}{dt} = -vy + xy,$$

where γ and v are positive constants.

9. Let P denote the carbon biomass of plants in an ecosystem, H be the carbon biomass of herbivores, and ϕ the rate of primary carbon production

in plants due to photosynthesis. A model of plant–herbivore dynamics is given by

$$P' = \phi - aP - bHP,$$
$$H' = \varepsilon bHP - cH,$$

where a, b, c, and ε are positive parameters.

(a) Explain the various terms and parameters in the model and determine the dimensions of each parameter.

(b) Find equilibrium solutions and nullclines, and identify two separate cases.

(c) Analyze the dynamics in the cases of high primary production ($\phi > ac/\varepsilon b$) and low primary production ($\phi < ac/\varepsilon b$). Explain, in terms of bifurcation, what happens in the system if primary production is slowly increased from a low value to a high value.

10. Show that period orbits for the system

$$x' = y,$$
$$y' = -ky - V'(x),$$

are possible only if $k = 0$. Here $V'(x)$ denotes dV/dx.

11. Analyze the following nonlinear system, which models cell differentiation.

$$x' = y - x, \quad y' = -y + \frac{5x^2}{4 + x^2}.$$

12. Consider the scaled predator–prey model

$$x' = \frac{2}{3}x\left(1 - \frac{x}{4}\right) - \frac{xy}{1 + x},$$
$$y' = ry\left(1 - \frac{y}{x}\right),$$

where the carrying capacity of the predator depends on the prey population. Examine the dynamics of the model.

13. Consider the "falling fruit" model,

$$R' = F - aRP, \quad P' = P(kaR - d),$$

where R represents resources (biomass) and P, predators (biomass).

(a) Interpret the model and state what the parameters mean. Reduce the model to dimensionless form.

(b) Analyze the dimensionless model completely, finding nullclines, equilibria, and stability. Draw a phase diagram.

14. In the model (4.23)–(4.24), analyze the two cases $a = L$ and $a > L$.

15. Comment on the meaning of and analyze the Leslie predator–prey model

$$\frac{dx}{dt} = ax - bx^2 - cxy, \quad \frac{dy}{dt} = \delta y \left(1 - \frac{fy}{\delta x} \right).$$

Criticize the assumptions of the model.

16. Verify that the conditions for the Kolmogorov theorem hold for the Rosenzweig–MacArthur system.

17. (SIRS disease) In some diseases, called SIRS diseases, the individuals who get over the disease become susceptible again after a short period. Formulate a set of governing equations that describe the evolution of an SIRS disease. What is R_0 for this disease? Analyze the dynamics in the SI phase plane and determine when there is an endemic state where the disease is always present. What happens when the endemic state is not present?

18. The Gause model for predator–prey interactions has the form

$$\begin{aligned} x' &= xg(x) - yp(x), \\ y' &= y(q(x) - \mu). \end{aligned}$$

Assume that $g(x) > 0$ for $x < K$, $g(x) < 0$ for $x > K$, $g(K) = 0$; $p(0) = 0$ and $p(x) > 0$ for $x > 0$; $q(0) = 0$ and $q'(x) > 0$ for $x > 0$.

(a) Explain what these conditions mean ecologically.

(b) In the case there is a nonzero coexistent state, show that $(0,0)$ and $(K,0)$ are saddle points.

(c) In this case, derive a condition for the coexistent case to be unstable and sketch the phase plane.

19. Consider a cooperation model of the form

$$u' = u(1 - u) + auv, \quad v' = cv(1 - v) + cbuv,$$

where a, b, and c are positive constants. Find a condition for which there is a positive coexistent equilibrium. In this case, determine all the equilibria, their stability, and sketch the phase plane. What does the phase plane look like if there is no positive coexistent state?

20. In Example 11 in Chapter 3 we presented a discrete model of the progression of an SIR disease, and in equations (4.5)–(4.7) we formulated a continous model. In this exercise we explore the properties of (4.5)–(4.7).

(a) By scaling the populations S, I, and R by the total population N, and time by r^{-1}, show that the model can be nondimensionalized to obtain

$$\frac{du}{d\tau} = -R_0 uv, \quad \frac{dv}{d\tau} = (R_0 u - 1)v, \quad \frac{dw}{d\tau} = v,$$

and determine R_0. Why do we need to examine only the first two equations?

(b) Draw a uv phase plane diagram in each of the two cases $R_0 < 1$ and $R_0 > 1$, with various initial conditions lying on the line $u + v = 1$. Find the nullclines, equilibria, and the stability of any equilibria.

(c) Determine conditions when an epidemic occurs, that is, when the number of infectives increases.

(d) Show that the disease dies out because of lack of infectives.

(e) Find an relation between u and v that describes the orbits. Are there individuals who never get the disease?

(f) If initially there are a very small number of infectives, show that

$$\frac{dw}{d\tau} = 1 - w - e^{-R_0 w},$$

and if $R_0 w$ is small (the epidemic is small), then, approximately,

$$\frac{dw}{d\tau} = (R_0 - 1)w \left(1 - \frac{w}{a}\right), \quad a = \frac{2(R_0 - 1)}{R_0^2}.$$

Obtain the solution in closed form by separation of variables. What is the shape of the solution?

21. There are many infectious disease models that prescribe the dynamics of susceptibles (S), infectives (I), exposed but not yet infective (E), and the removed class (R). Various models are denoted by $S \rightarrow I$, $S \rightarrow I \rightarrow S$, $S \rightarrow I \rightarrow R$, $S \rightarrow E \rightarrow I \rightarrow R$, and so on. Try to classify each of the following diseases as one of these types, or indicate why it may not fit any of them: a cold, the flu, head lice, chickenpox, tuberculosis, AIDS, gonorrhea, malaria, and Ebola.

22. A simple evolutionary model of the early stage of HIV infection, before viral mutations overwhelm the immune system, can be described as follows. Let v denote the virus load and x the magnitude of the specific immune response against the viral strain. Then

$$\frac{dv}{dt} = rv - pxv, \quad \frac{dx}{dt} = cv - bx,$$

where r, p, c, and b are parameters. Comment on the various terms in the model and what the constants mean. Investigate the dynamics. (See Nowak (2006) for a complete discussion of the model.)

MATLAB Note. The following m-file presents the Euler method for solving two differential equations in two unknowns. For example, we solve the competition equations

$$x' = x(8 - 4x - y), \quad y' = y(3 - 3x - y), \quad x(0) = 1, \ y(0) = 0.8.$$

```
function systemEuler
clear all
x=1; y=0.8; xlist=x; ylist=y; T=2; N=1000; h=T/N;
for n=1:N
u=x+h*x.*(8-4*x-y); v=y+h*y.*(3-3*x-y);
x=u; y=v; xlist=[xlist,x]; ylist=[ylist,y];
end
t=0:h:T;
plot(t,xlist,t,ylist)                          % plots time series
plot(xlist,ylist), xlim([0,3]),ylim([0,3])     % plots orbit
```

4.6 Reference Notes

A concise account of elementary phase plane analysis can be found in, for example, Ledder (2004), Logan (2006), or almost any sophomore-level differential equations text or any mathematical biology text. At the next level, where there is emphasis on the theory, we refer to Hirsch, et al (2004) and Strogatz (1994), both of which discuss applications to physics, biology, engineering, and chemistry.

Most intermediate books in mathematical biology have long chapters on higher-dimensional systems. We particularly mention Kot (2001), which is a very complete, and excellent, treatment of systems of differential equations. Pastor (2008) is one of the few books that treats ecosystems; it contains a large amount of material on resource, plant, consumer dynamics, and nutrient cycling. Murdoch et al (2003) is focused entirely on population dynamics, and it presents a large number of models.

The study of epidemics and diseases has a long history, even dating back to Daniel Bernoulli, in the mid-1700s, who investigated smallpox. In the early 1900s, work by Ross showed that malaria was transmitted by mosquitos, for which he won a Nobel prize. In 1927, based on some of the hypotheses put forth by Ross, Kermack and McKendrick published their classic paper on the SIR model. We can think of this paper in some sense as the beginning of the modern era of mathematical epidemiology. Since their work, many extensions have been made to include spatial issues, demography (births and deaths), stochasticity, macroparasites, vector-borne diseases, the evolution of virulence and resistance, and myriad other important effects. Many of the basic models (SI, SIR, SIRS, and so on) are discussed in Brauer & Castillo-Chavéz (2001), Anderson (1982), and Anderson & May (1991). A recent collection of articles is Brauer et al (2008). Also see Smith? (2008), Britton (2003), and an excellent chapter in Mangel (2006), who has an accessible pathway to the literature. HIV/AIDS is discussed in a chapter in Nowak (2006), and virus-immune system interactions can be found in Wodarz (2007).

References

Anderson, R. M. (ed.) 1982. *Population Dynamics of Infectious Diseases: Theory and Applications*, Chapman–Hall, New York.

Anderson, R. M. & May, R. M. 1991. *Infectious Diseases of Humans: Dynamics and Control*, Oxford University Press, Oxford, UK.

Brauer, F. & Castillo-Chavéz, C. 2001. *Mathematical Models in Population Biology and Epidemiology*, Springer-Verlag, New York.

Brauer, F, van den Driessche, P., & Wu, Jianhong (eds.) 2008. *Mathematical Epidemiology*, Springer-Verlag, Berlin.

Britton, N. F. 2003. *Essential Mathematical Biology*, Springer-Verlag, New York.

Hirsh, M. W., Smale, S., & Devaney. R. L. 2004. *Differential Equations, Dynamical Systems, & An Introduction to Chaos,* 2nd ed., Academic Press, San Diego.

Kot, M. 2001. *Elements of Mathematical Ecology*, Cambridge University Press, Cambridge, UK.

Ledder, G. 2004. *Differential Equations: A Modeling Approach*, McGraw-Hill, New York.

Logan, J. D. 2006. *A First Course in Differential Equations*, Springer-Verlag, New York.

Mangel, M. 2006. *The Theoretical Biologist's Toolbox*, Cambridge University Press, Cambridge, UK.

Murdoch, W. W., Briggs, C. J., & Nisbet, R. M. 2003. *Consumer–Resource Dynamics*, Princeton University Press, Princeton.

Nowak, M. A., 2006. *Evolutionary Dynamics: Exploring the Equations of Life*, Belknap Press of Harvard University Press, Cambridge, MA.

Pastor, J. 2008. *Mathematical Ecology of Populations and Ecosystems*, Wiley Blackwell, Chichester, UK.

Smith?, R. 2008. *Modelling Disease Ecology with Mathematics*, American Institute of Mathematical Sciences, Springfield, MO.

Strogatz, S. H. 1994. *Nonlinear Dynamics and Chaos*, Addison-Wesley, Reading, MA.

Wodarz, D. 2007. *Killer Cell Dynamics*, Springer-Verlag, New York.

5

Concepts of Probability

The concepts in probability theory form the underpinnings of statistics, or science of analyzing data. These concepts can be introduced in many ways. There is a formal, axiomatic approach that lays out the theory carefully and aesthetically, and sometimes this detailed level of understanding is necessary in setting up and analyzing problems. In the first section, however, our approach is intuitive and conversational, with less formality. We introduce through several examples the basic ideas of a sample space, event, probability measure, and random variable. Regardless of one's level of interest in probability, it is extremely important to understand the basic terminology used in the area.

5.1 Introductory Examples and Definitions

We begin with the very simple experiment of rolling a six-sided die. This is the quintessential example in probability and we can keep it in our heads to remind us always of the basic concepts. The experiment is a random experiment, the outcome of which is not known until the die is cast. The *set of all possible outcomes*, $1, 2, 3, 4, 5, 6$, is called the **sample space** and is denoted by S. An **event** is some subset of interesting outcomes to which we want to assign the likelihood of occurring. For example, $1, 2$ is the event that *a 1 or 2 is rolled*, 5 is the event that *a 5 is rolled*, and $2, 4, 6$ is the event that *an even number is rolled*. The **Event space** is the set of all relevant events. The actual assignment of a probability to each event is called a **probability measure**. One way to assign

Mathematical Methods in Biology,
By J. David Logan and William R. Wolesensky
Copyright © 2009 John Wiley & Sons, Inc.

a probability to an event is to perform the experiment many times and count the percentage of times that the event occurs. That percentage, or fraction, is an estimate of the probability of that event. Of course, some problems, such as ecological experiments, cannot be repeated and we must seek other methods to determine probabilities. Finally, a **random variable** is an assignment of a numerical value to each outcome; for example, in tossing a coin, the the player may be paid $2 if a 3 or 6 is rolled, but she may have to pay $4 if a 4 is rolled. If we are gambling on a horse race, the track may decide that the payoff is $5 if No. 3 is the winner, but the bettor pays $2 if No. 3 is not a winner. The assignment of a numerical value to each outcome leads to values for other events. Historically, random variables, the key idea in probability, were introduced by Laplace in the context of gaming over 200 years ago.

What we just described is the essence of all probability models, whether they are problems in gambling, ecology, financial markets, or whatever. However, sometimes the setup of a problem is subtle or difficult, and we need to take extreme care. Then a return to an axiomatic context, or quantitative approach, may be the only way to sort through the intricacies.

We present several examples that show the variety of contexts that can occur.

Example 5.1

Here are four simple examples of experiments and sample spaces. Let $S_1 = \{\omega_1, \omega_2, ..., \omega_n\}$ be the set of n turtles in a pond; the random experiment in the back of our minds is *pick a turtle*. Let $S_2 = \{\sigma_1, \sigma_2, ..., \sigma_N\}$ be the set of N cars registered in Lancaster County, Nebraska; *pick a car*. Let $S_3 = \{$(brown, blue), (blue, brown), (blue, blue), (brown, brown)$\}$ be the possible eye color genotypes for a child whose parents each possess genes for both brown and blue eyes, where the first member of each pair represents the gene received from the father; the random experiment is to observe the eye color of a child born to this couple. Finally, let $S_4 = \{ss, sf, fs, ff\}$ denote the possible outcomes of a predator entering a patch on two consecutive days, with s = success in making a kill and f = failure to make a kill. The underlying random experiment is to observe the success of the predator over a two-day period.   □

As stated previously, a subset of a sample space S is called an **event.** In some problems, all the subsets of S are events, whereas in others only some of the subsets are designated events. For some uncountable sets it is not possible to assign a probability to each subset; however, this is a technicality and will not concern us here. As we observe later, there are some rules for the sets that belong to event space.

More formally than discussed above, a **random variable** X on a sample space S is an assignment of a real number to each element of the sample space; that is, it is a real-valued function on the sample space. We write $X : S \rightarrow \mathbb{R}$, as in calculus. In terms of standard function notation, if $\omega \in S$, then $X(\omega)$ is the real number associated with ω; or $\omega \rightarrow X(\omega)$. We always use capital letters for random variables. Formally, what makes X a random variable is the requirement that the set

$$\{\omega | X(\omega) \leq y, \quad y \in \mathbb{R}\}$$

is an event and therefore has an associated probability. We discuss this key idea in a later section.

Example 5.2

For the sample spaces in Example 5.1, we can define the following random variables. First, on S_1,

$$\omega \rightarrow W(\omega),$$

where $W(\omega)$ is the weight (kilograms) of turtle ω. For S_2, define

$$\omega \rightarrow G(\omega),$$

where $G(\omega)$ is represents the average miles per gallon (mpg) of car ω, rounded off to the nearest integer. For S_3, define

$$\omega \rightarrow E(\omega),$$

where

$$E(\omega) = \begin{cases} 2, & \text{(brown, brown)}, \\ 1, & \text{(brown, blue) or (blue, brown)}, \\ 0, & \text{(blue, blue)}. \end{cases}$$

This random variable represents the number of inherited brown eye genes. Finally, for S_4, define

$$\omega \rightarrow C(\omega),$$

where

$$C(\omega) = \begin{cases} 2, & \text{if } \omega = ss, \\ 1, & \text{if } \omega = sf \text{ or if } \omega = fs, \\ 0, & \text{if } \omega = ff. \end{cases}$$

The random variable C counts the number of successes in the two-day foraging effort. $\square$

Random variables are classified as continuous or discrete. A random variable $X : S \to \mathbb{R}$ is called a **continuous random variable** if $X(\omega)$ can take on a continuum of values (e.g., all values in an interval of real numbers), and it is called a **discrete random variable** if $X(\omega)$ can take only a finite number of values or a countably infinite number of values. As ω varies over all of the sample space S, the values $X(\omega)$ form the **range** of X.

Example 5.3

Referring to Examples 5.1 and 5.2, the weight W of a turtle is continuous because W can take any real value between zero and some large fixed number which is larger than any possible turtle weight; usually, we just take the largest weight to be infinity, so $0 < W(\omega) < \infty$ is the range of W. On the other hand, G, E, and C are discrete random variables because their range is countable (remember that we defined G as an integer). The range of G can be taken as $\{1, 2, 3, ...\}$ while the range for E and C is $\{0, 1, 2\}$. Once a random variable is defined, we can then go about the task of finding the probability measure of particular events. □

The reader should take note that a very important perspective has occurred—events are usually determined by random variables on the sample space; it is the random variable that is of primary importance.

Example 5.4

If we pick a turtle at random, what is the chance that its weight is less than or equal to 0.25 kg? The set of turtles in the pond with weight less than or equal to 0.25 is a subset of S_1 and thus forms the event

$$A = \{\omega \in S_1 : W(\omega) \le 0.25\}. \tag{5.1}$$

To save writing, to simplify the notation, and because it is understood from the context, we almost always drop the symbol ω and write the event as

$$A = \{W \le 0.25\}.$$

In the second example (gas mileage), the event

$$E = \{15 < G \le 20\}$$

represents the event consisting of all cars whose average mpg is between 15 (not included) and 20, or $\{16, 17, 18, 19, 20\}$. For the sample space S_3, the event

$$B = \{E = 1 \text{ or } E = 2\}$$

represents the event that the child has brown eyes (because the gene for brown eyes is dominant). An event associated with S_4 is $F = \{ss, ff\}$, or

$$F = \{C = 2 \text{ or } C = 0\}, \qquad \square$$

Where do we get values for the probability measure? We can define probabilities experimentally or theoretically. We mentioned that we can perform a lot of experiments and determine the percentage of time that an event occurs. Consider the turtle event $A = \{W \leq 0.25\}$. Suppose that we go to the pond and catch 100 turtles. After weighing them we find that 35 have weight less or equal to 0.25 kg. Then we could assign, approximately, $\Pr(A) = 0.35$. So if we pick a turtle, about 35% of the time it will have weight less than 0.25 kg. Using frequencies of occurring events is an acceptable approach to finding probabilities experimentally. When the elements of the sample space have an equal chance of occurring we are able to compute the probability of an event E occurring by counting the number of sample points that satisfy the event E and dividing by the total number of elements in the sample space. That is,

$$\Pr(E) = \frac{n(E)}{n(S)},$$

where $n(E)$ denotes the number of ways E can occur and $n(S)$ is the total number of sample points in S.

It is important to note that there is a basic difference between the sample spaces S_3 (eye colors) and S_4 (foraging) given in Example 5.1. Because each parent is just as likely to contribute a gene for brown eyes as for blue eyes, the sample points of S_2 are equally likely. This allows us to compute the probability of E of the event that a child will have blue eyes by

$$\Pr(E) = \Pr(\text{blue, blue}) = \frac{n(E)}{n(S_3)} = \frac{1}{4}.$$

However, it is not correct to assume that the sample points of S_4 are equally likely. This would be true only if the chance of making a kill was equal to the chance of failing to make a kill. It may be, for example, that saturation plays a role and the probability of success on a given day depends on whether a kill was made the previous day. When the sample points are not equally likely, we are unable to compute $\Pr(F)$ simply by counting sample points. Instead, we have to use other counting arguments involving combinations and permutations and various properties of probability. We observe later that there is an intimate connection between a probability measure and the choice of a random variable.

Given two events on a sample space, we can form other events using the set operations of union and intersection. With respect to the turtle example, let

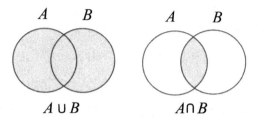

Figure 5.1 Venn diagram showing two events A and B, and their union $A \cup B$ (meaning that A or B occurs), and intersection $A \cap B$ (meaning A and B occur). The event A' means that A does not occur.

$A = \{W \le 0.25\}$ and $B = \{W > 0.5\}$. Then the event $A \cup B$ represents the event A or B, and $A \cap B$ is the event A and B. Here

$$A \cup B = \{W \le 0.25 \text{ or } W > 0.5\}.$$

Notice that there are no turtles in $A \cap B$, so we write $A \cap B = \phi$, where ϕ represents the empty, or null set. We say that two events A and B are **mutually exclusive** if $A \cap B = \phi$. The set of turtles with weight greater than 0.25 kg, or $A' = \{W > 0.25\}$, is said to be the **complement** of the event A in S, which is everything in S that is not in A. There is a simple geometric method to represent events. We draw a Venn diagram showing the sample space as a set S, as shown in Fig. 5.1. Then events A and B are subsets of S. The events $A \cup B$ and $A \cap B$ are the union and intersection of A and B, respectively.

In general, if X is a random variable on S, we want to associate a probability of the occurrence of any interesting event. For example, if we pick a car, what is the chance of event E occurring? That is, what is the probability that the car we pick has an average mpg between 15 and 20? We are asking about the probability of a property of the sample space of cars: namely, their gas mileage. This property is defined by a random variables. Again, the event is defined via the random variable.

Axiomatically, **event space** is a distinguished subset of events $\mathcal{E}$ from S that satisfies the rules: $S \in \mathcal{E}$ (i.e., S is an event); if $A \in \mathcal{E}$, then $A' \in \mathcal{E}$ (if A is an event, so is *not* A); if A_1, A_2, A_3, ... belong to $\mathcal{E}$, the union of A_1, A_2, A_3, ... belongs to $\mathcal{E}$ (i.e., A_1 or A_2 or A_3, or ... is an event). Also useful are the **DeMorgan laws**:

$$(A \cap B)' = A' \cup B', \quad (A \cup B)' = A' \cap B'.$$

These show that intersections of events are events.

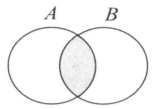

Figure 5.2 Conditional probability.

Formally, a **probability measure** is an association (function) that assigns a number $\Pr(A)$ to any event A in event space $\mathcal{E}$. It has the properties

$$0 \leq \Pr(A) \leq 1, \quad \Pr(S) = 1,$$

and if A and B are mutually exclusive, or $A \cap B = \phi$, the probabilities add:

$$\Pr(A \cup B) = \Pr(A) + \Pr(B).$$

It easily follows that

$$\Pr(A') = 1 - \Pr(A) \quad \text{and} \quad \Pr(\phi) = 0.$$

The additivity law extends to countable sums as well. That is, if $A_1, A_2, A_3, \ldots$ are **pairwise mutually exclusive**, or $A_i \cap A_j = \phi$ when $i \neq j$, then

$$\Pr(A_1 \cup A_2 \cup A_3 \cup \cdots) = \Pr(A_1) + \Pr(A_2) + \Pr(A_3) + \cdots.$$

If A and B are not mutually exclusive, one can show

$$\Pr(A \cup B) = \Pr(A) + \Pr(B) - \Pr(A \cap B).$$

Let A and B be events with $\Pr(B) > 0$. The **conditional probability** of A, given B, is defined by

$$\Pr(A|B) = \frac{\Pr(A \cap B)}{\Pr(B)}.$$

So the conditional probability of A given B is that part of A that belongs to B (Fig 5.2), because B has already occurred. Verbally, we say "$\Pr(A|B)$ is the probability that A occurs given that B has already occurred."

We say that two events A and B are independent if the occurrence of one event does not alter the probability of the other event happening. That is, A and B are **independent** if, and only if,

$$\Pr(A|B) = \Pr(A), \quad \Pr(B|A) = \Pr(B),$$

where $\Pr(A) \neq 0$, $\Pr(B) \neq 0$. It follows that

$$\Pr(A \cap B) = \Pr(A)\Pr(B). \tag{5.2}$$

We sometimes take the latter equation to be the definition of independence because it works equally well if the probabilities are zero.

Example 5.5

Assume that the success of a predator on a given search is *independent* of other hunting success. Using the sample space $S_4 = \{ss, sf, fs, ff\}$, which denotes the possible outcomes of a predator entering a patch on two consecutive days (s = success in making a kill, and f = failure to make a kill), the probabilities of the sample points in S_4 are given by

$$\Pr(ff) = \Pr(f)\Pr(f),$$
$$\Pr(sf) = \Pr(s)\Pr(f) = \Pr(fs),$$
$$\Pr(ss) = \Pr(s)\Pr(s). \qquad \square$$

When events are not independent, we can no longer use (5.2) to find $\Pr(A \cap B)$. Instead, we use conditional probability.

Example 5.6

We revisit the example of the success of a predator in two days of hunting. For this example we consider the simple case that a predator will make at most one kill on a given day. Assume that the probability that the predator is successful in making a kill is dependent on the success it experienced on the preceding day, as may be the case when satiation plays a role in determining foraging or search effort. Let's assume that if the predator was successful the first day, then the probability of success on the next day is 0.3, whereas if the predator fails to make a kill the first day, then the probability of success on the second day increases to 0.6. Further, assume that we know the probability of a kill on the first day is 0.7. We can express this information as $\Pr(s_2|s_1) = 0.3$, $\Pr(s_2|f_1) = 0.6$, and $\Pr(s_1) = 0.7$. We can then use the **multiplication rule**,

$$\Pr(A \cap B) = \Pr(B|A)\Pr(A) = \Pr(A|B)\Pr(B), \tag{5.3}$$

to determine the probability of making a kill on both days. The multiplication rule is applied when the events under consideration are determined to be dependent. Using (5.3), we find

$$\Pr(s_1 \cap s_2) = \Pr(s_2|s_1)\Pr(s_1) = 0.21. \qquad \square$$

Next, suppose that we are performing a field study and come upon a predator that is in the process of making a kill. Using probabilities given in Example 5.6, what is the probability that the predator made a kill the preceding day? When the given information is not readily apparent to answer the question at hand, we turn to **Bayes' theorem**. Bayes' theorem, a simple version of which is (5.3), is used to find $\Pr(B \mid A)$ when we know $\Pr(A|B)$.

Example 5.7

Using the information from Example 5.6, apply Bayes' theorem to find the probability that a predator made a kill on the first day given that the predator makes a kill on the second day. We are given that

$$\Pr(s_1) = 0.7, \quad \Pr(s_2|s_1) = 0.3,$$
$$\Pr(f_1) = 0.3, \quad \Pr(s_2|f_1) = 0.6.$$

To find $\Pr(s_1|s_2)$, the multiplication rule yields

$$\Pr(s_1|s_2) = \frac{\Pr(s_1 \cap s_2)}{\Pr(s_2)}. \tag{5.4}$$

The rules of probability also give

$$\Pr(s_2) = \Pr((s_2 \cap s_1) \cup (s_2 \cap f_1)) = \Pr(s_2 \cap s_1) + \Pr(s_2 \cap f_1)$$
$$= \Pr(s_2|s_1)\Pr(s_1) + \Pr(s_2|f_1)\Pr(f_1) = 0.21 + 0.18 = 0.39.$$

We again apply (5.3) to write

$$\Pr(s_1 \cap s_2) = \Pr(s_2|s_1)\Pr(s_1) = 0.21.$$

Substituting the last two equations into (5.4), we find that

$$\Pr(s_1 \mid s_2) = \frac{0.21}{0.39} = 0.5385. \quad \square$$

To prove a general version of Bayes' theorem, we need the following law of total probability, which is an important result in itself. It states that we can compute $\Pr(A)$ in terms of conditional probabilities of a partition of the sample space.

Theorem 5.8

(**Total Probability**) Let A be any event and B_i, $i = 1, 2, 3, ...$, be a set of events that partition S, meaning that $\Pr(B_i) > 0$, $B_i \cap B_j = \phi$, $\cup_i B_i = S$. Then

$$\Pr(A) = \sum_i \Pr(A|B_i)\Pr(B_i). \quad \square$$

Figure 5.3 shows the proof using a simple geometric illustration.

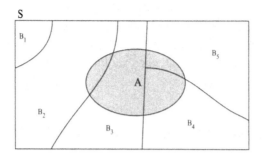

Figure 5.3 Proof of the theorem of total probability.

The general version of Bayes' theorem is:

Theorem 5.9

(**Bayes' theorem**). Let B_1, B_2, ..., B_n be a collection of events that partition a sample space S. Let A be an event such that $\Pr(A) \neq 0$. Then for any of the event B_k, $k = 1, 2, ..., n$,

$$\Pr(B_k|A) = \frac{\Pr(A|B_k)\,\Pr(B_k)}{\sum\limits_{i=1}^{n} \Pr(A|B_i)\,\Pr(B_i)}. \qquad \square$$

We leave this proof as an exercise for the reader.

EXERCISES

1. Use the fact that $A \cap A' = \phi$ and $A \cup A' = S$ to show that $\Pr(A') = 1 - \Pr(A)$.

2. Once every 20 years there is a flood on the Platte River that destroys the nesting habitat of a certain waterfowl, and the flood years are independent from one year to another.

   (a) What is the probability that in a given year there is no flood?

   (b) What is the probability that there is a flood in two consecutive years?

   (c) If there was a flood this year, what is the probability that there will be a flood next year?

   (d) What is the probability there will be a flood either this year or next year?

3. Verify equation (5.2).

4. What is the relation between $\Pr(A|B)$ and $\Pr(A'|B)$, if any?

5. Prove Bayes' theorem.

6. Assuming the blood type distribution to be A: 41%, B: 10%, AB: 3%, O: 46%, what is the probability that the blood of a randomly selected person will contain the A antigen? That it will contain the B antigen? That it will contain neither the A nor the B antigen?

7. Two birds are placed in captivity. The probability that the bird of species A will live 6 months is 0.5, and the probability that species B will live 6 months is 0.7. The probability that both die before 6 months is 0.4. What is the probability that neither will die within 6 months?

8. A lab test is 95% effective in detecting a virus and 1% of the time it gives a false positive. Overall, 0.5% of the population has the virus. If a random person tests positive, what is the probability that the person has the virus?

5.2 The Hardy–Weinberg Law

The question of how genes are passed on from parents to subsequent generations is a simple but nontrivial example of how probability models come into focus in evolutionary biology. If desired, this section may be omitted without loss of continuity.

To examine the simplest case in genetics, suppose that an organism has a locus (location) on a chromosome that has spots for two possible alleles, a or b. Therefore, the locus can have the pair aa, ab, or bb, which are called **genotypes**. The aa and bb are homozygous, and the ab is heterozygous. A **phenotype** is the expression of the trait that a pair dictates. One allele in the pair comes from one parent, and the other comes from the second parent. We are interested in understanding the dynamics of how genotypes change, generation through generation, with random breeding. To motivate a general model, we begin with an example.

Example 5.10

Consider a population of $N = 6$ individuals whose genotypes are aa, aa, aa, ab, bb, bb. This set is the gene pool, and there are $2 \times 6 = 12$ alleles. See Fig.

genotypes

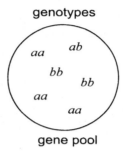

gene pool

Figure 5.4 Gene pool showing the genotypes of six people.

5.4. Let x, y, and z denote the fraction, or frequency, of each genotype, or

$$x = \text{frequency of } aa = \frac{3}{6},$$

$$y = \text{frequency of } ab = \frac{2}{6},$$

$$z = \text{frequency of } bb = \frac{1}{6}.$$

Note that $x+y+z = 1$. We can think of x, y, and z as being the probabilities of each genotype occurring in the gene pool. Knowing the frequencies of genotypes, we can easily calculate the frequency of each allele that occurs in the gene pool:

$$
\begin{aligned}
p &= \text{frequency of allele } a \\
&= \frac{\text{no. of } a \text{ alleles}}{\text{total no. of } a \text{ alleles}} \\
&= \frac{\text{no. of } a \text{ alleles in } aa \text{ genotype} + \text{no. of } a \text{ alleles in } ab \text{ genotype}}{\text{total number of } a \text{ alleles}} \\
&= \frac{\frac{3}{6} \times 2 \times 6 + \frac{2}{6} \times 6}{2 \times 6} = \frac{7}{12}.
\end{aligned}
$$

Similarly, we can compute

$$q = \text{frequency of allele } b = \frac{5}{12}.$$

Notice that we always have $p + q = 1$. □

Remark 5.11

To make general, summarizing statements based on Example 5.10, if there are N individuals in a gene pool, and if x, y, and z are the fraction of genotypes aa, ab, and bb that occur, respectively, then the fraction of alleles occurring is

$$p = x + \frac{1}{2}y, \quad q = z + \frac{1}{2}y, \tag{5.5}$$

where p = fraction of allele a, and q = fraction of allele b. Clearly,

$$x + y + z = 1, \quad p + q = 1.$$

Thus, the genotypes uniquely determine the allele frequencies p and q. However, p and q do not determine x, y, and z. (Can you give an example?) We interpret these quantities as probabilities and we write, with the obvious meaning, $x = \Pr(aa)$, $y = \Pr(ab)$, and $z = \Pr(bb)$. $\square$

Given the genotype and allele frequencies in a fixed generation, we ask, under conditions of random mating: what are the genotype and allele frequencies in the next generation. We now return to Example 5.10.

Example 5.12

In Example 5.10 we asked: What is the frequency of the genotype aa in the next generation? In terms of probability, the event aa in the next generation is equal to the event that parent 1 contributes a *and* parent 2 contributes a. Thus, by independence,

$$\begin{aligned}
x &= \Pr(aa) = \Pr(\text{parent 1 contributes } a) \cdot \Pr(\text{parent 2 contributes } a) \\
&= \frac{7}{12} \cdot \frac{7}{12} = \left(\frac{7}{12}\right)^2 = p^2.
\end{aligned}$$

Similarly,

$$\begin{aligned}
z &= \Pr(bb) = \Pr(\text{parent 1 contributes } b) \cdot \Pr(\text{parent 2 contributes } b) \\
&= \frac{5}{12} \cdot \frac{5}{12} = \left(\frac{5}{12}\right)^2 = q^2.
\end{aligned}$$

Finally, to get an ab in the next generation, parent 1 must contribute a and parent 2 contribute b, or, parent 1 contribute b and parent 2 contribute a. Thus, by independence,

$$\begin{aligned}
y &= \Pr(ab) = \Pr(\text{parent 1 contributes } a) \cdot \Pr(\text{parent 2 contributes } b) \\
&\quad + \Pr(\text{parent 1 contributes } b) \cdot \Pr(\text{parent 2 contributes } a) \\
&= \frac{7}{12} \cdot \frac{5}{12} + \frac{5}{12} \cdot \frac{7}{12} = 2\left(\frac{7}{12}\right)\left(\frac{5}{12}\right) = 2pq.
\end{aligned}$$

Knowing the genotype frequencies for the next generation, we can use (5.5) to compute the allele frequencies for that generation. We have, for the next generation,

$$p = \left(\frac{7}{12}\right)^2 + \frac{1}{2} \cdot 2 \left(\frac{7}{12}\right) \left(\frac{5}{12}\right) = \frac{7}{12},$$

$$q = \left(\frac{5}{12}\right)^2 + \frac{1}{2} \cdot 2 \left(\frac{7}{12}\right) \left(\frac{5}{12}\right) = \frac{5}{12}.$$

Therefore, the allele frequencies are the same as in the preceding generation! It is clear that if we continue the process to additional generations, the allele frequency will remain $\frac{7}{12}$ and $\frac{5}{12}$. $\square$

This example illustrates the famous **Hardy–Weinberg law**, proved in the 1920s. It states that under random mating, the allele frequencies p and q are constant through time, having the same values as in the initial generation. The genotypes, after the initial generation, remain at constant values

$$x = p^2, \quad y = 2pq, \quad z = q^2,$$

which are called the **Hardy–Weinberg ratios**. Our brief excursion into population genetics hardly does justice to the subject. Of course, specific influences can alter the Hardy–Weinberg equilibrium: for example, nonrandom mating, mutations, selection, limited population size, random genetic drift, and gene flow. An interesting second step is to see how selection and fitness affect allele frequencies over time; good elementary introductions to this theory may be found in Allman & Rhodes (2004) and Roughgarden (1998).

EXERCISES

1. In a population of 2000 caterpillars, let a and b be two alleles at a single locus. Suppose that two caterpillars have a rare genotype ab, none have bb, and the remaining caterpillars have a common genotype aa. Compute the frequency of genotypes of each type in the second generation.

2. In the United States, about 1 in 3700 people suffers from cystic fibrosis, which is caused by a recessive allele b occurring in the homozygous bb genotype. Estimate the percentage of the population that carries the recessive allele without showing signs of the disease.

5.3 Continuous Random Variables

In Section 5.1 we introduced in simple terms some of the elementary ideas in probability theory: the sample space, events, probability measure, and the concept of a random variable. However, to carry out more detailed analysis we require a more careful, quantitative formulation of these concepts. We are now ready to take up this task and show how all of these ideas fit together to describe a random system. Because all readers have been exposed in one way or another to the bell-shaped curve, or normal probability distribution, we begin with a general discussion of continuous random variables in the context of a familiar scenario in probability.

5.3.1 The Normal Distribution

Consider some measurable characteristic x of a large population of people (e.g., the height of an adult male, the gestation period for a birth, the IQ of eighth graders in a school system). If we choose a person at random from the population and take the measurement as x, we can consider x as a random variable X. So the sample space is the population and the random variable X is a mapping that associates to each person the measurement x. It comes with the caveat that $\{X \leq x\}$ is an event. If we take a large number of measurements and make a frequency plot, or the fraction of ipeople in the total population having the observation falling in the range x to $x + dx$, where dx is small, versus x, we get a histogram whose outline is a curve shaped like a bell, called a **bell-shaped curve** or **normal probability density**. There is some average value μ about which the frequency data are nearly symmetric, and there is some width σ that measures the variability about the average value. We have all seen this curve, shown in Fig. 5.5, but often, students in elementary courses do not realize that there is a mathematical formula for it:

$$f(x) = \frac{1}{\sqrt{2\pi\sigma^2}}e^{-(x-\mu)^2/2\sigma^2}, \quad -\infty < x < \infty. \tag{5.6}$$

It is a function of a real variable x, and it contains two parameters, μ and σ. The parameter μ, called the average or **mean**, measures the center of the graph on the x axis, and the parameter σ, called the **standard deviation**, measures the width of the graph from the mean to the inflection points (where the concavity changes). The square of σ, that is, σ^2, is called the **variance** (see Section 5.3.2).

We catalog two additional important properties of the function describing the bell-shaped curve: $f(x)$ is a nonnegative function and the area under its graph is 1, regardless of the values of the two parameters. In mathematical symbols,

$$f(x) \geq 0, \tag{5.7}$$

$$\int_{-\infty}^{\infty} f(x)\,dx = 1. \tag{5.8}$$

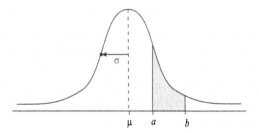

Figure 5.5 Normal probability density function with mean μ and standard deviation σ. The inflection points on the curve occur at $x \pm \sigma$. The shaded area is $\Pr(a < X \leq b)$.

It is an exercise in integral calculus (integration by parts) to show that the mean and variance can be calculated by the formulas

$$\mu = \int_{-\infty}^{\infty} x f(x)\, dx,$$

$$\sigma^2 = \int_{-\infty}^{\infty} (x - \mu)^2 f(x)\, dx.$$

The key connection to probability is that the area under the bell-shaped curve from a to b is the probability that the random quantity X, characterized by the bell-shaped curve, takes on a value between a and b. In symbols,

$$\Pr(a < X \leq b) = \int_a^b f(x) dx. \tag{5.9}$$

A good rule of thumb is the **68–95–97.5 rule**, which states that 68% of the data lies between $\mu - \sigma$ and $\mu + \sigma$ (within 1 standard deviation from the mean), 95% of the data lies between $\mu - 2\sigma$ and $\mu + 2\sigma$ (within 2 standard deviations from the mean), and 97.5% of the data lies between $\mu - 3\sigma$ and $\mu + 3\sigma$ (within 3 standard deviations from the mean).

Example 5.13

IQ scores for the 20–34 age group are normally distributed and have a mean of $\mu = 110$ and a standard deviation of $\sigma = 25$. By the rule of thumb, 68% of IQ scores for this group lie between 85 and 135. By symmetry, 2.5% have an IQ over 160 (greater than 2 standard deviations from the mean). □

If X is a random variable (RV) characterized by the normal probability density curve, we say X is **normally distributed** with mean μ and variance

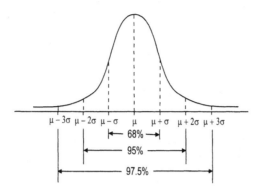

Figure 5.6 Normal probability density. The 68–95–97.5 rule: approximately 68% of the data lies between $x - \sigma$ and $x + \sigma$ (within 1 standard deviation of the mean), approximately 95% of the data lies within 2 standard deviations of the mean, and approximately 97.5% lies within 3 standard deviations of the mean.

σ^2, and we write $X \sim N(\mu, \sigma^2)$. If $X \sim N(\mu, \sigma^2)$, we define the **standard normal** RV Z by

$$Z = \frac{X - \mu}{\sigma} \quad \text{or} \quad X = \mu + \sigma Z.$$

One can show that $Z \sim N(0, 1)$, or that Z is a normal RV with mean 0 and variance 1. If x is an observation of X, then $z = (x - \mu)/\sigma$ is its associated **z-score**. Equations (5.3.1) allow us to go back and forth from observations x to z-scores. For the IQ scores from Example 5.13, if $x = 121$, then $z = (121 - 110)/25 = 0.44$, which is to say that an IQ score of 121 is 0.44 standard deviation above the mean. Generally, z is the fraction of the standard deviation that x deviates from the mean.

In tables of probabilities contained in books and in software packages, it is the standard normal Z that is tabulated. What data are actually given depends on the table. A common choice is to give values of the function $\Phi(z)$, where

$$\Phi(z) = \Pr(Z \leq z).$$

$\Phi(z)$, called the **cumulative distribution function** (cdf), is the area under the standard normal curve from $z = -\infty$ and z, or the area to the left of z (see figure 5.7).

Example 5.14

In the United States, the length X of a female's gestation period is normally distributed with mean $\mu = 266$ days and a standard deviation $\sigma = 16$ days.

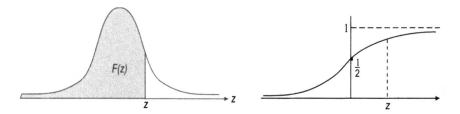

Figure 5.7 (a) Standard normal curve. The shaded area is the value of the cdf $\Phi(z)$. (b) Plot of the cdf $\Phi(z)$.

What percentage of pregnancies are longer than 290 days? We can proceed as follows.

$$
\begin{aligned}
\Pr(X > 290) &= \Pr(266 + 16Z > 290) \\
&= \Pr(Z > 1.5) \\
&= 1 - \Pr(Z \leq 1.5) \\
&= 1 - 0.9332 = 0.067,
\end{aligned}
$$

or 6.7%. The number $0.9332 \, [= \Pr(Z \leq 1.5)]$ was found from a table; MATLAB, and other computer algebra systems, also tabulate these values.   □

5.3.2 General Random Variables

Now we extrapolate ideas from the normal random variable case to any continuous random variable. A **continuous random variable** is a real variable, or quantity, that can take on an uncountable number of values, and the one it takes is based on the outcome of a random experiment. We can characterize a continuous random variable by its **probability density function** (pdf). A pdf is a nonnegative, real-valued function $f(x)$ defined for $-\infty < x < \infty$ that satisfies the two conditions (5.7)–(5.8), and a condition that links the RV to the pdf, namely,

$$
\Pr(a < X \leq b) = \int_a^b f(x) \, dx. \tag{5.10}
$$

So the probability of the event $a < X \leq b$, or that X lies in a certain range, is the area under the pdf curve from $x = a$ to $x = b$ (Fig. 5.4). Notice that if Δx is a very small quantity, then

$$
\Pr(x < X \leq x + \Delta x) = \int_x^{x + \Delta x} f(y) \, dy \approx f(x) \, \Delta x.
$$

Thus

$$f(x) \approx \frac{\Pr(x < X \le x + \Delta x)}{\Delta x},$$

which states that the pdf is a probability per unit length, or a density-like quantity. In the last equation we make an error, called $O(\Delta x)$, that goes to zero as $\Delta x \to 0$,[1] and so we can write

$$f(x) = \frac{\Pr(x < X \le x + \Delta x)}{\Delta x} + O(\Delta x).$$

An alternative function that may also be used to characterize a RV is the **cumulative distribution function** (cdf) $F(x)$, defined by

$$F(x) = \Pr(X \le x) = \int_{-\infty}^{x} f(z)\, dz.$$

The cdf "accumulates" probability as shown in the plot in Fig. 5.7 for the normal distribution. Clearly, $F(x) \to 0$ as $x \to -\infty$ and $F(x) \to 1$ as $x \to +\infty$. It need not be a continuous function; but we will assume that it is a nondecreasing, piecewise continuous function. At points x where f is continuous, it follows from the fundamental theorem of calculus[2] that

$$f(x) = F'(x).$$

Therefore, If one knows either the cdf or pmf, one can compute the other.

Example 5.15

(**Exponential random variable**) Suppose that an animal is foraging for food in a given region. How long will the animal search before it finds a food item?

---

[1] There are two useful **order symbols** (also called Landau symbols) that quantify the size of a term that depends on a small quantity h. In the discussion, $h = \Delta x$. The symbols are **little oh** of h, denoted by $o(h)$, and **big oh** of h, denoted by $O(h)$. We say that a quantity is $o(h)$ if the quantity goes to zero *faster than* h; symbolically, $o(h)$ is defined by

$$\lim_{h \to 0} \frac{o(h)}{h} = 0.$$

The order symbol $O(h)$ denotes a term that is bounded by $|h|$ for small h. For example, $h^{3/2} = o(h)$, but it is not bounded by h, so it is not $O(h)$. On the other hand, $h^{1/3} = O(h)$ because $h^{1/3} \le |h|$ for sufficiently small h; yet $\lim h^{1/3}/h$ does not go to zero as $h \to 0$, so $h^{1/3}$ is not order $o(h)$.

[2] One version of the fundamental theorem of calculus is that $\frac{d}{dx} \int_{a}^{x} f(y)\, dy = f(x)$ at a points x where f is continuous.

This searching time is a random variable T. Ecologists often model this search time with an **exponential RV** whose density is given by

$$f(t) = \begin{cases} 0, & t < 0, \\ \lambda e^{-\lambda t}, & t \geq 0, \end{cases}$$

where λ is a positive parameter. See Fig. 5.8. Later we show that $1/\lambda$ equals the average search time. The cdf is

$$F(t) = \int_0^\infty \lambda e^{-\lambda t}\, dt = 1 - e^{-\lambda t}. \qquad \square$$

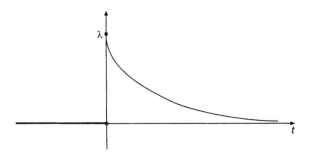

Figure 5.8 The pdf of an exponential RV. We refer to the pdf as the exponential distribution and we say that T is exponentially distributed.

Example 5.16

(**Uniform random variable**) Let X be a RV with density

$$f(x) = \begin{cases} \frac{1}{b-a}, & a \leq x \leq b, \\ 0, & \text{otherwise.} \end{cases}$$

This is the uniform pdf, and if a RV has this density, we say that X is uniformly distributed (constant) on the interval $a \leq x \leq b$. Clearly, this density models the case where all values in an interval are equally likely. $\square$

Example 5.17

(**Gamma distribution**) A random variable has the gamma distribution with positive parameters α and λ if its pdf is

$$f(x) = \frac{\lambda^\alpha e^{-\lambda x} x^{\alpha-1}}{\Gamma(\alpha)}, \qquad x \geq 0,$$

and zero otherwise, where $\Gamma(\alpha)$ is the gamma function:

$$\Gamma(\alpha) = \int_0^\infty e^{-x} x^{\alpha-1} \, dx, \quad \alpha > 0.$$

When $\alpha = n$, a positive integer, the distribution models the amount of time that one has to wait until n events have occurred. When $n = 1$ the gamma distribution reduces to the exponential distribution. When $\alpha = n/2$ and $\lambda = 1/2$, the gamma distribution is called the **chi-squared distribution**, written χ_n^2, with n degrees of freedom. The chi-squared distribution is a key distribution that arises in statistical inference studied in Chapter 6. We leave it to the reader to show, using integration by parts, that the gamma function behaves like a generalized factorial function in that

$$\Gamma(\alpha + 1) = \alpha\Gamma(\alpha)$$

for any number $\alpha > 0$, not just integers. For an integer n, we have $\Gamma(n) = (n-1)!$. □

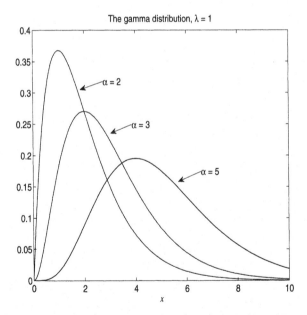

Figure 5.9 Gamma distribution for $\lambda = 1$ and $\alpha = 2, 3, 5$.

Given an RV X with pdf $f(x)$, we define the **expected value** (or **mean**) of X by

$$EX = \int_{-\infty}^{\infty} x f(x)\,dx.$$

[We use both notations $E(X)$ and EX.] Thinking of the integral as a limit of a sum, as in calculus, the right side looks like a weighted average (a sum) of the x values, where each x is weighted by the pdf at that point. The expected value measures the central tendency, or average, of the random variable. Often we denote $EX = \mu$. A measure of the spread of the distribution is the **variance**, defined by

$$\operatorname{var} X = \int_{-\infty}^{\infty} (x - \mu)^2 f(x)\,dx,$$

which can be regarded as a weighted average of the squared distances of x's from the mean. An alternative notation is $\operatorname{var} X = \sigma^2$. The square root of this, σ, is called the **standard deviation**. For a function of the RV X, say $g(X)$, we also have

$$Eg(X) = \int_{-\infty}^{\infty} g(x) f(x)\,dx.$$

The special choice of $g(X) = X^n$ ($n = 1, 2, 3, ...$) gives

$$E(X^n) = \int_{-\infty}^{\infty} x^n f(x)\,dx,$$

which are called the **moments** of the RV X. A very important result, and useful as an alternative way to calculate the variance, is

$$\operatorname{var}(X) = E(X^2) - (E(X))^2.$$

Example 5.18

The mean of an exponential RV with parameter λ is

$$\mu = \int_{-\infty}^{\infty} x\lambda e^{-\lambda} f(x)\,dx = \int_{0}^{\infty} x\lambda e^{-\lambda} f(x)\,dx = \frac{1}{\lambda}.$$

The integral was done using integration by parts, and we leave that calculation as an exercise. Similarly, one can show that the variance of an exponential RV is $1/\lambda^2$. □

Example 5.19

(**Approximating means and variances**) Let X be a random variable with mean μ and variance σ^2, and let g be a function. If the variance is small, we can use Taylor's theorem to approximate the expected value of g and its variance; this is called the **small variance approximation**. We have

$$
\begin{aligned}
E(g(X)) &= \int g(x)f(x)\,dx \\
&= \int \left(g(\mu) + g'(\mu)(x-\mu) + \frac{1}{2}g''(\mu)(x-\mu)^2 + \cdots\right)f(x)\,dx \\
&= g(\mu)\int f(x)dx + g'(\mu)\int (x-\mu)f(x)\,dx \\
&\quad + \frac{1}{2}g''(\mu)\int (x-\mu)^2 f(x)dx + \cdots = g(\mu) + \frac{1}{2}g''(\mu)\sigma^2 + \cdots.
\end{aligned}
$$

Now we can calculate the variance of $g(X)$ using the formula $\mathrm{var}(g(X)) = E(g(X)^2) - E(g(X))^2$. The first term, $E(g(X)^2)$, can be calculated from the previous formula by replacing $g(X)$ by $g(X)^2$. Using $(g^2)'' = 2gg'' + 2g'^2$, we get

$$
E(g(X)^2) = g(\mu)^2 + (g(\mu)g''(\mu) + g'^2(\mu))\sigma^2 + \cdots.
$$

Then

$$
\begin{aligned}
\mathrm{var}(g(X)) &= E(g(X)^2) - E(g(X))^2 \\
&= g(\mu)^2 + (g(\mu)g''(\mu) + g'^2(\mu))\sigma^2 - (g(\mu) + \frac{1}{2}g''(\mu)\sigma^2)^2 + \cdots \\
&= g'^2(\mu)\sigma^2 + \cdots. \quad \square
\end{aligned}
$$

Example 5.20

Let X be $N(\mu,\sigma^2)$ and $Y = g(X) = e^X$. Then Y is called a **lognormal** random variable. From the preceding calculation,

$$
E(Y) = e^\mu + \frac{1}{2}e^\mu\sigma^2 + \cdots, \qquad \mathrm{var}(Y) = e^{2\mu}\sigma^2 + \cdots.
$$

The exact values are given by

$$
E(Y) = e^{\mu+\sigma^2/2}, \qquad \mathrm{var}(Y) = e^{2\mu+\sigma^2}(e^{\sigma^2} - 1).
$$

Expanding these last two expression in powers of σ and retaining the lower-order terms shows the correctness of the approximation. $\square$

We now summarize key properties of continuous RVs. The reader should make certain that he or she understands these definitions and connections. A random variable X is characterized by its pdf $f(x)$ or its cdf $F(x)$. Then:

1. $F(x) = \Pr(X \le x) = \int_{-\infty}^{x} f(y)\,dy.$

2. $f(x) = F'(x)$ for x where f is continuous.

3. $E(X) = \int_{-\infty}^{\infty} x f(x)\, dx$ and $E(g(X)) = \int_{-\infty}^{\infty} g(x) f(x)\, dx$.

4. $\mathrm{var}(X) = E(X^2) - (E(X))^2$.

5.3.3 Predation with a Random Search

The exponential distribution plays a key role in foraging theory in ecology, and a brief excursion into this subject illustrates the salient points and will give the reader a good feel for this important distribution and its properties. Consider a single prey, which we call a **victim**, subject to a constant rate of predation a (measured in units of time^{-1}). If T is the victim's time of death, which is a random variable, the probability of T being less than time t is the cumulative distribution function $F(t) = \Pr(T \leq t)$. A sensible model for this distribution is

$$F(t) = \Pr(T \leq t) = 1 - e^{-at}, \quad t \geq 0,$$

and zero otherwise. The reader should graph this cdf. It follows, by differentiation, that the pdf for T is the exponential density

$$f(t) = a e^{-at}, \quad t \geq 0,$$

with $f(t) = 0$ for negative t. Notice that $\Pr(T \leq t) = 1 - \Pr(T > t)$, so $\Pr(T > t) = e^{-at}$. Simply, the probability of surviving to a time greater than t decreases exponentially. This distribution has the property of being **memoryless** in the sense that

$$\Pr(\text{victim survives to time } t + s \mid \text{it survives to time } t)$$
$$= \frac{\Pr(\text{victim survives to time } t + s \text{ and it survives to time } t)}{\Pr(\text{it survives to time } t)}$$
$$= \frac{e^{-a(t+s)}}{e^{-at}} = e^{-as}.$$

So the probability that it survives to time $t + s$ does not depend on t, the starting time, but rather only on s, the length of the interval. This says, for example, that the probability of surviving from day 5 to day 8 is the same as the probability of surviving from, say, day 10 to day 13. This implies that the predator, or the victim, does not learn from previous history. This type of search is called a **random search**.

Looking at this in a different way, let us assume that

$$\Pr(\text{victim is killed in the next time interval } dt) = a\,dt + \mathrm{o}(dt).$$

Taking

$$q(t) = \Pr(T > t) = \Pr(\text{victim survives to time } t),$$

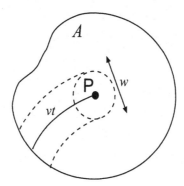

Figure 5.10 Area searched by a predator in time t/N with visual acuity w and speed v.

we have

$$q(t + dt) = q(t)(1 - a\,dt + o(dt)).$$

This means that to survive to time $t + dt$ the victim has to survive to time t and not be killed in the next dt (we can multiply probabilities because the events are assumed to be independent). Rearranging this last statement to get a difference quotient, we have

$$\frac{q(t + dt) - q(t)}{dt} = -aq(t) + \frac{o(dt)}{dt}.$$

Taking the limit as $dt \to 0$ gives a differential equation (a decay equation)

$$q'(t) = -aq(t).$$

We know that $q(0) = 1$, and therefore we can solve the differential equation to obtain

$$q(t) = e^{-at}.$$

Again we obtain the survivorship function for the victim.

A practical question is to ask how we can obtain the predation rate a. It should be clear that it could be estimated from survival data. But here is a theoretical model of predation that fits well into theoretical ecology. Imagine that we take the perspective of a single predator. Let A be the area of a region where the victim resides, and let v be the speed that the predator searches (distance per time). See fig. 5.10. Further, suppose that w is the width that the predator can detect as it searches (its visual acuity). We can break the total time t of its search into N small time increments of length t/N. Then, in this small interval the predator travels a distance vt/N and therefore covers an area

$$\frac{vtw}{N}.$$

Because the victim can reside anywhere in A, the probability of detecting the victim in this small time increment is the ratio

$$\frac{vtw}{NA}.$$

So the probability of not detecting the victim is

$$1 - \frac{vtw}{NA}.$$

Therefore, the probability of the victim surviving the search is

$$\Pr(T > t) = \left(1 - \frac{vtw}{NA}\right)^N.$$

Again, to survive each of the N time increments, the victim must survive the first, the second, and so on; so we multiply probabilities (by independence). Finally, taking the limit as $N \to \infty$ gives

$$\Pr(T > t) = e^{-at},$$

where

$$a = \frac{vw}{A}.$$

Therefore, by a theoretical argument, we have related the predation rate a to the search speed, detection width, and total area of the region. The number $1/a$ is called the characteristic time of the search process.

Example 5.21

(**Nicholson–Bailey equations**) Parasitoids are abundant insects, including some wasps and flies, that lay their eggs inside the eggs or larva of other insects. The victims, or hosts, therefore provide a nutritious environment for the growth and development of the parasitoid eggs, which eventually hatch into larva and escape. Nicholson and Bailey, in the mid-1930s, studied host–parasitoid dynamics and developed a model for predation that is central in theoretical ecology. Let H_t and P_t be the population of hosts and parasitoids at the beginning of season t. The usual geometric growth model for hosts is

$$H_{t+1} = rH_t,$$

where $r > 1$ is the growth rate. However, only a fraction of the hosts survive to the next season because of parasitoid predation. We can argue as in Example 5.20 that the probability of survival of a host in the presence of a single predator is e^{-a}, where a is the predation rate; we have taken the search time to be one season. If there are P_t predators, the survival probability is e^{-aP_t} because a host must survive attack by each predator (independence again). Therefore, the model equation for the host is

$$H_{t+1} = rH_t e^{-aP_t}. \tag{5.11}$$

Assuming that parasitoids lay one egg per host, the number of parasitoids in the next generation is

$$P_{t+1} = H_t(1 - e^{-aP_t}), \tag{5.12}$$

where $1 - e^{-aP_t}$ is the fraction of hosts H_t that survive. Equations (5.11)–(5.12), a pair of coupled difference equations, form the **Nicholson–Bailey model**. This model can be analyzed by the methods in Chapter 3. (See the Exercises.)   □

Example 5.22

(**Hazard rates**) All living creatures suffer random events in their environments that cause death. These hazards include predation, weather events, random accidents, and so on. Humans are subject to additional health hazards based on risky behavioral choices such as smoking. Here is how we model death rates due to these types of hazards. Let T be the time of death, a random variable, and let $f(t)$ and $F(t)$ be its pdf and cdf, respectively. If one lives to time t, what is the probability of being alive at time $t + dt$? That is, what is the conditional probability

$$\Pr(T \geq t + dt \,|\, T > t)?$$

By the definition of conditional probability, this can be evaluated as

$$\frac{\Pr(T \geq t + dt)}{\Pr(T > t)} = \frac{1 - \Pr(T < t + dt)}{1 - \Pr(T \leq t)}$$
$$= \frac{1 - F(t + dt)}{1 - F(t)}.$$

Therefore, if one lives to time t, the probability of *not* surviving to $t + dt$ is 1 minus this quantity, or

$$1 - \frac{1 - F(t + dt)}{1 - F(t)} = \frac{F(t + dt) - F(t)}{1 - F(t)}$$
$$\approx \frac{F'(t)}{1 - F(t)} dt$$
$$= \frac{f(t)}{1 - F(t)} dt,$$

where we have used the fact that $F(t + dt) = F(t) + F'(t)dt + \cdots = F(t) + f(t) \, dt + \cdots$. We define the **hazard rate** $\lambda(t)$ by

$$\lambda(t) = \frac{f(t)}{1 - F(t)}.$$

Therefore, in summary, the interpretation of the hazard rate is: if one lives to time t, the probability of not surviving to time $t + dt$ is $\lambda(t)dt$. □

Example 5.23

Given a hazard rate $\lambda(t)$, the equation

$$\lambda(t) = \frac{F'(t)}{1 - F(t)}$$

determines, by integration, the cdf of the time of death T. In the simplest case, if $\lambda(t) = \lambda$ is constant, we can integrate both sides from 0 to t to obtain

$$\lambda t = \int_0^t \frac{F'(s)}{1 - F(s)} ds = \int_1^{1-F(t)} \frac{-1}{r} dr = -\ln(1 - F(t)).$$

[In the second inequality we made the substitution $r = 1 - F(s)$, $dr = -F'(s) \, ds$.] Therefore, $F(t) = 1 - e^{-\lambda t}$, which is the cdf for an exponential distribution. □

5.3.4 Central Limit Theorem

If you have had any experience with probability or statistics, you have noticed that the normal probability distribution plays an essential role. The reason is the **central limit theorem** (CLT), which states that the average (mean) of n independent RVs tends to a normal distribution as n tends to infinity. It was first shown by de Moivre in the early 1700s that the normal distribution approximates a binomial distribution (binomial RVs are discrete RVs, discussed in the next section), and then Laplace and Gauss noted that measurement errors, which are the average of many smaller errors, are approximately normally distributed. In the 1800s, Galton showed that weights and heights, and other qualities, are approximately normal. There are many variations of the CLT, depending on the types of RVs being averaged.

To be precise, let $X_1, X_2, ..., X_n$ be n independent random variables with identical distributions, and let μ and σ^2 denote the mean and variance of each. We know that (shown in Section 5.6)

$$\begin{aligned} \mathrm{E}(X_1 + X_2 + \cdots + X_n) &= n\mu, \\ \mathrm{var}(X_1 + X_2 + \cdots + X_n) &= n\sigma^2. \end{aligned}$$

Let the sum be denoted by

$$S_n = X_1 + X_2 + \cdots + X_n.$$

The central limit theorem tells us about the properties of S_n as $n \to \infty$. To set it up, let Z_n be the z-score of S_n, defined by

$$Z_n = \frac{S_n - n\mu}{\sqrt{n\sigma^2}}.$$

[Recall that if X is an RV, its z-score is $Z = (X - \text{mean})/(\text{standard deviation})$.] Note that $E(Z_n) = 0$ and $\text{var}(Z_n) = 1$. Then:

Theorem 5.24

(**Central limit theorem**) Under the stated assumptions, the sequence Z_n of z-scores converges in distribution to a standard normal distribution $N(0,1)$. This means that

$$\lim_{n\to\infty} \Pr(Z_n \le z) = \int_{-\infty}^{z} \frac{1}{\sqrt{2\pi}} e^{-x^2/2}\, dx. \qquad \Box$$

This implies, of course, that S_n converges to an $N(n\mu, n\sigma^2)$ random variable. We will not prove the CLT, but refer to other texts. A simple straightforward proof using moment generating functions is given in Grimmett & Welsh (1986) as well as in other places.

Example 5.25

Let S_n be the sum of n RVs X_i, each with an exponential distribution $f(x) = \lambda e^{-\lambda t}$. Then the sum $S_n = X_1 + X_2 + \cdots + X_n$ approaches a normal distribution with mean and variance $1/\lambda$. So

$$\Pr(S_n \le z) \simeq \int_{-\infty}^{z} \frac{1}{\sqrt{2\pi/\lambda}} e^{-(x-1/\lambda)^2/(2/\lambda)}\, dx. \qquad \Box$$

EXERCISES

1. Referring to the gestation period of mothers in Example 5.14, answer the following questions.

   (a) What fraction of pregnancies are less than or equal to 275 days?

   (b) What percentage of pregnancies are between 275 and 290 days?

(c) How many days must a woman carry a child to be in the longest 10% of pregnancies?

2. The height X of adult women in the United States, given in inches, is normally distributed, or $X \sim N(64.5, 2.5^2)$. What is the probability that a randomly chosen woman is at least 62 inches tall?

3. Let $f(x) = cx^2$, $0 \le x \le 2$, where c is a constant.

   (a) Determine the value of c that makes f a pdf on $0 \le x \le 2$.

   (b) If X is an RV with pdf f, determine $E(X)$, the second moment $E(X^2)$, and $\text{var}(X)$.

   (c) Find the cdf $F(x)$ and plot it on $\mathbb{R}$.

   (d) Find $\Pr(X > 1)$.

4. Let X be an exponential RV with parameter $\lambda = 0.1$. Compute $\Pr(X > 15)$ and $\Pr(X > 20 | X > 5)$.

5. Let X be a continuous RV with pdf $f(x)$. The **moment-generating function** (mgf) $M_X(t)$ of X is defined by

$$M_X(t) = E(e^{tX}) = \int_{-\infty}^{\infty} f(x)e^{tx}\, dx$$

for values of t for which the improper integral exists. The mgf sometimes provides an easy way to calculate moments of an RV.

   (a) Show that $E(X) = M'_X(0)$.

   (b) Show that $\text{var}(X) = M''_X(0) - (M'_X(0))^2$.

   (c) Show that the mgf for an exponential RV is $M_X(t) = \frac{\lambda}{\lambda - t}$, $\lambda > t$.

   (d) Find the mgf for X if $X \sim N(0, 1)$.

   (e) Show that $\text{var}(aX + b) = a^2 \text{var} X$.

6. Let $X_1, X_2, ..., X_{25}$ be a random sample from a uniform distribution on the interval $[0, 1]$. Use the central limit theorem to estimate $\Pr(X_1 + X_2 + \cdots + X_{25} < 12.5)$.

7. If $F(x)$ is the cdf for a nonnegative continuous RV X, show that

$$E(X) = \int_{-\infty}^{\infty} (1 - F(x))dx.$$

8. Show that the mean and variance of a uniform RV on the interval $[a, b]$ are

$$\mu = \frac{b-a}{2}, \qquad \sigma^2 = \frac{(b-a)^2}{12},$$

and find the cdf. Sketch the density and the cdf function.

9. If $X \sim N(\mu, \sigma^2)$ is a normal RV with pdf $f(x)$, $Y = e^X$ is called a **lognormal RV**. The following steps compute the pdf $f_Y(y)$ of the lognormal RV Y. Give a precise reason for each equality:

$$\begin{aligned}
\Pr(Y \le y) &= \Pr(X \le \ln y) \\
&= \int_{-\infty}^{\ln y} f(x)\, dx \\
&= \int_0^y f(\ln w) \frac{1}{w}\, dw.
\end{aligned}$$

Therefore, $f_Y(y) = f(\ln y) \frac{1}{y}$, or

$$f_Y(y) = \frac{1}{y\sqrt{2\pi\sigma^2}} e^{-(\ln y - \mu)^2/2\sigma^2}.$$

10. Often a random effect Y is the result of a large number of multiplicative factors $X_1, X_2, \ldots, X_n$. That is, $Y = X_1 X_2 \cdots X_n$, where the X_i are identically distributed random variables.

   (a) If n is large, explain why Y should be log-normally distributed.

   (b) For a population of animals, assume a geometric growth law where the growth rate is a random variable. That is, the population growth law is

   $$N(t+1) = R(t)N(t), \quad t = 0, 1, 2, 3 \ldots, \quad N(0) = n_0,$$

   where $N(t)$ is the population at time t and $R(1), R(2), \ldots$ are identically distributed RVs. Explain why the total population $N(t)$ at time t, as t gets large, is approximately lognormally distributed.

11. Let X be a continuous RV with pdf and cdf $f(x)$ and $F(x)$, respectively, and let $g(x)$ be a differentiable, strictly increasing function of x. If Y is a random variable defined by $Y = g(X)$, show that the pdf for Y is

$$f_Y(y) = f(g^{-1}(y)) \frac{d}{dy} g^{-1}(y),$$

where g^{-1} is the inverse function of g. [Hint: Begin with $\Pr(Y \le y)$.]

12. Using Exercise 11, show that the pdf $f_Y(y)$ of $Y = X^2$ is

$$f_Y(y) = \frac{1}{2\sqrt{y}} f(\sqrt{y})$$

if $X \geq 0$ and has density $f(x)$.

13. If the hazard rate is a linear function of t, or $\lambda(t) = a + bt$, show that the density function for the time of death T is

$$f(t) = (a + bt)e^{-(at+bt^2/2)}.$$

When $a = 0$, this density is called the **Rayleigh density**.

14. An article in the newspaper states that a recent study shows that the death rate of a person who smokes is twice that of a nonsmoker. This means, in terms of hazard rates, that $\lambda_s(t) = 2\lambda_n(t)$, with the obvious notation. Conclude that of two people the same age, one a smoker and one not, the probability that the smoker survives to any given age is the *square* of the corresponding probability for a nonsmoker.

15. Show that the expected value of a gamma-distributed RV X is $E(X) = a/\lambda$ and its variance is $\text{var}(X) = a/\lambda^2$.

16. Sketch a graph of the density for a Cauchy RV given by

$$f(x) = \frac{1}{\pi} \frac{1}{1 + x^2}.$$

From the plot, guess the expected value of X. Now compute the expected value. What is your conclusion?

5.4 Discrete Random Variables

The preceding definitions and results for continuous RVs go over to discrete random variables in an obvious way. Let X be a discrete RV that assumes the values $x_1, x_2,...,x_k, ...$ The analog of the pdf is the discrete **probability mass function**

$$p_k = \Pr(X = x_k), \quad k = 1, 2, 3, ...,$$

which gives the probabilities of X taking the various values. Clearly, $\sum_k p_k = 1$. The **expected value** of X is the weighted average

$$\mu = E(X) = \sum_k x_k p_k,$$

and more generally,

$$E(g(X)) = \sum_k g(x_k)p_k.$$

The **variance** is

$$\mathrm{var}(X) = E((X - \mu)^2)$$

and

$$\mathrm{var}(X) = E(X^2) - (E(X))^2.$$

Note that the mean and variance do not have the same units, but the mean and standard deviation do. Sometimes the dimensionless **coefficient of variation** CV, defined by

$$\mathrm{CV} = \sigma/\mu,$$

is used as a measure of variation in a sample; for example, CV= 0.40 carries the rough interpretation that on average X varies 40% from its mean value.

Some important discrete random variables are listed below, and the reader should plot the pmf in each case for several values of the parameters. For a discrete RV, the pmf plots as a histogram, or vertical bar graph, with the values x_i on the horizontal axis and the corresponding probabilities p_i on the vertical axis. The MATLAB command: bar([x1, x2,...,xn],[p1, p2,...,pn]), ylim([0 1]) produces the desired plot.

Now we present several important examples of discrete RVs and distributions that arise in biological problems.

Example 5.26

(**Bernoulli**) The simplest discrete RV is the Bernoulli RV. Here $X \in \{0, 1\}$ and the probabilities are defined by

$$p_1 = \mathrm{Pr}(X = 0) = p, \quad p_2 = \mathrm{Pr}(X = 1) = 1 - p,$$

where $0 \le p \le 1$. This models a single random experiment, called a Bernoulli trial, where one chooses between two items that have been labeled or assigned the numbers 0 and 1. In MATLAB the command rand gives a value of a uniformly distributed continuous random variable on the interval $[0, 1]$, so the following MATLAB commands generate a value of a Bernoulli RV; the user inputs the value of p.

```
r=rand;
if r<p
X=1;
else
X=0;
```

end

X   □

Example 5.27

(**Binomial**) In a "binomial experiment" we perform N independent, identical trials of an experiment that has only two outcomes, success S and failure F. We denote by X the random variable giving the number of successes in N trials. It is important that the trials be exactly the same; therefore, if selections are made from a group of items, they must be done with replacement of the item chosen before performing another trial. (The issue of performing experiments with replacement or without replacement is an important one.) If $\Pr(S) = p$ and $\Pr(F) = 1 - p$, the probability mass function for a binomial RV is

$$\Pr(X = k) = \binom{N}{k} p^k (1-p)^{N-k},$$

which is the probability of obtaining k successes in the N trials. The symbol $\binom{N}{k}$ is called a **binomial coefficient** and is defined by

$$\binom{N}{k} = \frac{N!}{k! \, (N-k)!}.$$

Combinatorially speaking, the binomial coefficient is the number of ways to choose k items out of N items, ignoring the order of the items chosen. The binomial coefficient is also, recalling the binomial theorem, the coefficient of x^k in the expansion of $(x + 1)^N$; that is,

$$(x+1)^N = \sum_{k=0}^{N} \binom{N}{k} x^k.$$

For a binomial RV, the mean and variance are

$$\mu = Np, \qquad \sigma^2 = Np(1-p).$$

The binomial RV defines the binomial distribution. We write $X \sim \mathrm{bin}(N, p)$ to say that X is binomial-distributed. Observe that a binomial RV X can be written as a sum of N Bernoulli RVs $Y_1, Y_2, ..., Y_N$:

$$X = Y_1 + Y_2 + \cdots + Y_N.$$

That is, X counts and adds up the number of successes out of N independent attempts. We remark that if the population is very large, there will be little difference in the result if we replace the items or if we don't.

How can we choose, or simulate, a random value from a binomial distribution, say bin(100,0.95)? We can use the fact that X is the sum of 100 Bernoulli RVs, each of which gives the value 0 or 1, depending on failure or success. The MATLAB code is

```
function BinomialSim
N=100; p=0.95; X=0;
for i=1:N
r=rand;
if r < p
Y=1;
else Y=0;
end
X=X+Y;
end
X
```

Some random population models are based on a binomial process. That is, if the population is N at some time t, and if p is the survival rate, or fraction that survive, the population at the next time $t + 1$ is a value chosen from the distribution bin(N, p). In other words, we pick out a random number of survivors. □

Example 5.28

(**Geometric**) Again consider successive independent trials where an event S (success) occurs with probability p. Let X be the number of trials until a success occurs the first time. So, X is the **waiting time** for success. Then

$$\Pr(X = k) = p(1 - p)^k, \ k = 1, 2, 3, \ldots$$

is called the **geometric distribution** with parameter p. Again, experiments are done with replacement. The expected value and variance are

$$E(X) = \frac{1}{p}, \quad \text{var} = \frac{1 - p}{p^2}. \quad \square$$

Example 5.29

(**Negative binomial distribution**) Again in the binomial setting with replacement, let X be the number of trials required until the rth success occurs. Then

$$\Pr(X = k) = \binom{k - 1}{r - 1} p^r (1 - p)^{k-r}, \quad k = r, r + 1, \ldots$$

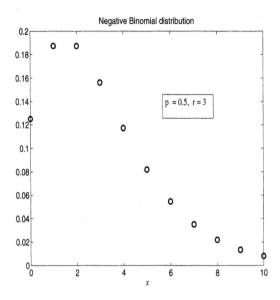

Figure 5.11 Negative binomial distribution when $p = 0.5$ and $r = 3$.

is the negative binomial distribution. Its mean and variance are

$$\mu = \frac{r}{p}, \quad \sigma^2 = \frac{r(1-p)}{p^2}. \quad \square$$

Example 5.30

(**Poisson distribution**) The Poisson distribution with parameter $\lambda > 0$ is

$$\Pr(X = k) = \frac{\lambda^k e^{-\lambda}}{k!}, \quad k = 0, 1, 2, \dots$$

A Poisson RV models the probability of unpredictable or rare events over a time period, where λ is the average occurrence of the event in that period. For example, it counts the events witnessed by an observer during a given period, such as the number of seedlings that germinate in one week. Its mean and variance are

$$\mu = \lambda, \quad \sigma^2 = \lambda.$$

If you know only the rate r that events occur (events per time), then $\lambda = rt$, where t is the time period. For large λ the density is bell-shaped. $\square$

Example 5.31

(**Multinomial distribution**) Recall that the binomial experiment has two outcomes, S and F (success and failure). A multinomial experiment has n trials, *with* replacement, and more outcomes. Specifically, assume that

1. There are n trials with replacement.

2. There are k outcomes $E_1, ..., E_k$ with associated probabilities $p_1, ..., p_k$.

3. Let X_i be the number of outcomes of E_i, $i = 1, 2, 3, ...$. Then

$$\Pr(X_1 = x_1, X_2 = x_2, ..., X_k = x_k) = \begin{pmatrix} n \\ x_1 \quad x_2 \quad ... \quad x_k \end{pmatrix} p_1^{x_1} p_2^{x_2} \cdots p_k^{x_k},$$

where

$$\sum_{i=1}^{k} x_i = n, \quad \sum_{i=1}^{k} p_i = 1, \quad \begin{pmatrix} n \\ x_1 \quad x_2 \quad ... x_k \end{pmatrix} = \frac{n!}{x_1! x_2! \cdots x_k!}.$$

The latter symbol is called the **multinomial coefficient**, and the distribution is the **multinomial distribution**.

As an example, suppose that a pond's turtle population is 75% painted turtles and 25% snapping turtles. If you pick two turtles from the pond, what is the probability that you will get one of each? Here there are two ($k = 2$) outcomes (painted turtle and snapper) and two trials ($n = 2$)—pick 2 with replacement. We have

$$\Pr(X_1 = 1, X_2 = 1) = \begin{pmatrix} 2 \\ 1 \quad 1 \end{pmatrix} (0.75)^1 (0.25)^1 = 0.375. \quad \square$$

Example 5.32

(**Hypergeometric distribution**) The hypergeometric RV models an experiment with two outcomes (S and F) *without* replacement. Suppose that there are N items in a bin and k of them represent successes and $N - k$ represent failures. We pick n of them without replacement. What is the probability that we will obtain x successes? Letting X denote the random number representing successes, we have

$$\Pr(X = x) = \frac{\binom{k}{x} \binom{N-k}{n-x}}{\binom{N}{n}}, \quad x = 0, 1, 2, ..., n < N.$$

Combinatorially, $\binom{N}{n}$ is the number of ways to choose the n items, $\binom{k}{x}$ is the number of ways to choose the x's, and $\binom{N-k}{n-x}$ is the number of ways to choose the failures. The mean and variance are

$$\mu = np, \quad \sigma^2 = np(1-p)\frac{N-n}{N-1}.$$

Notice that if the sample size n is much smaller that the population size N, then $(N-n)/(N-1) \approx 1$ and the mean and variance are nearly the same as for the binomial distribution.

For example, in a pond there are 30 turtles, 20 of them being painted turtles ($C.\ picta$). We choose, without replacement, 12 turtles. What is the probability that at least half of them will be $C.\ picta$? Here we want

$$\Pr(X \geq 6) = \sum_{x=6}^{k=12} \frac{\binom{k}{x}\binom{N-k}{n-x}}{\binom{N}{n}} = \sum_{x=6}^{k=12} \frac{\binom{20}{x}\binom{10}{12-x}}{\binom{30}{12}}.$$

We leave this calculation to the reader.   □

5.4.1 Likelihood

A common problem in ecology is to census an animal population and determine from the census how many animals are present in a given region. Of course, we usually cannot catch them all. One way to make an estimate is to perform a **mark-and-recapture** experiment. We catch a number of animals, mark them in some way, and then release them back into the environment. Later, we capture or sight another group of animals and count how many are marked. Can that procedure give us an accurate estimate of the population?

Example 5.33

(**Mark and recapture**) A simple analogous situation is to estimate the crowd size at a baseball game. As fans enter, we give out 100 free hats at random. Then, when the game begins, we pick a section of size 200, say, and count the number of hats in that section. If we get 4 hats in that section, we conclude that there are $100/4 = 25$ sections. Hence, there are $200 \times 25 = 5000$ attendees at the ball game.

We proceed similarly for the animal experiment. Suppose that we mark $k = 100$ fish in a large lake, and then later we catch $n = 200$ fish and get 4 successes (marked animals). Therefore, we conclude that there are

$$N = 25 \times 200 = \frac{100}{4} \times 200 = \frac{k}{x}n = 5,000$$

fish in the lake. The number N, defined by

$$N = \frac{k}{x}n,$$

is called the **Lincoln index**. Here N is the unknown population, k is the number originally marked, n is the recaptured sample size, and x is the number found marked out of the recaptured sample. □

Example 5.34

(**Maximum likelihood estimate**) Consider a discrete RV X with pmf $f_\theta(x)$, where θ is some parameter in the density. Now suppose that we perform an experiment and $X = x$ occurs. Then the function of θ defined by

$$L(\theta) = f_\theta(x),$$

where x is now fixed, is called the **likelihood function**. We can think of asking what is the most likely value of θ given that x actually occurred. In different notation, thinking of finding θ, given x, we often write $L(\theta) = L(\theta|x)$. The value $\hat\theta$ where L attains its maximum value is called the **maximum likelihood estimator** of θ. (Because products often appear, we often maximize the logarithm of L).

The mark-and-recapture experiment should remind us of the hypergeometric distribution, where we count the number of successes x in n draws from a population of size N. Therefore, we take the likelihood function as a function of N, or

$$L(N) = \frac{\binom{k}{x}\binom{N-k}{n-x}}{\binom{N}{n}}, \quad x = 0, 1, 2, ..., n < N.$$

Suppose that we mark $k = 25$ and recapture $n = 60$. Of those, we find $x = 5$ marked. So the likelihood function is

$$L(N) = L(N|x = 5) = \frac{\binom{25}{5}\binom{N-25}{60-5}}{\binom{N}{60}}.$$

The Lincoln estimate gives $N = 300$; the maximum likelihood estimator of L is $\hat N \approx 300$. [To obtain this estimate we can plot $L(N)$, or $\ln L(N)$, for values of N and observe where the maximum occurs.] □

Example 5.35

Let T be a continuous random variable representing the time that a cell dies, and suppose that T is exponential with unknown parameter λ. Thus, the pdf is

$$f(t; \lambda) = \lambda e^{-\lambda t}, \quad t > 0.$$

To determine λ, we observe n independent cells and measure their deaths at times $t_1, t_2, \ldots, t_n$. The likelihood function is defined by

$$
\begin{aligned}
L(\lambda | t_1, \ldots, t_n) &= \Pr(T = t_1, \ldots, T = t_n) \\
&= \Pr(T = t_1) \cdots \Pr(T = t_n) \\
&= \lambda e^{-\lambda t_1} \cdots \lambda e^{-\lambda t_n} \\
&= \lambda^n \exp\left(-\sum t_i\right).
\end{aligned}
$$

Here the sum is over $i = 1$ to $i = n$. To maximize the likelihood function L, we first take its logarithm to get

$$
\mathcal{L} = \ln L = n \ln \lambda - \lambda \sum t_i.
$$

Then

$$
\frac{d\mathcal{L}}{d\lambda} = \frac{n}{\lambda} - \sum t_i = 0.
$$

Therefore, the maximum likelihood estimator is

$$
\hat{\lambda} = \frac{n}{\sum t_i}.
$$

Notice that

$$
\frac{1}{\hat{\lambda}} = \frac{\sum t_i}{n},
$$

which is the average of the t_i. Recall that for the exponential distribution, $E(T) = 1/\lambda$. □

Example 5.36

If $x_1, \ldots x_n$ is a sample from a normal distribution $N(\mu, \sigma^2)$, find the maximum likelihood estimator of μ and σ. Now the likelihood function is

$$
L(\mu, \sigma | x_1, \ldots x_n) = \prod_{k=1}^{n} \frac{1}{\sigma \sqrt{2\pi}} e^{(x_k - \mu)^2 / 2\sigma^2}.
$$

Because products are involved, to maximize L we maximize its logarithm. A straightforward calculation, which we leave as an exercise, gives maximum likelihood estimators

$$
\hat{\mu} = \overline{x}, \quad \hat{\sigma}^2 = \frac{1}{n} \sum_{k=1}^{n} (x_k - \overline{x})^2,
$$

where $\overline{x}$ is the mean of the sample data. □

Example 5.37

(**Data fitting**) One of the most important problems in ecology is to fit a model to an observed time series, with stochasticity included. To fix the idea, consider the Ricker model for a population N_t,

$$N_{t+1} = bN_t e^{-cN_t} e^{W_t},$$

where $W_t \sim N(0, \sigma^2)$, and b, c, and σ^2 are not known. Suppose that we have $q+1$ observations y_0, y_1, ...,y_q of the population at times $t = 0, 1, 2, ..., q$. How do we find b, c, and σ^2 to fit the model to these data in the best possible way? Here we show how to do this using likelihood methods.

The first step is to change variables so that the noise, or randomness, is additive. Let

$$X_t = \ln N_t.$$

Then the model can be written

$$X_{t+1} = \ln b + X_t - ce^{X_t} + W_t = g(X_t, b, c) + W_t,$$

which defines $g(X_t, b, c) = \ln b + X_t - ce^{X_t}$. Next, let the logarithms of the observations be denoted by Y_0, Y_1, Y_2,..., Y_q.

Now we ask: What is the probability of observing Y_{t+1} given that Y_t was already observed? If $X_t = Y_t$, we have

$$Y_{t+1} - X_{t+1} = W_t$$

or

$$Y_{t+1} - g(Y_t, b, c) = W_t.$$

Think of it this way. We observed a *deviation* D_{t+1} given by

$$D_{t+1} = Y_{t+1} - g(Y_t, b, c),$$

and this deviation is therefore normally distributed with mean 0 and variance σ^2. Thus, the likelihood of observing this deviation, or the probability of it occurring, is the likelihood function

$$\frac{1}{\sqrt{2\pi\sigma^2}} e^{-(Y_{t+1}-g(Y_t,b,c))^2/2\sigma^2}.$$

Consequently, the probability of observing all the q values Y_1, Y_2,...,Y_q is given by the likelihood function

$$L(b, c, \sigma) = \prod_{t=0}^{q-1} \frac{1}{\sqrt{2\pi\sigma^2}} e^{-(Y_{t+1}-g(Y_t,b,c))^2/2\sigma^2}.$$

We want to maximize this. As we noted earlier, it is often easier to maximize $l(b, c, \sigma) = \ln L(b, c, \sigma)$. Easily,

$$l(b, c, \sigma) = \frac{q}{2} \ln 2\pi - \frac{q}{2} \ln \sigma^2 - \sum_{t=0}^{q-1} \frac{(Y_{t+1} - g(Y_t, b, c))^2}{2\sigma^2}.$$

[To use software, for example the Nelder–Mead package in MATLAB, it is better to minimize $-l(b, c, \sigma)$.] Defining

$$S(b, c) = \sum_{t=0}^{q-1} (Y_{t+1} - g(Y_t, b, c))^2, \tag{5.13}$$

we maximize

$$\Lambda(b, c, \sigma) = -q \ln \sigma - \frac{1}{\sigma^2} S(b, c).$$

From calculus

$$\frac{\partial \Lambda}{\partial \sigma} = -\frac{q}{\sigma} + \frac{1}{\sigma^3} S(b, c) = 0$$

or

$$\sigma^2 = \frac{1}{q} S(b, c). \tag{5.14}$$

As an aside, note that

$$\frac{\partial \Lambda}{\partial b} = -\frac{1}{\sigma^2} \frac{\partial S(b, c)}{\partial b} = 0, \qquad \frac{\partial \Lambda}{\partial c} = -\frac{1}{\sigma^2} \frac{\partial S(b, c)}{\partial c} = 0.$$

Therefore, b and c maximize, after simplification,

$$\Lambda \left(b, c, \sqrt{\frac{1}{q} S(b, c)} \right) = -\frac{q}{2} \left(1 + \ln \left(\frac{S(b, c)}{q} \right) \right).$$

We conclude that the maximum likelihood estimates *minimize* the sums of the squares of the deviations (5.13). So the algorithm is to minimize (5.13) and then find σ^2 from (5.14). This calculation can be performed easily on a computer to find b, c, and σ^2 using the method of least squares. □

EXERCISES

1. Determine the mean and the variance of a Bernoulli RV and a geometric RV.

2. In a family with five children, what is the probability that three of them re girls?

3. Fifty-five percent of a physician's patients are women. In a given afternoon, the physician sees 12 patients. What is the probability that there were at most two women?

4. In a region of a forest, an ecologist counted 100 suitable nest sites for a certain species of bird. Thirty nests were safe, but 70 were subject to predation. Of 20 birds that came into the region, 11 chose safe nests and 9 chose unsafe nests. Are the birds randomly choosing their nesting sites, or are they making preferences of safe sites?

5. Male bower birds construct elaborate nests to attract females for mating. Females that visit a nest either mate or they leave; they mate on 25% of their visits. What is the expected number of visits by females to a male's nest before mating occurs? What is the variance? What is the probability that mating will occur on the first or second visit?

6. A predator has a 20% chance of a kill on a given hunt. On how many hunts, on average, must it make to get 10 kills? What is the probability that it will make its tenth kill on the twentieth hunt?

7. A small forest has three species of trees. The percentage of each species, 1, 2, and 3, is 60%, 38%, and 2%, respectively. A botanist examines 30 of the trees in the forest. What is the probability that at least three of the trees are the rare species 3? (Set up this problem but do not calculate.)

8. Use the exponential RV with parameter λ to compute the probability that an event occurs the first time between times $n - 1$ and n. Compare your answer to a value obtained using the geometric distribution, taking $p = 1 - e^{-\lambda}$.

9. Water foul make their nests in wetland regions. Suppose that a major flooding event occurs in a wetland ecosystem about once in every 13 years, causing the destruction of many nests. Find the probability of the following events:

   (a) No floods occurring in a given decade.

   (b) Two consecutive floods occur at least five years apart.

   (c) In a 13-year period there is exactly one flood.

   (d) Three consecutive decades have exactly one such flood each.

10. Let $X \sim \text{bin}(5, p)$, and let $B = \{X \geq 3\}$. Compute the conditional probability $\Pr(X = k | B)$.

11. Let Z_n, $n = 1, 2, 3, \ldots$ be a sequence of discrete random variables, each with range $1, 2$ and $\Pr(Z_n = 1) = 1/n$, $\Pr(Z_n = 2) = 1 - 1/n$ for all n. Show that

$$\lim_{n \to \infty} \text{E}|Z_n - 2|^2 = 0.$$

12. (Probability generating functions) A **probability generating function** (pgf) of a discrete RV X "generates" its probabilities in the following way. If $p_k = \Pr(X = k)$ are the probabilities, the pgf is defined to be that function $G(s)$ for which

$$p_0 + p_1 s + p_2 s^2 + \cdots = G(s) \quad \text{or} \quad G(s) = \sum_{k=0}^{\infty} p_k s^k.$$

Here s is a dummy variable in a range where the series converges.

(a) Show that
$$G(s) = \frac{ps}{1 - qs}, \quad q = 1 - p,$$
is the pgf for the geometric distribution.

(b) Find the pgf for the binomial distribution.

(c) If X is a discrete RV, show that $E(X) = G'(1)$ and $E(X^2) = G''(1) + G'(1)$. These formulas sometimes provide an easy way to calculate the mean and variance of a RV.

(d) Use parts (b) and (c) to find the mean of a binomial RV.

13. Complete the details in Example 5.36.

14. Let $x_1, \ldots, x_n$ be a sample of data points taken from a Poisson distribution with parameter λ. Find the maximum likelihood estimator $\hat{\lambda}$.

15. Let $X_1, X_2, ..., X_{25}$ be a random sample from a Poisson distribution with parameter λ. If $\overline{X} = \frac{1}{25}(X_1 + X_2 + \cdots + X_{25})$ is the sample mean, use the central limit theorem to approximate $\Pr(\overline{X} < \lambda)$ and $\Pr(\overline{X} < \lambda + \sqrt{\lambda}/5)$.

16. Let $X_1, X_2, ..., X_{15}$ be a random sample from a binomial distribution $\text{bin}(6, 1/3)$. Approximate the sample mean $\overline{X}$.

17. If you toss a coin 10 times and get 8 heads, what is the probability p that you will get heads on a single toss? Use likelihood methods; that is, from the binomial distribution write the likelihood function and maximize it to get p. Sketch a graph of the likelihood function vs. p.

5.5 Joint Probability Distributions

In many situations several random variables come into play, and it is important to understand their joint behavior, for example, what is the probability of joint outcomes? Examples of myriad applications are:

1. The weight and carapace length of a turtle in a pond.

2. Records of wind velocity, air temperature, and solar radiation at a meteo-rological station.

3. Two traits possessed by a plant species.

4. The population of an animal species at time t and its population at a later time $t + s$.

We limit the discussion to the case of two random variables X and Y. The careful reader will see how the ideas easily extend to several random variables. Our goal is to present enough ideas so that the reader can go to Chapter 7 on stochastic processes, where several random variables is the norm, rather than to be exhaustive. The following example with discrete RVs illustrates the key ideas.

Example 5.38

A randomly chosen member of a group of freshwater turtles is classified according to which pond it was caught, numbered 1 or 2, and its size class, numbered 1, 2, 3, 4. To each turtle, let $X = 1, 2$ and $Y = 1, 2, 3, 4$ be random variables for its pond and for its size, respectively. The following table shows the relative frequencies:

$X \backslash Y$	1	2	3	4	$p_X(x)$
1	$\frac{1}{8}$	$\frac{1}{16}$	$\frac{3}{16}$	$\frac{1}{8}$	$\frac{1}{2}$
2	$\frac{1}{16}$	$\frac{1}{16}$	$\frac{1}{8}$	$\frac{1}{4}$	$\frac{1}{2}$
$p_Y(y)$	$\frac{3}{16}$	$\frac{2}{16}$	$\frac{5}{16}$	$\frac{3}{8}$	1

The eight central entries in the table, $p(x, y) = \Pr(X = x, Y = y)$, $x = 1, 2$, $y = 1, 2, 3, 4$, define the **joint probability mass function** of X and Y. For example, $p(1, 3) = 3/16$ is the probability that a selected turtle comes from pond 1 and is in the third size class. Note that the sum of all is 1. The conditions for a joint pmf are

$$\sum_{x,y} p(x, y) = 1, \quad p(x, y) \geq 0.$$

The last column is the row sum, and the last row is the column sum. Row sums form the probabilities that $X = 1$ and $X = 2$, and column sums form the probabilities that $Y = 1, 2, 3$, and 4. For example, we have

$$\Pr(X = 1) = \frac{1}{8} + \frac{1}{16} + \frac{3}{16} + \frac{1}{8} = \frac{1}{2},$$

and similarly, $\Pr(X = 2) = \frac{1}{2}$. So the row sums form a probability density $p_X(x)$, which is called the **marginal density** of X; similarly, the column sums define the marginal density $p_Y(y)$ of Y. Therefore, the joint pmf $p(x, y)$ determines the marginal densities of X and Y. Generally stated, the marginal densities are

$$p_X(x) = \sum_{y=1}^{4} p(x, y), \ x = 1, 2; \quad p_Y(y) = \sum_{x=1}^{2} p(x, y), \ y = 1, 2, 3, 4.$$

One can understand the use of the word marginal because the densities are calculated in the margins of the table. Conversely, knowledge of the pmfs of X and Y (i.e., the marginal densities) does not well-define the joint pmf of X and Y. That is, the row and column sums do not uniquely determine the entries. A concrete example is requested in the Exercises.

Correspondingly, the joint cdf is the function

$$P(x, y) = \Pr(X \le x, Y \le y) = \sum_{x_i \le x, \, y_j \le y} p(x_i, y_j),$$

where the sum is a double sum ranging over all discrete, double indices (x_i, y_j) with $x_i \le x$, $y_j \le y$. For example, in the turtle data,

$$P(2, 2) = \frac{1}{16} + \frac{1}{16} + \frac{1}{16} + \frac{1}{8} = \frac{5}{16}. \quad \square$$

These ideas for discrete RVs extend in a natural way to continuous RVs. Again, the most important instrument is the joint pdf. Specifically, the relationship between two continuous random variables X and Y is characterized by the **joint probability density function** (jpdf) $f(x, y)$, having the properties

1. $f(x, y) \ge 0$.

2. $\int_{-\infty}^{\infty} \int_{-\infty}^{\infty} f(x, y) \, dx dy = 1$.

Similar to the case of a single RV, we regard $f(x, y) dx dy$ as the probability (approximately) that (X, Y) lies in the rectangle $x < X \le x + dx$, $y < Y \le y + dy$. The basic probability law can be expressed in terms of the **joint cumulative distribution function** (jcdf) of X and Y, defined by

$$F(x, y) = \Pr(X \le x, Y \le y) = \int_{-\infty}^{x} \int_{-\infty}^{y} f(\xi, \eta) \, d\xi d\eta. \tag{5.15}$$

In theory, all information about the random variables X and Y can be answered in terms of the density or distribution function. By our definition the

joint distribution is the (double) integral of the density, and, assuming sufficient smoothness of the functions involved, the density is the derivative of the
distribution function; that is,

$$f(x,y) = \frac{\partial^2 F(x,y)}{\partial x \, \partial y}.$$

The properties of the jcdf are similar to those of a cdf for a single random
variable.

1. $F(x,y)$ is nonnegative and nondecreasing in both variables.

2. $F(-\infty, y) = F(x, -\infty) = 0, \quad F(y, \infty) = F(x, \infty) = 1.$

If the jcdf's for X and Y are known, we can recover the cdf's of the single
random variables X and Y as follows. Note that

$$F_X(x) = \Pr(X \leq x) = \Pr(X \leq x, Y < \infty) = F_{XY}(x, \infty).$$

Similarly,

$$F_Y(y) = F(\infty, y).$$

In this context, when the single distributions are found from the joint distributions, $F_X(x)$ and $F_Y(y)$ are called the **marginal distribution functions**.
The **marginal density functions** are

$$f_X(x) = \int_{-\infty}^{\infty} f(x,y)\, dy, \quad f_Y(y) = \int_{-\infty}^{\infty} f(x,y)\, dx.$$

On the other hand, the two single cdf's (pdf's) do not always determine the
jcdf (jpdf).

When considering two random variables X and Y we can think of an arbitrary event as a subset A of the xy plane (with the caveat that there may be
some unusual sets that are not events). Then

$$\Pr((X,Y) \in A) = \int\int_A f_{XY}(x,y)dxdy,$$

where the double integral is taken over the two-dimensional domain A. Computing this probability requires some facility in calculating double integrals.

Example 5.39

Let X and Y be random variables with joint density $f(x,y)$. Then the probability that $X + Y < 1$ is

$$\Pr(X + Y < 1) = \int\int_{x+y<1} f(x,y)\, dxdy = \int_{-\infty}^{\infty}\int_{-\infty}^{1-y} f(x,y)\, dxdy. \quad \square$$

Example 5.40

Let

$$f(x,y) = \begin{cases} x+y, & 0 \le x \le 1, 0 \le y \le 1, \\ 0, & \text{otherwise,} \end{cases}$$

be a joint pmf on the unit square. Then the marginal density for X is

$$f_X(x) = \int_0^1 (x+y)\, dy = x + \frac{1}{2}.$$

The calculation of the integral for the jcdf $F(x,y)$ depends on the location of the point (x,y). Notice, for example, that $F(x,y) = 0$ if (x,y) lies in the third quadrant. For (x,y) in the square,

$$F(x,y) = \int_0^x \int_0^y (u+v)\, dvdu = \int_0^x \left(yu + \frac{1}{2}y^2 \right) du = \frac{1}{2}yx^2 + \frac{1}{2}xy^2.$$

It is easily checked by partial differentiation that

$$\frac{\partial^2 F}{\partial x\, \partial y} = x + y,$$

verifying the connection between the jpdf and the jcdf.   □

The **expected value** of a function $g(X,Y)$ is given by

$$E(g(X,Y)) = \int\int g(x,y)f(x,y)\, dxdy,$$

where the double integral is over all (x,y) in the plane. It follows that

$$E(X+Y) = E(X) + E(Y).$$

Similar results hold for discrete RVs. For more than one RV, it is not clear what we mean by the variance; it will take a special section to discuss this case.

To recall, two events A and B, which are both in the same sample space, are **independent events** if

$$\Pr(A \cap B) = \Pr(A)\Pr(B). \tag{5.16}$$

Two RVs X and Y are **independent random variables** if, and only if,

$$\Pr(X \le x, Y \le y) = \Pr(X \le x)\Pr(Y \le y).$$

For continuous RVs this means that the cdfs and pdfs are multiplicative, that is, they break up into a product of a function of x and a function of y. Thus, X and Y are independent if, and only if,

$$F(x,y) = F_X(x)F_Y(y),$$

or, in terms of densities,

$$f(x, y) = f_X(x)f_Y(y).$$

Similar equations hold in the discrete RV case. For independent RVs,

$$E(XY) = E(X)E(Y),$$

but the converse does not hold.

EXERCISES

Some of these exercises use the unit step function h, or **Heaviside function**, defined by $h(x) = 0$ if $x < 0$, and $h(x) = 1$ if $x \geq 0$.

1. Concoct an example of two discrete random variables, showing that the marginal densities do not uniquely determine the joint probability mass function.

2. Can

$$F(x, y) = h(x)h(y)(1 - e^{-x-y})$$

   be a jcdf for two random variables X and Y?

3. Let X and Y be jointly distributed random variables with joint density $f(x, y) = 8xy$ if $0 \leq x \leq 1$, $0 \leq y \leq x$, and zero otherwise.

   a) Find the marginal densities $f_X(x)$ and $f_Y(y)$.

   b) Are X and Y independent?

   c) Find the probability that $X + Y \geq 1$.

   d) Compute $E(X)$, $E(Y)$ and $E(XY)$.

4. If X and Y are independent random variables, show why

$$\Pr(Y \leq X) = \int_{-\infty}^{\infty} f_X(x)F_Y(x)\, dx.$$

5. Let $f(x, y) = e^{-(x+y)}$ for $x \geq 0$, $y \geq 0$, and $f(x, y) = 0$ otherwise, be a joint pdf for continuous RVs X and Y. Find the pdf of the random variable X/Y. [Hint: Begin with $\Pr(Z \leq z)$, where $Z = X/Y$.]

6. Let X and Y be independent exponential random variables with parameter λ, and let $Z = X + Y$. Show that Z is gamma-distributed. [Hint: Calculate $\Pr(Z \leq z)$.]

7. Two independent random variables X_1 and X_2 are uniformly distributed
   with common density
   $$\frac{1}{a}(h(x) - h(x - a)).$$
   Find the joint pdf of X_1 and X_2. If $W = X_1 + X_2$, find $E(W)$ and find
   $E(X_1 X_2)$ in two ways.

8. If X and Y are independent continuous random variables, show why
   $E(XY) = E(X)E(Y)$.

5.6 Covariance and Correlation

First, by way of motivation and review, consider two *data sets* X and Y defined
by the lists of numbers $x_1, ..., x_N$ and $y_1, ..., y_N$, respectively. We recall from
elementary statistics (see Chapter 1) that the correlation coefficient is defined
by

$$r = \frac{1}{\sigma_X \sigma_Y} \sum_{i=1}^{N} (x_i - \mu_X)(y_i - \mu_Y). \qquad (5.17)$$

The correlation coefficient, which takes values between -1 and 1, inclusively, is
a measure of how much the two data sets are related. The closer the value to 1
or to -1, the more the set of data points (x_i, y_i) lies near a line (the regression
line) in the xy plane, predicting a linear relationship between X and Y—the
closer the value of r to zero, the more scattered the data. The value r^2 is a
measure of what fraction of the data is explained by a linear relationship.

5.6.1 Covariance

As the name implies, the **covariance** is a measure of the strength of the dependency between two random variables. Given two RVs X and Y, two extreme
circumstances can occur:

1. There is no link at all between X and Y, and knowing the value of X gives
   no clue to the value of Y. In this case, the two variables are independent.

2. The dependency is so strong that there is a functional relationship between
   them where one determines the other without any uncertainty.

   Most often, however, the link between X and Y is somewhere in between
these two extremes, and knowing one of the values of one of the RVs reduces

the uncertainty about the other, but not completely. Covariance is one way to define this intermediate situation.

We can think of it like this. If X and Y are strongly (positively) linked, then:

– Whenever X is positive, Y is also likely to be positive.

– Whenever X is negative, Y is also likely to be negative too.

Because covariance should be unchanged if both probability distributions are translated by arbitrary amounts (e.g., to their respective means), instead of measuring the values of X and Y from the origin, as above, we measure them relative to their means μ_X and μ_Y. Therefore:

Whenever $X - \mu_X$ is positive, then $Y - \mu_Y$ is also likely to be positive.

Whenever $X - \mu_X$ is negative, then $Y - \mu_Y$ is also likely to be negative.

Thus, if X and Y are strongly (positively) related, then more often than not, $X - \mu_X$ and $Y - \mu_Y$ are simultaneously positive or simultaneously negative, meaning that the product $(X - \mu_X)(Y - \mu_Y)$ is very likely to be positive. Yet the product $(X - \mu_X)(Y - \mu_Y)$ is a random variable, and we want a fixed number. But a random variable that spends most of its time taking positive values is likely to have a positive expectation. So we will consider the expectation of $(X - \mu_X)(Y - \mu_Y)$, and call it the **covariance** of X and Y:

$$\mathrm{cov}(X, Y) = \mathrm{E}((X - \mu_X)(Y - \mu_Y)).$$

Clearly, a large positive value of the covariance is an indication that $X - \mu_X$ and $Y - \mu_Y$ often take large positive or large negative values simultaneously, which leads us to the conclusion that X and Y are strongly linked, whereas a small positive value of the covariance indicates that one of the variables may be close to its mean while the other may take large values. When there is a strong link, we can show that the link is close to linear relationship.

The argument we made above on the basis of a positive link between X and Y applies equally well to the case of a negative link. If $X - \mu_X$ takes large positive values makes it likely that $Y - \mu_Y$ takes large negative values, the covariance is a large negative number.

Now we follow the quantitative consequences. Consider two discrete RVs X and Y that take values (x_i, y_j) with probability p_{ij}. In the case of a single RV X, the variance is is $\mathrm{var}\, X = \mathrm{E}((X - \mu)^2)$, the average of the squares of the distances of the values from the mean. How do we define the variance for two random variables? Taking account of the discussion above, our definition is $\mathrm{E}((X - \mu_X)(Y - \mu_Y))$; then, if $X = Y$, it reduces to the single-variable case of

$\text{var} X = \text{E}((X - \mu)^2)$. If we expand the right side of (5.6.1) using the additivity rules for expectation, we easily get

$$\text{cov}(X, Y) = \text{E}(XY) - \text{E}(X)\text{E}(Y). \qquad (5.18)$$

Again, this reduces to the one-variable formula when $X = Y$, or $\text{cov}(X, X) = \text{Var}(X)$. Observe that if X and Y are independent, then

$$\text{cov}(X, Y) = 0 \qquad (X, Y \text{ independent}).$$

The converse, however, is not true. Further, it is straightforward to expand out the variance of a sum (using expected values) to obtain the following formula:

$$\text{var}(X + Y) = \text{var } X + \text{var } Y + 2\,\text{cov}(X, Y). \qquad (5.19)$$

Therefore, if X and Y are independent, the variance of a sum is the sum of the variances:

$$\text{var}(X + Y) = \text{var } X + \text{var } Y \qquad (X, Y \text{ independent}).$$

In summary, the covariance of two random variables is a measure of how much they are related, or dependent. In the random pair (X, Y), if most of the X and Y values lie on the same sides of their respective means (above or below), the quantities $X - \mu_X$ and $Y - \mu_Y$ will have the same sign most of the time and the product will be positive; hence, the covariance will be positive. If X values and Y values are on opposite sides of their means, $X - \mu_X$ and $Y - \mu_Y$ will have opposite signs most of the time and the product, and hence covariance, will be negative. In these two instances, X and Y are positively or negatively related. On the other hand, if there is no relation between X and Y and their values scatter so that the product $(X - \mu_X)(Y - \mu_Y)$ is sometimes positive and sometimes negative, the expected value may be close to zero because terms will cancel in computing the expected value (a sum).

The covariance is not scaled and it will give different answers depending on the units selected to measure values of the random variables. Suppose, for example, that X and Y are in millimeters and we calculate the value $\text{cov}(X, Y)$. If we change units to meters, we get new random variables $\widetilde{X}$ and $\widetilde{Y}$ with $\widetilde{X} = 10^3 X$ and $\widetilde{Y} = 10^3 Y$, where we have multiplied by the conversion factor. Then, using formula (5.18),

$$\text{cov}(\widetilde{X}, \widetilde{Y}) = \text{cov}(10^3 X, 10^3 Y) = 10^6\,\text{cov}(X, Y).$$

Therefore, we get a covariance 1000000 times more! We would like to have a dimensionless quantity that measures the covariance. The key is to examine the correlation coefficient r in (5.17) and note that each quantity in the sum is a z–score: namely, $(x - \mu_X)/\sigma_X$ and $(y - \mu_Y)/\sigma_Y$. Remember that the standard deviation has the same units as the random variable and the mean. Thus, we define the dimensionless **correlation coefficient** between two random variables X and Y by

$$\rho(X, Y) = \frac{1}{\sigma_X \sigma_Y} \text{cov}(X, Y).$$

We leave the following important properties as exercises.

1. If $Y = aX + b$ with $a > 0$, then $\rho(X, Y) = +1$.

2. If $Y = aX + b$ with $a < 0$, then $\rho(X, Y) = -1$.

3. $-1 \leq \rho(X, Y) \leq +1$.

4. $\text{E}(XY)^2 \leq \text{E}(X)\text{E}(Y)$ and $\text{cov}(X, Y)^2 \leq \text{var}(X)\text{var}(Y)$.

5.6.2 Covariance Matrix

In Chapter 7 we show how to construct a system of stochastic differential equations for a dynamic model involving two random variables. We need to know how to represent normal random variables in terms of standard normals, or z-scores. This leads in a natural way to the covariance matrix.

We recall from elementary statistics that if X is any normal RV with mean μ and variance σ^2 [i.e., $X \sim \text{N}(\mu, \sigma^2)$], it can be represented as

$$X = \mu + \sigma Z,$$

where $Z \sim \text{N}(0, 1)$. How does this work for two random variables X_1 and X_2? (We use this notation rather than X and Y because it fits better with vector and matrix representations.) Let

$$X_1 \sim \text{N}(\mu_1, \sigma_1^2), \quad X_2 \sim \text{N}(\mu_2, \sigma_2^2)$$

be RVs. At first we are tempted to write

$$X_1 = \mu_1 + \sigma_1 Z_1, \quad X_2 = \mu_2 + \sigma_2 Z_2, \quad Z_1, Z_2 \sim \text{N}(0, 1),$$

but it is easy to show, taking expectations, that X_1 and X_2 must be independent and therefore the covariance is zero; thus, there can be no dependencies between X_1 and X_2. Therefore, let us define the combinations

$$\begin{aligned} X_1 &= \mu_1 + a_1 Z_1 + a_2 Z_2, \\ X_2 &= \mu_2 + b_1 Z_1 + b_2 Z_2, \end{aligned}$$

where $Z_1, Z_2 \sim N(0, 1)$, and where a_1, a_2, b_1, and b_2 are to be determined. In vector–matrix notation, if we define

$$\mathbf{X} = \begin{pmatrix} X_1 \\ X_2 \end{pmatrix}, \quad \mu = \begin{pmatrix} \mu_1 \\ \mu_2 \end{pmatrix}, \quad \mathbf{Z} = \begin{pmatrix} Z_1 \\ Z_2 \end{pmatrix}, \quad B = \begin{pmatrix} a_1 & a_2 \\ b_1 & b_2 \end{pmatrix},$$

the last pair of equations can be written

$$\mathbf{X} = \mu + B\mathbf{Z}.$$

We want to determine the matrix B. Taking the transpose of both sides yields

$$\mathbf{X}^{\mathrm{T}} = \mu^{\mathrm{T}} + \mathbf{Z}^{\mathrm{T}} B^{\mathrm{T}}.$$

Then

$$\begin{aligned}
\mathbf{X}\mathbf{X}^{\mathrm{T}} &= (\mu + B\mathbf{Z})(\mu^{\mathrm{T}} + \mathbf{Z}^{\mathrm{T}} B^{\mathrm{T}}) \\
&= \mu\mu^{\mathrm{T}} + \mu\mathbf{Z}^{\mathrm{T}} B^{\mathrm{T}} + B\mathbf{Z}\mu^{\mathrm{T}} + B\mathbf{Z}\mathbf{Z}^{\mathrm{T}} B^{\mathrm{T}}.
\end{aligned}$$

Notice that each term is a matrix. Taking expected values (the expected value of a matrix is the matrix of expected values),

$$\begin{aligned}
E(\mathbf{X}\mathbf{X}^{\mathrm{T}}) &= E(\mu\mu^{\mathrm{T}}) + E(\mu\mathbf{Z}^{\mathrm{T}} B^{\mathrm{T}}) + E(B\mathbf{Z}\mu^{\mathrm{T}}) + E(B\mathbf{Z}\mathbf{Z}^{\mathrm{T}} B^{\mathrm{T}}) \\
&= \mu\mu^{\mathrm{T}} + BE(\mathbf{Z}\mathbf{Z}^{\mathrm{T}})B^{\mathrm{T}},
\end{aligned}$$

The second and third terms on the right side of the first equality are zero because $E(\mathbf{Z}) = 0$. If the reader does not have good facility with matrix notation, he or she should work out these formulas. Now,

$$E(\mathbf{Z}\mathbf{Z}^{\mathrm{T}}) = \begin{pmatrix} E(Z_1^2) & E(Z_1 Z_2) \\ E(Z_2 Z_1) & E(Z_2^2) \end{pmatrix} = \begin{pmatrix} 1 & 0 \\ 0 & 1 \end{pmatrix} = I,$$

the identity matrix. Therefore,

$$E(\mathbf{X}\mathbf{X}^{\mathrm{T}}) = \mu\mu^{\mathrm{T}} + BB^{\mathrm{T}},$$

or

$$\begin{aligned}
BB^{\mathrm{T}} &= E(\mathbf{X}\mathbf{X}^{\mathrm{T}}) - \mu\mu^{\mathrm{T}} \\
&= \begin{pmatrix} E(X_1^2) - \mu_1^2 & E(X_1 X_2) - \mu_1\mu_2 \\ E(X_2 X_1) - \mu_2\mu_1 & E(X_2^2) - \mu_2^2 \end{pmatrix} \\
&= \begin{pmatrix} \text{cov}(X_1, X_1) & \text{cov}(X_1, X_2) \\ \text{cov}(X_2, X_1) & \text{cov}(X_2, X_2) \end{pmatrix}.
\end{aligned}$$

This matrix is the **covariance matrix,** and we denote it by C; basically, it is the matrix of covariances. In summary, we have the important result and representation

$$\mathbf{X} = \mu + B\mathbf{Z}, \quad BB^{\mathrm{T}} = C. \tag{5.20}$$

This means that we can represent a normal vector $\mathbf{X}$ in terms of standard normal random variables using the covariance matrix.

Of course, to find B we must solve the equation $BB^T = C$. Because C is a real symmetric matrix, we can make use of eigenvalue theory to determine B. But this is beyond the scope of our discussion. However, in the case that B is symmetric, we have $BB^T = BB = B^2 = C$, and the question boils down to asking whether B has a square root. If so, $B = C^{1/2}$, and we have determined B. In Chapter 7 we discuss how to determine the square root of a matrix so that we can find the representation (5.20).

EXERCISES

1. In the example with the turtle data (Example 5.38), compute $E(X)$, $E(Y)$, $E(XY)$, $E(X^2)$, $E(Y^2)$, $\text{cov}(X, Y)$, and $\rho(X, Y)$. What is the conclusion?

2. If X and Y are random variables and $Y = aX + b$, show that the correlation coefficient is $\rho(X, Y) = +1$ if $a > 0$.

3. Let X and Y be random variables with joint density

$$f(x, y) = \frac{(x + y)^2}{40}, \quad -1 < x < 1, \, -3 < y < 3,$$

and zero otherwise.

   a) Find $E(XY)$, $E(Y)$, and $E(X)$, $\text{var}(X)$, and $\text{var}(Y)$.

   b) Compute the correlation coefficient.

4. Verify (5.18) and (5.19).

5.7 Reference Notes

Concepts in probability had a long development period prior to A. Kolmogorov, who, in the early 1930s, formalized the theory of probability and set down basic definitions and postulates about sample spaces, event space, and probability measures. Probability itself arose centuries ago out of games of chance. There is evidence that Egyptians, as early as 3500 B.C., gambled with a four-sided die, and with a six-sided die in 1600 B.C. Much later, dice were used in China during the period A.D.600 − 1000, with playing cards arising around 900. The Chinese also constructed paper dominos. It is believed that these gaming items moved from China to Central Asia, and then to Europe throughout the Middle Ages. During the Renaissance period, Cardano and Tartaglia (whose names come up in the history of elementary algebra), as well as Galileo, studied games of chance. The theory of probability, or at least serious investigations, began in the mid-1600s when the Chevalier de la Mere posed to Fermat a question about probabilities associated with rolling dice. That spawned correspondence with Pascal (the Fermat–Pascal letters), which can be regarded as the origin of probability theory. Not long after that, books by C. Huygens, J. Bernoulli, and de Moivre, in the late 1600s to the mid-1700s, spread the information across Europe, where serious mathematics was beginning to flourish. Laplace, Poisson, and Gauss made further contributions. Then, in the 1800s, the Russian school (Chebyshev, Markov, and Lyapunov), made significant contributions. In his famous presentation in 1900, David Hilbert named the "axiomatic development of probability" one of the central mathematical problems of the time. Borel, Berstein, and von Mises made further advances before Kolmogorov put together his axiomatic, aesthetic treatment of probability. Advanced probability theory was introduced to English-speaking mathematicians in the 1940s and 1950s, mainly by Feller and Doob.

Many, many books and papers have been written on pure and applied probability, and several research journals in all fields of mathematics, statistics, science, economics, and so on, document that work. A very large body of work appears in biostatistics. Because most events in the world seem random, it is not surprising that probability plays such a central role; most of our deterministic models in science seem only to represent an approximation to the real, random world—this is especially true in ecology.

Because this book is designed for students and focused around models in biology, and in particular, ecology, we make a few remarks about resources for further reading. Many elementary textbooks are accessible on probability theory. We mention a few books we have found useful, in the order of slightly increasing difficulty: Stirzaker (2005), Olofsson (2005), Grimmett & Welsh (1986), and Ross (2006); the latter has a wealth of examples and exercises. But the reader should be advised that these are mathematics texts and do not have many examples or motivations from the biological sciences. The general mathematical biology text by Otto & Day (2007) has several chapters on probability written at about the same level as the present text, and they are all focused around biology models. Mooney & Smith's (1999) excellent text on modeling has a strong stochastic and biology content, and it is elementary enough for a first-year calculus student. An advanced perspective on probability and statistics in biology is that of Mangel (2006); in fact, it is tempting to say that anyone, student or researcher, who considers herself or himself to be an ecologist or evolutionary biologist should own and know this important book by heart. Add to that the book of Hilborn & Mangel (1997), which should be read by everyone doing ecology.

The brief section on the Hardy–Weinberg law is just a tiny scratch on the overall theory of the subject of population genetics, or how natural selection enters the population cycles and can dictate evolution of the genetic characteristics of species. This topic is often considered one of the most mathematical in all of biology. There are many books solely on this topic. Two of the very best general mathematical biology texts that treat this subject are Allman & Rhodes (2004) and Roughgarden (1998); the former is a very elementary discussion suitable for readers with a limited mathematical background, and the latter has an excellent lengthy chapter on the central issues.

References

Allman, E. S. & Rhodes, J. A. 2004. *Mathematical Models in Biology*, Cambridge University Press, Cambridge, UK.

Grimmett, G. & Welsh, D. 1986. *Probability: An Introduction*, Oxford Science Publications, Oxford, UK.

Hilborn, R. & Mangel, M. 1997. *The Ecological Detective: Confronting Models with Data*, Princeton University Press, Princeton.

Mangel, M. 2006. *The Theoretical Biologist's Toolbox*, Cambridge University Press, Cambridge, UK.

Mooney, D. & Smith, R. 1999. *A Course in Mathematical Modeling*, Mathematical Association of America, Washington, DC.

Olofsson, P. 2005. *Probability, Statistics, and Stochastic Processes*, Wiley-Interscience, Hoboken, NJ.

Otto, S. P. & Day, T. 2007. *A Biologist's Guide to Mathematical Modeling in Ecology and Evolution*, Princeton University Press, Princeton, NJ.

Ross, S. 2006. *A First Course in Probability*, 7th ed., Prentice Hall, Upper Saddle River NJ.

Roughgarden, J. 1998. *Primer of Ecological Theory*, Prentice-Hall, Upper Saddle River, NJ.

Stirzaker, D. 2005. *Probability and Random Variables: A Beginner's Guide*, Cambridge University Press, Cambridge, UK.

6
Statistical Inference

Statistics is the mathematical science of organizing, analyzing, and understanding data sets. In Section 1.6 we introduced some elementary ideas in descriptive statistics. Terms such as *mean, standard deviation*, and *median* were introduced and defined to characterize simple data sets. (The reader should review that short section.) Now we want to take up a more serious investigation of statistical concepts. Many of these rely on ideas in probability theory, the subject of Chapter 5. We focus on the subject of statistical inference, which ties together probability models and data collection to make conclusions about basic characterizations of large populations where the properties of every individual cannot be measured. Voting is a classical example——we cannot know how every person will vote, so we survey a finite number of people and use that sample to draw inferences about the voting implications in the entire population. The topics of interest in our discussion include sampling distributions, confidence intervals, hypothesis testing, and bootstrap techniques.

Statistics is of such importance in the biological sciences that some biologists think that "mathematical biology" means statistics! Certainly, however, in ecology, medicine, and all the other biological sciences, statistical analysis plays a crucial role.

Mathematical Methods in Biology,
By J. David Logan and William R. Wolesensky
Copyright © 2009 John Wiley & Sons, Inc.

6.1 Introduction

Consider a population Ω of unknown size and a random variable X defined on Ω. To take an ecology example, Ω could be all the *C. picta* turtles in a pond and X could be the weight of a turtle. What else do we know? Really, nothing. We don't know how the weights are distributed (the pdf), we don't know the average weight μ, and we don't know any of the other population parameters (e.g., the standard deviation σ of the weights, the median weight, the maximum weight). Clearly, a problem of great interest and importance to ecologists is estimation of the population parameters.

How would we obtain these estimates? The only way that comes to mind is to take a random sample of the population and estimate the parameters from the sample. In particular, if we sample the population and get values $x_1, x_2, \ldots, x_n$, we could compute the **sample mean** $\overline{x}$ and the **sample variance** s^2 by

$$\overline{x} = \frac{x_1 + x_2 + \cdots + x_n}{n},$$

$$s^2 = \frac{(x_1 - \overline{x})^2 + (x_2 - \overline{x})^2 + \cdots + (x_n - \overline{x})^2}{n - 1},$$

and use these to approximate the true mean μ and variance σ^2. If we make these approximations, we are obliged to make some statements about their validity or accuracy.

The types of evidence for their accuracy can be found using the following methods:

1. (**Interval Analysis**) Because $\overline{x}$ is random, depending on the selection of the sample, we can say something about the probability that the true mean μ lies inside some interval $[\overline{x} - \mathrm{SE}, \overline{x} + \mathrm{SE}]$, called a **confidence interval**. The number SE is called the *standard error*.

2. (**Hypothesis Tests**) We can set up a hypothesis about the mean (or some other parameter) and ask if the sample data collected validates or invalidate the hypothesis.

3. (**Bootstrap Methods**) Using a single sample, we can create a large number of new simulated samples by drawing from the original single sample (e.g., with replacement). We can then study the statistical properties of the samples simulated.

To understand specifically how probability theory enters the analysis, note that if we took a second random sample of size n, we would probably get different values of the sample mean and variance. Looked at in this way, the values $\overline{x}$ and s^2 are values of random variables $\overline{X}$ and S^2 for the sample mean and sample variance. In fact, we can regard x_1 as a value of X_1, x_2 as a value of X_2, and so on, where X_1, $X_2, \ldots$ are independent, identically distributed (iid) random variables representing our first draw, our second draw, and so on. Regardless of the distribution, we have $E(X_i) = \mu$ and $\text{var}(X_i) = \sigma^2$. Therefore,

$$\overline{X} = \frac{X_1 + X_2 + \cdots + X_n}{n},$$
$$S^2 = \frac{(X_1 - \overline{X})^2 + (X_2 - \overline{X})^2 + \cdots + (X_n - \overline{X})^2}{n - 1},$$

define the random variables. The pdfs for these random variables are called **sampling distributions**, and specific values such as $\overline{x}$ and s^2 are called **point estimators**.

First let us turn to the sample mean. Immediately we can find the mean and variance of the sampling distributions, regardless of the distribution of the original population. We have

$$E(\overline{X}) = E\left(\frac{X_1 + X_2 + \cdots + X_n}{n}\right) = \frac{1}{n}\sum E(X_i) = \frac{1}{n} \cdot n\mu = \mu$$

and

$$\text{var}(\overline{X}) = \text{var}\left(\frac{X_1 + X_2 + \cdots + X_n}{n}\right) = \frac{1}{n^2}\sum \text{var}(X_i) = \frac{1}{n^2} \cdot n\sigma^2 = \frac{\sigma^2}{n}.$$

Notice that the standard deviation of the sample mean is $\sigma/\sqrt{n}$. But how is the sample mean $\overline{X}$ distributed? If X is normally distributed, each X_i is normally distributed and hence the sum is normally distributed. Therefore, in this case, the sampling distribution for the mean has a normal distribution $N(\mu, \sigma^2/n)$. Summarizing, the key result is:

Theorem 6.1

Let X_1, $X_2, \ldots, X_n$ be a random sample of size n taken from a normal distribution with mean μ and variance σ^2. Then the sampling distribution $\overline{X}$ is normally distributed with mean μ and variance σ^2/n. Therefore,

$$Z = \frac{\overline{X} - \mu}{\sigma/\sqrt{n}}$$

is a standard normal random variable $N(0, 1)$. $\square$

What if the underlying distribution is unknown? One of the important theorems in probability theory, stated in Chapter 5, answers the question in this case. We repeat it here.

Theorem 6.2

(**Central Limit Theorem**) Let $X_1,\ X_2,\ldots,X_n$ be independent, identically distributed, random variables with mean μ and variance σ^2, and denote $S_n = X_1 + X_2 + \cdots + X_n$. Then the sequence of random variables

$$Z_n = \frac{S_n/n - \mu}{\sigma/\sqrt{n}}, \quad n = 1, 2, 3, \ldots$$

approaches a standard normal random variable Z as $n \to \infty$. By this we mean that

$$\lim_{n \to \infty} \Pr\left(a < \frac{S_n/n - \mu}{\sigma/\sqrt{n}} \le b\right) = \int_a^b \frac{1}{\sqrt{2\pi}} e^{-x^2/2}\, dx.$$

Therefore, for large n the distribution of the sample mean $\overline{X} = S_n/n$ is approximately the standard normal distribution $N(0,1)$. Stated differently, the random variable

$$Z = \frac{\overline{X} - \mu}{\sigma/\sqrt{n}} \tag{6.1}$$

is an approximately standard normal random variable. □

With the two results above, Theorems 6.1 and 6.2, we can now find an interval estimate for μ in the case that the underlying distribution is either normal or not normal. Thus, these theorems provide a framework for finding an interval estimate for the population mean μ under the assumption that σ^2 is known.

EXERCISES

1. Let $Z \sim N(0,1)$.

    (a) Find $z_{\alpha/2}$ for which $\Pr(-z_{\alpha/2} < Z < z_{\alpha/2}) = 1 - \alpha$ when $\alpha = 0.10$.

    (b) Find z for which $\Pr(-z < Z < z) = 0.75$.

2. Let X denote the amount of radiation that can be absorbed by an animal before death ensues. Assume that X is normal with a mean of 300 roentgen and a standard deviation of 80 roentgen. Above what dosage level will only 5% of those exposed survive?

6.2 Interval Analysis

6.2.1 Known Variance

To estimate the mean of a population, we take a random sample from that population and use its sample mean as s statistic to create an interval estimate.

When selecting a random sample from a *normal distribution*, Theorem 6.1 applies; when the underlying distribution is not normal, Theorem 6.2 applies. In either case we end up examining a standard normal RV. Example 6.3 illustrates the general case.

Example 6.3

(**Sampling from a normal distribution**) Freshwater turtles occupy a local pond. Twenty turtles are caught and weighed, and the sample mean is $\overline{x} = 310$ grams. The distribution of turtle weights is assumed to be normal, the standard deviation σ is presumed to be 40 grams, and the population mean μ is unknown. We want to find statistics L_1 and L_2 so that $\Pr(L_1 \leq \mu \leq L_2) = 0.90$. The interval $(L_1 \leq \mu \leq L_2)$ is called a 90% **confidence interval**. To proceed, we first find numbers a and b such that

$$\Pr(a \leq Z \leq b) = 0.90,$$

where $Z = (\overline{X} - \mu)/(\sigma/\sqrt{n}) \sim N(0,1)$. Using software (or a table) we find that $a = -1.64$ and $b = 1.64$. That is,

$$\Pr\left(-1.64 < \frac{\overline{X} - \mu}{\sigma/\sqrt{n}} < 1.64\right) = 0.90.$$

To find L_1 and L_2, we isolate μ algebraically as follows:

$$\Pr\left(\frac{-1.64\sigma}{\sqrt{n}} \leq \overline{X} - \mu \leq \frac{+1.64\sigma}{\sqrt{n}}\right) = 0.90,$$

$$\Pr\left(-\overline{X} - \frac{-1.64\sigma}{\sqrt{n}} \leq -\mu \leq -\overline{X} + \frac{1.64\sigma}{\sqrt{n}}\right) = 0.90,$$

$$\Pr\left(\overline{X} - \frac{-1.64\sigma}{\sqrt{n}} \leq \mu \leq \overline{X} + \frac{1.64\sigma}{\sqrt{n}}\right) = 0.90.$$

Therefore, $L_1 = \overline{X} - 1.64\sigma/\sqrt{n}$ and $L_2 = \overline{X} + 1.64\sigma/\sqrt{n}$. Substituting in the point estimate $\overline{x} = 310$, $\sigma = 40$, and $n = 20$ yields the 90% confidence interval $[310 - 14.67, 310 + 14.67] = [295.33, 324.67]$. $\square$

Theorem 6.4

Let X_1, $X_2, \ldots, X_n$ be a random sample of size n taken from a normal distribution with mean μ with known variance σ^2. Then the lower bound, L_1, and upper bound, L_2, for a $100(1 - \alpha)\%$ confidence interval for μ are given by

$$L_1 = \overline{X} - \frac{z_{\alpha/2}\sigma}{\sqrt{n}} \quad \text{and} \quad L_2 = \overline{X} + \frac{z_{\alpha/2}\sigma}{\sqrt{n}}.$$

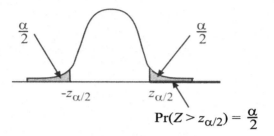

Figure 6.1 For the symmetric normal distributions, finding the value of z that solves $\Pr(-z_{\alpha/2} < Z < z_{\alpha/2}) = 1 - \alpha$ is equivalent to finding the z for which $\Pr(Z > z_{\alpha/2}) = \alpha/2$ is true.

That is,

$$\Pr\left(\overline{X} - \frac{z_{\alpha/2}\sigma}{\sqrt{n}} \le \mu \le \overline{X} + \frac{z_{\alpha/2}\sigma}{\sqrt{n}}\right) = 1 - \alpha,$$

where $z_{\alpha/2}$ denotes the value from the Z distribution where $\Pr(Z > z_{\alpha/2}) = \alpha/2$. □

Observe that by symmetry of the normal distribution, $\Pr(-z_{\alpha/2} < Z < z_{\alpha/2}) = 1 - \alpha$ is equivalent to $\Pr(Z > z_{\alpha/2}) = \alpha/2$ (Fig. 6.1), which is often the form we use in calculations to compute the critical value $z_{\alpha/2}$.

The number

$$\mathrm{SE} = \frac{z_{\alpha/2}\sigma}{\sqrt{n}}$$

is called the **standard error**. In Example 6.3, SE $= 14.67$.

Now an important remark regarding interpretation. For example, if $\alpha = 0.10$, this does *not* mean that there is a 90% chance that the interval contains the value of μ. Rather, it means that if we repeat the random experiment many times, 90% of the intervals we compute will contain μ (Fig. 6.2). There is an important distinction between these two interpretations, and care must be taken to use the correct one.

Example 6.5

A random sample of size 8 from a N$(\mu, 50)$ distributed population yielded $\overline{x} = 57$. Find a 90% confidence interval for the mean. Here $\alpha = 0.10$. We need to find the value $z_{.05}$ for which $\Pr(Z > z_{.05}) = \alpha/2 = 0.05$. From software, $z_{.05} = 1.645$. Therefore, the confidence interval is

$$\left(57 - 1.645\sqrt{50}/\sqrt{8}, 57 + 1.645\sqrt{50}/\sqrt{8}\right) = (52.8875, 61.1125). \quad \Box$$

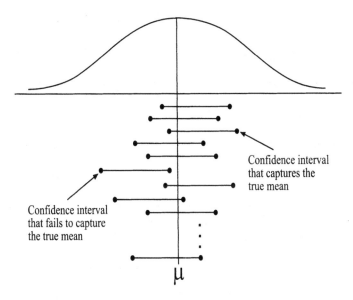

Figure 6.2 If the random experiment is conducted n times, we would expect $(1 - \alpha)100\%$ of the $(1 - \alpha)100\%$ confidence intervals to capture the true value of the parameter being estimated.

Example 6.6

If we want a smaller confidence interval, we must take a larger sample size n. In Example 6.3 we found that SE $= 14.67$. If we want $m = $ SE $= 7$, the confidence interval is $[303, 317]$ and

$$n = \frac{z_{\alpha/2}^2 \sigma^2}{m^2} = \frac{1.64^2 \times 40^2}{7^2} \simeq 88.$$

So the sample size should be about 88. □

Notice that all the preceding calculations are the same if the original population is not normal.

EXERCISES

1. A random sample of size 15 from a normal distribution $N(\mu, 25)$ yielded $\bar{x} = 70.2$. Find a 95% confidence interval for μ.

2. Wildlife managers were queried about how long they have had a job with their current company or organization. One hundred fourteen responded and the average was 11.8 years. From past employment data, it is believed that the standard deviation is $\sigma = 3.2$ years. Find an interval in which you will be 95% confident that the true mean lies.

3. Referring to Exercise 2, suppose that you want a margin of error of SE = 0.2. How many wildlife managers should you query?

4. Sixty-six percent in a poll of 1664 adults near a wildlife sanctuary in Alaska said they favor allowing oil exploration in that area. The local newspaper reported that the margin of error was 3% at a 95% confidence level. What does that mean?

6.2.2 The Gamma Distribution

The next question, which is the obvious one, is: How do we estimate the mean if the variance σ^2 of the underlying population from which we are sampling is unknown? This is almost always the case. From previous sampling with large sample sizes, we may have a good idea of what the variance is, but we never know it exactly.

If we assume that the underlying population is normal, which occurs frequently, it appears reasonable to replace σ^2 by the statistic s^2. Then, from the derivations in Section 6.1 [see (6.1)], we are led to consider the random variable

$$T = \frac{\overline{X} - \mu}{S/\sqrt{n}},$$

where S^2 is the random variable for sample variance and S is the sample standard deviation. It is not clear a priori what the distributions of S, S^2, or T are. We do know T is a jointly distributed random variable depending on the two random variables $\overline{X}$ and S, but we do not know if they are independent. These questions are answered by key results from mathematical statistics which state that $\overline{X}$ and S are, in fact, independent random variables, and S^2 has a chi-squared distribution. The random variable T, given by

$$T = \frac{\overline{X} - \mu}{\sqrt{S^2/n}}, \tag{6.2}$$

is a t-distribution with $n - 1$ **degrees of freedom** (degrees of freedom is denoted by df). We now explain what we mean by these distributions using additional detail.

Both of these distributions (the t- and chi-squared distributions) can be defined in terms of a fundamental distribution called the **gamma distribution**. The pdf for the gamma distribution depends on two positive parameters α and β and is given by

$$f(x) = \frac{1}{\beta^\alpha \Gamma(\alpha)} x^{\alpha-1} e^{-x/\beta}, \quad x > 0,$$

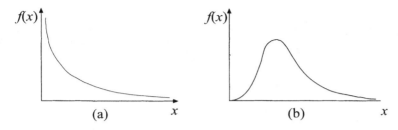

Figure 6.3 The gamma distribution for various values of α and β: (a) an exponential distribution. (b) a chi-squared distribution.

and $f(x) = 0$ for $x \leq 0$. Here df $= \alpha$, and $\Gamma = \Gamma(z)$ is the **gamma function**, a special function in mathematics defined by

$$\Gamma(z) = \int_0^\infty t^{z-1} e^{-t} \, dt, \quad z > 0.$$

You can regard the gamma function as just another common function in mathematics like ln, exp, sin, erf, and so on, but it is defined by an integral. This is not unusual; witness, for example, that many calculus texts define the natural logarithm function by the integral $\ln x = \int_1^x (1/u) \, du$. Values of the gamma function are tabulated in computer algebra systems. But some values are easy to calculate. By simple integration we can find that $\Gamma(1) = 1$ and $\Gamma(2) = 1$. Using integration by parts we can obtain, for any value z, integer or not,

$$\Gamma(z + 1) = z\Gamma(z).$$

Therefore, the Γ function acts like a generalized factorial function. In fact, $\Gamma(n) = (n - 1)!$ for positive integers n.

The gamma distribution has a variety of shapes, depending on the parameter values (Fig. 6.3). The exponential distribution, which we encountered in Chapter 5, is a gamma distribution with $\alpha = 1$ (Fig. 6.3a). A **chi-squared distribution with ν degrees of freedom** is a gamma distribution with $\alpha = \nu/2$ and $\beta = 2$ (Fig. 6.3b). Thus, a chi-squared distribution, denoted by $\chi^2(\nu)$, has density $f(x) = 0$ for $x \leq 0$, and

$$f(x) = \frac{1}{2^{\nu/2}\Gamma(\nu/2)} x^{\nu/2-1} e^{-x/2}, \quad x > 0.$$

Finally, a **t-distribution with ν degrees of freedom** (df $= \nu$) has a pdf given by

$$f(t) = \frac{\Gamma((\nu + 1)/2)}{\sqrt{\nu\pi}\,\Gamma(\nu/2)} \left(1 + \frac{t^2}{\nu}\right)^{-(\nu+1)/2}, \quad -\infty < t < \infty.$$

The t-density $f(t)$ is symmetric about the origin and has a bell shape, like the normal distribution. For $\nu \geq 30$ the graph of f very closely approximates the normal distribution. But for all degrees of freedom the t-distribution has "fatter" tails than the normal; that is, it doesn't decay as fast and there is more area under its tails (near $\pm\infty$) than does the normal distribution (Fig. 6.4). This means that confidence intervals are larger than when the variance is known exactly. One can show that if T is t-distributed with ν degrees of freedom, then

$$\mathrm{E}(T) = 0, \quad \mathrm{var}(T) = \frac{\nu}{\nu - 2}, \quad \nu > 2.$$

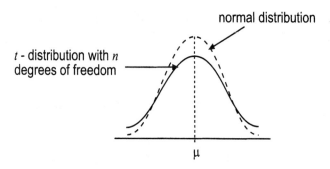

Figure 6.4 The t-distribution compared to the normal distribution.

The t-distribution has an interesting story that is part of the folklore of statistics. It is credited to the chemist W. Gossett (1876–1936), who worked for the Guinness Brewery and published under the pseudonym "Student." So the t-distribution is sometimes called Student's distribution. Conjectures abound as to why he published in secret. Was it an industrial secret, or did he not want his employer to know he was spending his time doing mathematics?

EXERCISES

1. Show that the chi-squared distribution with two degrees of freedom is the same as the exponential distribution with parameter $1/2$.

2. Sketch plots of the t-distribution for $\nu = 1$ and $\nu = 4$ degrees of freedom.

3. Let $X \sim \mathrm{N}(0, 1)$. Show that X^2 is chi-squared distributed with one degree of freedom. [*Hint*: Begin with $\Pr(Y \leq y) = \Pr(-\sqrt{y} \leq X \leq \sqrt{y})$, and so on, next using the standard normal density.]

6.2.3 The Sample Variance

The general process of finding confidence intervals for any parameter closely mirrors the process of finding confidence intervals for μ when σ^2 is known, and μ when σ^2 is unknown. The key element was knowledge of the distribution of the particular random variable. In an interval estimation, an estimate for a population parameter θ is given as an interval in which we expect the actual value of the population parameter to lie. In the case of a confidence interval, we are able to assign a measure of the accuracy for this interval. Formally, a $100(1 - \alpha)\%$ **confidence interval** for a parameter θ is a random interval $[L_1, L_2]$ such that

$$\Pr(L_1 \leq \theta \leq L_2) = 1 - \alpha,$$

regardless of the value of θ. We call the probability $1 - \alpha$ the confidence coefficient, and the random variables L_1 and L_2 the lower and upper confidence limits, respectively. As noted previously, the confidence coefficient gives the proportion of confidence intervals constructed that will actually contain the population parameter. That is, if we take 100 different samples from a population, using each of these samples to construct a confidence interval for a particular parameter θ, a confidence coefficient of 0.95 (i.e., 95% confidence interval) means we are confident that 95 of the 100 constructed intervals will actually contain the true value for the parameter θ.

We now discuss the distribution of the sample variance S^2. It is a direct calculation using the properties of expectation to show that

$$E(S^2) = \sigma^2.$$

We can obtain a confidence interval using the random variable

$$\chi^2 = \frac{(n-1)S^2}{\sigma^2},$$

which is a chi-squared random variable with $n - 1$ degrees of freedom. Then, given a number α, there exists constants a and b for which

$$\Pr\left(a < \frac{(n-1)S^2}{\sigma^2} < b\right) = 1 - \alpha.$$

There are many choices of a and b. We select a and b such that $\Pr(\chi^2 \leq a) = \Pr(\chi^2 \geq b) = \alpha/2$ (see Fig. 6.5). Upon rearranging the inequalities in the last equation, we have

$$\Pr\left(\frac{(n-1)S^2}{b} < \sigma^2 < \frac{(n-1)S^2}{a}\right) = 1 - \alpha,$$

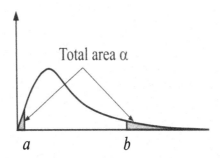

Figure 6.5 The chi-squared distribution showing the choice of a and b.

and we have the variance trapped between two endpoints that depend on a random experiment. Once the experiment is performed and we obtain the sample statistic s^2, the $100 \times (1 - \alpha)$ confidence interval for the variance is

$$\left(\frac{(n-1)s^2}{b}, \frac{(n-1)s^2}{a} \right).$$

As before, the correct interpretation is that if the experiment were performed many times, the true variance of σ^2 would be trapped in the confidence interval about $100 \times (1 - \alpha)$ percent of the time.

6.2.4 The Mean with Unknown Variance

Finally, we return to the question of estimating the mean if the variance is unknown. For unknown variance the confidence interval for the mean μ is computed exactly as in the case of known variance, but with S replacing σ and $t_{\alpha/2}$ replacing $z_{\alpha/2}$, where $t_{\alpha/2}$ is the t-statistic, with $n-1$ degrees of freedom, for which

$$\Pr\left(-t_{\alpha/2} < \frac{\overline{X} - \mu}{S/\sqrt{n}} < t_{\alpha/2} \right) = 1 - \alpha.$$

Then the $(1 - \alpha) \times 100\%$ confidence interval for the mean is

$$\left(\overline{x} - \frac{t_{\alpha/2}s}{\sqrt{n}} < \mu < \overline{x} + \frac{t_{\alpha/2}s}{\sqrt{n}} \right),$$

where $t_{\alpha/2}$ is $\Pr(T > t_{\alpha/2}) = \alpha/2$, and where T is a t-distribution with $n-1$ degrees of freedom.

Example 6.7

To measure toxicity of a harmful chemical in a lake biology students sampled the water at 15 different locations. Measured in parts per million, they found the concentrations:

$$259 \quad 260 \quad 206 \quad 265 \quad 284 \quad 291 \quad 229 \quad 216$$
$$232 \quad 251 \quad 225 \quad 225 \quad 242 \quad 238 \quad 250$$

The sample mean is $\bar{x} = 244.867$ and the sample variance $s^2 = 583.124$. To find a 95% confidence interval for the mean we set $\alpha = 0.05$ and find the value $t_{.025}$ for which $\Pr(T > t_{.025}) = \alpha/2 = 0.025$. We know that T is t-distributed with $n - 1 = 14$ degrees of freedom. From software, $t_{.025} = 2.145$. Therefore, the confidence interval is

$$\left(244.867 - \frac{2.145 \times \sqrt{583.124}}{\sqrt{14}}, \ 244.867 + \frac{2.145 \times \sqrt{583.124}}{\sqrt{14}} \right)$$

or

$$[231.024, 258719]. \quad \square$$

EXERCISES

1. Let T be a t-distributed random variable with 5 degrees of freedom.

   (a) Find the value a for which $\Pr(T > a) = 0.025$.

   (b) Find $\Pr(-1.5 < T \leq 1.5)$.

2. The number of days required for seeds of a certain plant to mature is $N(\mu, \sigma^2)$. A random sample of 13 seeds yielded $\bar{x} = 18.9$ and $s^2 = 10.7$. Find a 90% confidence interval for σ^2.

3. The number of days required for seeds of a certain plant to mature is $N(\mu, \sigma^2)$. A random sample of 13 seeds yielded $\bar{x} = 18.9$ and $s^2 = 10.7$. Find a 90% confidence interval for σ^2.

4. At a biological station ozone levels were measured on seven randomly chosen days. The levels were (in ppm): 0.06, 0.07, 0.08, 0.11, 0.12, 0.14, 0.21. Assuming a normal distribution, find a 95% confidence interval for the mean μ.

5. Using the data in Exercise 4, find a 95% confidence interval for the variance σ^2, again assuming a normal distribution.

6.3 Estimating Proportions

Exactly as we proceeded for the mean and variance, confidence intervals can be constructed for a population proportion p, the difference between population proportions p_1 and p_2 (i.e., $p_1 - p_2$), and various other quantities involving proportions. The classic example of estimating a proportion occurs in elections. Pollsters continually sample the population to determine if voters are going to choose one candidate over another.

But here we examine ecological problems. Suppose that a large pond has a turtle population and we want to estimate the fraction p of females in the pond. The obvious strategy is to catch a random sample of n turtles and determine their sex (e.g., males have longer claws and a greater distance along the tail from the shell to the coecal openning). If x of them are females, the ratio

$$\widehat{p} = \frac{x}{n}$$

is an obvious **point estimator** of the value of p. Notice that p is the probability of selecting a female on one random selection from the pond. The obvious question, therefore, is to determine how close our estimate is to the true value. To fix the idea, suppose that we found $x = 25$ female turtles in a random sample of $n = 50$ turtles. Then $\widehat{p} = x/n = 25/50 = 0.5$. How confident are we that this value is close to the actual value of p, which we do not know?

The first thing to notice is that $\widehat{p}$ is determined by the sample we took. If we took another sample of size 50, we may find $x = 20$ females, giving the estimate $\widehat{p} = 0.4$. Therefore, we can regard the fraction of female turtles as a random variable $\widehat{P}$, and the two values of 0.5 and 0.4 are two values of $\widehat{P}$. To answer questions about $\widehat{P}$, we need to know how it is distributed. The following result is basic to the analysis.

Theorem 6.8

Let $X_1, X_2, X_3, \ldots, X_n$ be a random sample where X_i is a Bernoulli random variable (taking on the value 0 or 1, depending on the presence of a given trait). The random variable

$$\widehat{p} = \frac{X}{n} = \frac{\text{number in the sample with particular trait}}{\text{sample size}}$$

is approximately normally distributed with mean p and variance $\sigma^2 = p(1 - p)/n$. $\square$

The pdf for $\widehat{P}$ is the **sampling distribution** of $\widehat{P}$. Let X be the random variable representing the number of females chosen in a sample (with replacement) of size n. It is clear that X is binomially distributed with $E(X) = np$ and $\text{Var}(X) = npq$, where $q = 1 - p$. Therefore, $\widehat{P}$ is binomial and

$$E(\widehat{P}) = E\left(\frac{1}{n}X\right) = \frac{1}{n}E(X) = p,$$

$$\text{var}(\widehat{P}) = \text{var}\left(\frac{1}{n}X\right) = \frac{1}{n^2}\text{var}(X) = \frac{pq}{n}.$$

With this information we can evaluate the accuracy of a point estimate $\widehat{p}$.

For large n it is known from the central limit theorem that the random variable

$$Z = \frac{\widehat{P} - p}{\sqrt{pq/n}}$$

is $N(0,1)$. This fact helps us obtain an interval estimate that assesses the error in taking $\widehat{p}$ for p. For a given probability $1 - \alpha$, we find the value of Z, $z_{\alpha/2}$, for which

$$\Pr\left(-z_{\alpha/2} < \frac{\widehat{P} - p}{\sqrt{pq/n}} < z_{\alpha/2}\right) = 1 - \alpha.$$

For example, if $\alpha = 0.05$, and so $1 - \alpha = 0.95$, we can find from a table or software that $z_{\alpha/2} = 1.96$. Notice that $z_{\alpha/2}$ depends on the selected value of α. Using some algebra to isolate and sandwich p, we can write this in the form

$$\Pr\left(\widehat{P}_L < p < \widehat{P}_R\right) = 1 - \alpha,$$

where $(\widehat{P}_L, \widehat{P}_R)$ is a random interval for p depending on $z_{\alpha/2}$, n, and $\widehat{P}$. The endpoints are given by

$$\widehat{P}_L = \frac{\left(\widehat{P} + z_{\alpha/2}^2/2n\right) - z_{\alpha/2}\sqrt{\widehat{P}(1 - \widehat{P})/n + z_{\alpha/2}^2/4n^2}}{1 + z_{\alpha/2}^2/n},$$

$$\widehat{P}_R = \frac{\left(\widehat{P} + z_{\alpha/2}^2/2n\right) + z_{\alpha/2}\sqrt{\widehat{P}(1 - \widehat{P})/n + z_{\alpha/2}^2/4n^2}}{1 + z_{\alpha/2}^2/n}.$$

Now, when we take a random sample and compute $\widehat{p}$, which is a value for $\widehat{P}$, we will have values $\widehat{p}_L$ and $\widehat{p}_R$ of $\widehat{P}_L$ and $\widehat{P}_R$. The interval $(\widehat{p}_L, \widehat{p}_R)$ forms a $(1 - \alpha) \times 100\%$ confidence interval for the population proportion p.

If n is large, we can approximate $\widehat{p}_L$ and $\widehat{p}_R$ by simpler expressions. Using the fact that $z_{\alpha/2}^2/n$, $z_{\alpha/2}^2/4n^2$, and $z_{\alpha/2}^2/2n$ are small, the confidence interval is closely approximated by

$$\left(\widehat{p} - z_{\alpha/2}\sqrt{\frac{\widehat{p}\widehat{q}}{n}}, \widehat{p} + z_{\alpha/2}\sqrt{\frac{\widehat{p}\widehat{q}}{n}}\right), \tag{6.3}$$

with $\widehat{q} = 1 - \widehat{p}$.

Example 6.9

In the turtle example introduced earlier, $n = 50$ and $\hat{p} = 0.5$. A 95% confidence interval requires that $1 - \alpha = 0.95$, or $\alpha = 0.05$. From a table of standard normal values, we find $z_{\alpha/2} = 1.96$. Therefore, using (6.3), a 95% confidence interval is $(0.36, 0.64)$. □

EXERCISES

1. The problem of comparing two proportions arises frequently in mathematical biology. For example, suppose that we want to see if the proportion of female turtles in two different ponds is different. We can compare two population proportions by looking at the difference, $p_1 - p_2$, between the two population proportions. A natural point estimate for the difference of two proportions is

$$\hat{p}_1 - \hat{p}_2 = \frac{X_1}{n_1} - \frac{X_2}{n_2},$$

where X_1 is the count of individuals with the particular trait of interest in a sample of size n_1 taken from the first population, and X_2 is the count of individuals with the particular trait in a sample of size n_2 taken from the second population. It can be shown for large sample sizes that the random variable

$$Z = \frac{(\hat{p}_1 - \hat{p}_2) - (p_1 - p_2)}{\sqrt{p_1(1 - p_1)/n_1 + p_2(1 - p_2)/n}}$$

is approximately the standard normal distribution. Use this information to derive a $100(1-\alpha)\%$ confidence interval for the difference of two population proportions $p_1 - p_2$.

2. Due to differing environmental conditions, the proportion of female turtles in two ponds is believed to differ. To check this, a random sample of 40 turtles is taken from the first pond, and it is noted that 17 of these turtles are females. In the second pond, a random sample of 54 turtles are captured, and it is noted that 28 of the turtles in this sample are female. Construct a 90% confidence interval for the difference of the proportion of female turtles in these two ponds. Using this confidence interval, can we safely conclude that there is a difference between the proportion of female turtles in these two ponds? Provide support for your conclusion.

3. A geneticist is interested in the proportion of males and females in a region of Africa that have a minor blood disorder. In a random sample of 1000 males, 250 are found to be afflicted, while 275 of 1000 females are afflicted.

   (a) Find a 95% confidence interval for the proportion of males with the blood disorder.

(b) Find a 95% confidence interval for the proportion of females with the blood disorder.

(c) Compute a 95% confidence interval for the difference between the proportion of males and females that have the blood disorder.

4. Residents in a Midwest county are asked whether they favor stopping the construction of a bridge across a local river where the endangered tiger beetle has its habitat. How large a sample is required if one wants to be at least 95% confident that the estimate of the proportion who favor stopping construction is within 0.04 of the true proportion?

5. A botanist found that 10 seeds out of 20 germinated at 5°C, whereas at 15°C he found that 15 out of 20 seeds germinate. Is there a significant difference in germination at these two different temperatures?

6.4 The Chi-Squared Test

In this section, in the context of a chi-squared test, we introduce some terminology of hypothesis testing, a topic taken up in detail in the next section.

Suppose that we toss a coin 160 times and it comes up heads 86 times, and therefore tails 74 times. Is this a fair coin, or is it "loaded"? Does this experiment allow us to reject a hypothesis that it is a fair coin? One might argue that the results of the experiment is within chance. Yes or no? If the result was 120 heads and 40 tails, we would have much stronger evidence that the coin is not fair. The chi-squared test is a statistical test that allows us to assess experiments like this and make a conclusion about the validity of hypothesis that we might pose. Let us arrange the data in a table where the values in parentheses are the values expected:

$$\underline{\text{Heads}} \quad \underline{\text{Tails}}$$
$$86(80) \quad 74(80)$$

The expected values are based on our hypothesis that it is a fair coin. In all chi-squared tests we first make a hypothesis and then we compute expected values based on that hypothesis. Next we ask how to measure the discrepancy, or difference, between the two observed and expected values. There are several ways that one might think of to do this. The method we select, which should remind us of the least squares method, is to sum the squares of the relative differences in each cell. That is, we define the test statistic X^2 by the formula

$$X^2 = \sum \frac{(\text{observed value} - \text{expected value})^2}{\text{expected value}},$$

where the sum is taken over the two cells in the table. Thus, in the present case,

$$X^2 = \frac{(86-80)^2}{80} + \frac{(74-80)^2}{80} = 1.2.$$

This number X^2 is a statistic; if we took a different collection of turtles, the value of X^2 would be different. If X^2 is small, we expect that the hypothesis has some validity; if it is large, we would question the hypothesis that the coin is fair. Is the number 1.2 large enough to reject the hypothesis that the coin is fair? Is it within chance (i.e., the random error of the experiment)? We need a quantitative way to decide.

X^2 is a random variable because we get a different result each time we repeat the random experiment. Its probability density function is **chi-squared** with 1 degree of freedom. The number of degrees of freedom df is, as we see shortly, related to the number of cells in the table. Therefore, let us ask: What is the probability that we would get a value at least as large as 1.2? That is, what is $\Pr(X^2 \geq 1.2)$, or the probability that the number we obtained for X^2 is 1.2 or more? From software, or a table of values, we find that, approximately,

$$\Pr(X^2 \geq 1.2) = 0.28,$$

which means that X^2 would take a value equal to or larger than 1.2 about 28% of the time. This does not give us an indication that we can reject the hypothesis; getting a value of 0.28 or larger happens relatively often. The value $p = 0.28$ is called the **p-value**, and it gives some indication of how valid the hypothesis is; it is the area under the chi-squared distribution curve for $X^2 \geq 1.2$. See Fig. 6.6. If p had turned out to be a very small value, then we would have evidence to reject the hypothesis. For example, suppose that we found $X^2 = 5$. Then, from software, $\Pr(X^2 \geq 5) = 0.025$. This statement implies that getting a value of at least 5 is fairly rare, occurring only about 2.5% of the time. So there is evidence to reject the hypothesis. Where is the cutoff? We explain this after one more example.

Example 6.10

Biology students working with turtles conjectured that small male and female turtles were in the same ratio as large turtles. To test this hypothesis the students caught 48 turtles from a large pond and measured their length. There were 23 males and 25 females. They formed the following **contingency table** to display the data:

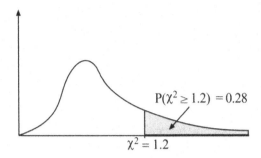

Figure 6.6 The probability $\Pr(X^2 \geq 1.2) = 0.28$ for a chi-squared distribution with 1 degree of freedom. The area is the p-value $p = 0.28$

	Length ≤ 15	Length > 15	Totals
Males	**11**	**12**	23
Females	**7**	**18**	25
Totals	18	30	48

Now, if the student's hypothesis is correct, then we would expect certain values w, x, y, z, in the four data positions:

	Length ≤ 15	Length > 15	Totals
Males	w	x	23
Females	y	z	25
Totals	18	30	48

Because the hypothesis is that the ratios are the same, we must have

$$\frac{w}{18} = \frac{23}{48}, \quad \frac{x}{30} = \frac{23}{48}$$

and

$$\frac{y}{18} = \frac{25}{48}, \quad \frac{z}{30} = \frac{25}{48}.$$

Solving for w, x, y, and z gives $w = 8.6$, $x = 14.3$, $y = 9.4$, $z = 15.7$. Therefore, the values expected may be displayed as

	Length ≤ 15	Length > 15	Totals
Males	**8.6**	**14.3**	23
Females	**9.4**	**15.7**	25
Totals	18	30	48

It is common to display the values expected in parentheses in a single table next to the values observed. The values expected are based on the hypothesis we made: namely, that the ratios are the same. They differ from the data values in the first table above. But are they different enough to reject the hypothesis? After all, there was some randomness in our collection of turtles.

Again, the question we ask is how we measure the discrepancy, or difference, between the four observed and expected values. As in the first example we define the test statistic X^2 by the formula

$$X^2 = \sum \frac{(\text{observed value} - \text{expected value})^2}{\text{expected value}},$$

where the sum is taken over the four cells in the table. Thus,

$$X^2 = \frac{(11 - 8.6)^2}{8.6} + \frac{(12 - 14.3)^2}{14.3} + \frac{(7 - 9.4)^2}{9.4} + \frac{(18 - 15.7)^2}{15.7} = 2.019.$$

A different collection of turtles would give a different value of X^2. Clearly, if X^2 is small, we expect that the hypothesis has some validity; if it is large, we would really question the hypothesis. Is the number 2.019 large enough to reject the hypothesis? Or, is it within the random error of the experiment? The random variable X^2 is chi-squared distributed with 1 degree of freedom. The number of degrees of freedom is, as we observe shortly, related to the number of cells in the contingency table. So we may ask what $\Pr(X^2 \geq 2.019)$ is, or the probability that the number we obtained for X^2 is 2.019 or more. From software we find that

$$\Pr(X^2 \geq 2.019) = 0.155,$$

which means that X^2 would take a value equal to or larger than 2.019 about 15.5% of the time. The value $p = 0.155$ is the p-value, and it gives the area under the chi-squared distribution curve for $X^2 \geq 2.019$. But do we reject the hypothesis?

There is some standard language in this regard that helps formulate our conclusion precisely. We compare the p-value we obtained to other, standard p-values: for example, $p = 0.01$, $p = 0.05$, $p = 0.10$ $p = 0.25$. (1%, 5%, 10%, and 25%).

We say that our value of $p = 0.155$, which is between $p = 0.10$ and $p = 0.25$, is **significant** at the 25% level but **not significant** at the 10% level. Suggestive of the terminology to be used in the next section, we **reject** the hypothesis at the 25% level and **do not reject** the hypothesis at the 10% level. We usually do not say we "accept the hypothesis" but rather, we "do not reject it." This is suggestive of the fact that one example does not prove anything, but one example can disprove a statement. The higher the chi-squared statistic (value), the smaller the p-value and the more it is that likely the hypothesis can be rejected at some level. Many researchers often use the 5% significance ($p = 0.05$) level as a benchmark for rejecting a hypothesis. □

Finally, we return to the issue of the number of degrees of freedom df. It is related to the number of cells in the table of values. If there is one row and n columns, df $= n-1$. If there are r rows and c columns, then df $= (r-1)\times(c-1)$. For each degree of freedom, the chi-squared density has a different shape.

EXERCISES

1. Let X^2 be $\chi^2(5)$. Find the value a for which $\Pr(X^2 \geq a) = 0.01$.

2. A hypothesis is tested and the chi-squared statistic is $X^2 = 4.50$. There is 1 degree of freedom. Can the hypothesis be rejected at the 5% level? The 10% level?

3. A hypothesis is tested and the chi-squared statistic is $X^2 = 10.0$. There are 5 degrees of freedom. Can the hypothesis be rejected at the 5% level? The 10% level?

4. A fishing area below a dam is contaminated with a fungus that infects the gills of fish. Over a period of a few weeks several fish were caught and examined for the fungus. The counts for three species are as follows:

	Drum	White bass	Sunfish
No fungus	50	47	56
Fungus	5	14	8

Test the hypothesis that there is no difference in the species of fish with regard to becoming infected with the fungus.

5. An ecologist is testing the mortality of flour beetles in a severe environment. She placed 100 adult beetles in the environment and for four consecutive weeks measured the number of survivors at the end of each week; she found that the populations were $53, 27, 15$, and 9. From these data she postulated that a good model for the dynamics is that the weekly mortality rate is 50%. Discuss the validity of her hypothesis.

6.5 Hypothesis Testing

In Section 6.4 we introduced the basic idea of testing a hypothesis. Now we embark on a more detailed analysis that fine tunes the preceding ideas. In a typical estimation problem there is some population parameter whose value is to be approximated based on sample data. Usually, there is no predetermined notion concerning the actual value of this parameter. We are simply attempting to ascertain its value by using gathered data. In *hypothesis testing* we must have a preconceived idea of the parameter value. This preconceived notion of the value of the parameter implies *two hypotheses*: the hypothesis being proposed by the researcher, which is called the **alternative hypothesis**, and the negation of this hypothesis, which is called the **null hypothesis**. We use H_0 to denote the null hypothesis and H_1 (or H_a) to denote the alternative hypothesis. When performing a hypothesis test, the purpose of the experiment is to determine if the data gathered tend to disprove the null hypothesis. The following guidelines can be used to help decide how to state H_0 and H_1 correctly.

1. For a hypothesis test concerning the value of some parameter (i.e., μ, σ), the null hypothesis will always be the statement that includes equality. That is, the statement included in H_0 *must* contain one of the binary relations $\leq$, $\geq$, or $=$.

2. The alternative hypothesis, or research hypothesis, contains the statement that the experimenter is trying to support. This hypothesis contains the binary relation that negates the null hypothesis.

3. Because H_1 is the research hypothesis, it is hoped that the data leads us to reject H_0, and thus accept H_1.

Example 6.11

The Food and Drug Administration has stated that the mean level μ of mercury in canned tuna is 0.12 ppm. It is feared that this figure has increased as a result of increased atmospheric mercury releases. Data are gathered to determine if the level of mercury has indeed increased. The hypotheses for this study are

$$H_0 : \mu \leq 0.12 \text{ ppm},$$
$$H_1 : \mu > 0.12 \text{ ppm}. \quad \square$$

Example 6.12

Studies of the habits of white-tailed deer indicate that these deer live and feed within very limited ranges. To determine whether the ranges of deer in two different geographical areas differ, researchers caught, tagged, and fitted 40 deer with small radio transmitters. Define μ_1 as the mean range of deer in the first geographic location and μ_2 as the mean range of deer in the second geographic location. It is believed by the researcher that the mean ranges for deer in these two geographic regions is statistically different. The hypotheses for this study are

$$H_0 : \mu_1 = \mu_2,$$
$$H_1 : \mu_1 \neq \mu_2. \quad \square$$

Example 6.13

Linear regression (see Section 1.7) is a form of regression analysis that uses a linear model to describe the relationship between one or more independent variables, $x_1, x_2, x_3, \ldots, x_p$, and a dependent variable y. This relationship is described by the model

$$y = \beta_0 + \beta_1 x_1 + \beta_2 x_2 + \cdots + \beta_j x_j + \cdots + \beta_p x_p + \varepsilon.$$

There are p parameters $\beta_0, \beta_1, \beta_2, \ldots, \beta_p$ to be determined (typically by the method of least squares) and ε is an error term. Part of the analysis of a linear regression model is to determine if a *fixed* independent variable x_j, $j = 1, 2, \ldots, p$, is important, or useful, in describing, predicting, or controlling y when using this regression model. To determine if an independent variable x_j is related significantly to the dependent variable y, we can perform the hypothesis test

$$H_0 : \beta_j = 0,$$
$$H_1 : \beta_j \neq 0. \quad \square$$

For additional terminology, Example 6.11 is a **one-tailed hypothesis test** as H_1 contains a strict inequality, whereas Examples 6.12 and 6.13 are **two-tailed hypothesis tests** as H_1 contains "does not equal."

After the hypotheses have been determined and the sample data collected, a decision is made. We either reject H_0 or fail to reject H_0. Using the collected data, we compute a **test statistic** θ^*. The test statistic is a random variable whose probability distribution is known under the assumption that the null hypothesis H_0 is true. If the test statistic assumes a value that is rare when H_0 is true, the data would tend to support the alternative hypothesis, and the decision would be to "reject H_0 in favor of H_1" (see Fig. 6.7). On the other hand, if the value of the test statistic is reasonable under the assumption that H_0 is true, our decision would be to "fail to reject H_0." After the decision is made, there are four possible outcomes:

1. We reject H_0 in favor of H_1, but H_0 is actually true. In this case we have made an error. This type of error is called a **type I error**.

2. We reject H_0 in favor of H_1, and H_1 is actually true. In this case we have made the correct decision.

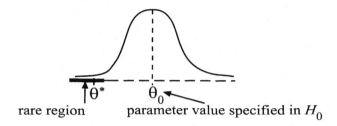

rare region   parameter value specified in H_0

Figure 6.7 If the test statistic, θ^*, falls in the rejection region, we reject H_0 in favor of H_1. In this figure, the rejection region shown is representative of a one-tailed test to the left.

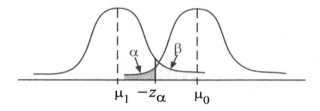

Figure 6.8 μ_1 represent the true parameter value. If the test statistic falls to the right of $-z_\alpha$, we will make a type II error.

3. We fail to reject H_0 in favor of H_1, but H_1 is actually true. This case also results in an error, called a **type II error**.

4. We fail to reject H_0 in favor of H_1, and H_0 is actually true. In this case we have made the correct decision.

As these four outcomes show, in the process of hypothesis testing there is the possibility of making an error. It is standard to let α denote the probability of making a type I error and β the probability of making a type II error. The probability α is called the **level of significance** of the test and the probability β is called the **power** of the test. Figure 6.8 shows the relationship between α and β. It is important to note that computing β, the probability of a type II error, is a little harder to handle than α, the probability of a type I error, which is usually dictated by the researcher beforehand. In fact, β can be determined only if a particular value of the alternative hypothesis is specified, which is seldom the case. For the purposes of hypothesis testing, we restrict our focus to type I errors.

Example 6.14

In Example 6.11, a type I error occurs if we reject H_0 when H_0 is true. In this case we would conclude incorrectly that the mean level of mercury in canned tuna is greater 0.12 ppm. This error could lead to making unnecessary recommendations for people to lower their consumption of canned tuna. A type II error occurs if we fail to reject H_0 when H_0 is actually false. In this case we would conclude that it is not necessary to encourage reduced canned tuna consumption, when in fact we should be implementing a program to do so. □

Example 6.15

In a certain population, we expect 52% of the population to get a cold in a 3-month period. We give 196 volunteers 1000 mg of vitamin C per day for 4 months. At the end of the period, 96 people (49%) have gotten colds. So the proportion of our sample that have gotten colds is $\hat{p} = 0.49$. Let p be the proportion of people being treated with vitamin C who will get colds. We would like to know if there is evidence that p is smaller than 0.52. □

We now discuss how the data gathered are used to determine if we "reject H_0" or "fail to reject H_0." The decision can be made using either *classical hypothesis testing* or *significance testing*. The two approaches are very similiar in that both depend on the value of the test statistic, but there is one key difference. With classical hypothesis testing a *rejection region* is decided beforehand. If the test statistic θ^* falls in the rejection region, we reject H_0 in favor of H_1. The rejection region depends on three things: the level of significance α that the researcher is using, the distribution of the sample statistic $\hat{\theta}$, and the relation stated in the alternative hypothesis H_1. Figure 6.9 shows the three types of rejection regions: one-tailed test to the left [part (a)], two-tailed test [part (b)], and a one-tailed test to the right [part (c)].

With significance testing, the researcher uses θ^* to compute the p-value. The size of the p-value is then used to make the decision to reject H_0. If the p-value, which is the probability of making a type I error, is "small", the decision will be made to reject H_0. The determination of what is small depends on what the ramifications of making a type I error are. If a type I error is very costly, then to reject H_0 the researcher will generally require a p-value < 0.01. Otherwise, the researcher may deem a p-value smaller than 0.05 sufficient to reject H_0. With the advent of statistical software packages, which generally return p-values, significance testing is now the preferred method for performing hypothesis testing. The next two examples illustrate the process of significance testing.

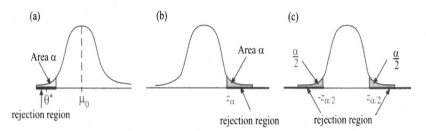

Figure 6.9 The three types of rejection regions. (a) the rejection region for a one-tailed test to the left; (b) the rejection region for a one-tailed test to the right; (c) the rejection region for a two-tailed test.

Example 6.16

Significance testing can be thought of as a four-step process. We describe these steps for the question posed in Example 6.15.

1. Determine null and alternative hypotheses. We are interested in testing if the proportion of people treated with vitamin C who get colds is smaller than the population proportion of $p = 0.52$. In this case we have a one-tailed test to the left, and the hypotheses are

$$H_0 : p \geq 0.52,$$
$$H_1 : p < 0.52.$$

2. Use the data to compute the test statistic. By Theorem 6.8 we see that $\hat{p}$ is normally distributed; thus, by the central limit theorem the random variable $z = (\hat{p} - p)/\sigma$ is a standard normal random variable. We see in Theorem 6.8 that $\sigma = \sqrt{0.52(1 - 0.52)/196} = 0.357$. The test statistic is then given by

$$z^* = \frac{0.49 - 0.52}{0.357} = -0.8403.$$

3. Compute the p-value using the test statistic. The p-value is simply the probability that we would see a sample statistic more extreme (in this case to the left of z^*) than the test statistic computed (Fig. 6.10). That is,

$$p\text{-value} = Pr(z < z^*) = 0.2033.$$

4. State the conclusion with regard to the stated hypotheses. The computed p-value of 0.2033 implies that if we were to reject H_0 in favor of H_1, the probability that we would have made a type I error is 0.2033, which is quite large. Thus, it would seem prudent that we fail to reject H_0. By failing to

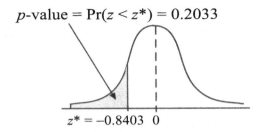

$$p\text{-value} = \Pr(z < z^*) = 0.2033$$

$z^* = -0.8403 \quad 0$

Figure 6.10 For a one-tailed test to the left, the p-value is the area under the density function to the left of the test statistic.

reject H_0, we are not implying that H_0 is true, but rather, that the data collected do not provide significant support to disprove the null hypothesis. In general, p-values less than 0.05 are said to be statistically significant. □

Example 6.17

Refineries, steel mills, food-processing plants, and other industries separate oil and water using polyelectrolytes. These work better when pH is closely controlled. This wastewater must be neutralized before it is released into the environment. The following data are obtained on the pH of wastewater samples that have been treated: 6.4, 7.3, 7.7, 6.9, 7.1, 7.8, 8.2, 6.6, 7.4, 6.9. Based on these data, is there evidence that the treatment process does not yield an average pH of 7 as desired?

1. Determine null and alternative hypotheses. We are interested in determining if the average pH of treated wastewater yields a pH level that is different than 7. This leads to a two-tailed hypothesis test with

$$H_0 : \mu = 7,$$
$$H_1 : \mu \neq 7.$$

2. Use data to compute the test statistic. Equation (6.2) shows that the random variable $t = (\overline{X} - \mu)/(s/\sqrt{n})$ follows a T-distribution with $n-1$ degrees of freedom. Using our data, we have $\overline{x} = 7.23$, $s = 0.5618$, and $n = 10$. Thus, the test statistic t^* is given by

$$t^* = \frac{\overline{x} - \mu}{s/\sqrt{n}} = \frac{7.23 - 7}{0.5618/10} = 4.09.$$

3. Compute the p-value using the test statistic. For a two-tailed test the p-value is given by

$$p\text{-value} = 2P(t \geq 4.09) < 0.01.$$

4. State your conclusion with regard to the stated hypotheses. A p-value that is less than 0.01 is very small and is generally classified as highly significant. Small p-values lead us to reject H_0 in favor of H_1. We would conclude by stating that the data gathered provide evidence that the treatment process does not yield a desired average pH of 7. $\square$

The general procedure of significance testing can be summarized as follows:

1. Determine the null and alternative hypotheses.

2. Use the data to compute the test statistic.

3. Compute the p-value using the test statistic.

4. State the conclusion with regard to the stated hypotheses.

EXERCISES

1. Opponents of drilling for oil in a previous restricted area claim that less than half the residents living in this area are in favor of drilling. A survey was conducted to gain support for this point of view.

   (a) Set up the appropriate null and alternative hypotheses.

   (b) Of the 500 people surveyed, 230 are in favor of allowing this drilling. Is this sufficient evidence to justify the claim of the opponents of this drilling?

   (c) To what type of error are you now subject? Discuss the practical consequences of making such an error.

2. A procedure used to produce identical twins in cattle entails the microsurgical division of the embryo into two groups of cells, followed by immediate embryo transfer. This procedure is thought to be more than 50% effective. ("Embryo Transfer Technology for the Enhancement of Animal Reproduction," R. Maplesoft, *Biotechnology*, February 1984, pp. 149–159.)

   (a) Set up the null and alternative hypotheses needed to support this claim.

   (b) When the experiment is conducted 120 times, 55 of the transplants result in the birth of twins. Can H_0 be rejected? Interpret your results in the context of this problem.

3. In a particular region of the Chesapeake Bay area, the average number of blue crab that a fisherman caught in a crab pod three years ago was 22.6 crabs. In the current year, it is believed that the yield is lower. The first 12 pods produced the following number of crabs: 24, 18, 12, 26, 21, 24, 17, 13, 24, 22, 16, 20.

(a) Set up the appropriate null and alternative hypotheses.

(b) Compute the p-value and state the appropriate conclusion.

4. Ozone is a component of smog that can be hazardous to plants even at low levels. In 2008 a federal ozone standard of 0.075 ppm was set. It is thought that the ozone level in a particular region exceeds this level. To verify this contention, air samples are obtained from 30 monitoring stations in this region.

(a) Set up the appropriate hypotheses for verifying this contention.

(b) When the data are analyzed, a sample mean of 0.082 and a sample standard deviation of 0.03 are obtained. Use these data to test H_0. Using the p-value, determine if H_0 can be rejected.

(c) What assumption are you making concerning the distribution of the random variable X, the ozone level in the air?

5. In a clinical trial for an appetite suppressing medication, the weight was measured for 10 people before and after taking the medication; the differences (before minus after): -8, 0, 2, 4, 9, 14, 19, 22, 32, 35. Did the people experience weight loss while taking the medication? (Test the hypothesis $H_0: \mu = 0$.)

6.6 Bootstrap Methods

A probability distribution is distinguished by its density function. In general, the density function depends on various numerical parameters that are related to the population. Common parameters that are used in many distribution functions include μ and σ. The object of statistical inference is to provide estimates for these unknown parameters. To make these estimates, we begin by taking a sample, or samples, from the population of interest. Next, we use these samples to compute statistics, for example $\bar{x}$ and s, that we use to provide estimates for the unknown probability distribution parameters. Often, these estimates take the form of a confidence interval. In constructing a confidence interval, we generally assume that the *sampling distribution of the sample statistic*, or simply the sampling distribution, is known. In other words, the sampling distribution is the frequency distribution that would result for the statistic of interest if we would sample the population repeatedly, taking all possible samples. The best known sampling distribution is given by the central limit theorem, which describes the sampling distribution of the sample mean. Often confused with the sampl*ing* distribution is the term *sample distribution*. The **sample distribution** refers to the relative frequency distribution of the sample data.

Example 6.18

(**Sampling Distribution vs. Sample Distribution**) To illustrate the idea, suppose you know that in a particular lake there are a total of 180 turtles, and in addition, you know the exact weight of each turtle (remember, we use this only to illustrate the difference). Suppose that you capture 40 of these turtles (your sample), and record the weight for each turtle in your sample. Using your sample data you compute the sample mean $\bar{x}$ and create a frequency distribution for your sample data. This represents the sample distribution. Now suppose that you were able to take ALL possible samples of size 40 from the population [there would be $180!/(40!\,140!)$ possible samples], and for each sample you record the sample mean. You could then create a frequency distribution for the sample means. This frequency distribution represents the sampling distribution of the statistic $\bar{x}$. In this example, since our sample size is fairly large (40), the central limit theorem applies and tells us that the sampling distribution for the sample means is approximately normal./*quad*/*square*

Bootstrap methods allow us to approximate an unknown sampling distribution for a statistic from a *single* random sample. The term *bootstrap* implies that one relies on the resources that are available (the old adage to "pull yourself up by your bootstraps"). In our case we view the single random sample as the available resource, and we use it to make statistical inferences. A bootstrap method is actually a simple Monte Carlo method that uses repeated sampling (with replacement) of an empirically generated data set, $X = \{x_1,\ x_2,\ x_2, \ldots, x_n\}$, in order to create new, simulated data sets. In essence, we treat the set X as the population of interest from which we draw repeated samples. In many ecological studies, it is often costly or impractical to gather multiple data sets; thus, using bootstrapping to draw inferences can be very cost-effective. In addition, bootstrapping allows us to use a single data set to draw statistical inferences without *any* theoretical assumptions.

In classical hypothesis testing and interval estimation, it is essential that we know the underlying sampling distribution for the statistic of interest. For example, when computing a confidence interval for the mean μ of a population, we generally take a sample and then use the sample data to create a confidence interval for μ. It is often assumed that the population is normally distributed about the mean μ, and thus the sampling distribution of $\bar{x}$ is approximately normal with standard error $\sigma/\sqrt{n}$. These assumptions about the distribution of the population, and in turn the sampling distribution, allow us to use $Z = (\bar{X} - \mu)/(\sigma/\sqrt{n})$ to construct a confidence interval for μ. However, in many cases we are unsure of the actual sampling distribution. For example, we may wish to find a confidence interval for the *median* weight of turtles in a pond. It is likely that we are unsure of the sampling distribution for the sample median. In general, when we create confidence intervals using a parametric approach, we assume that we know the sampling distribution for the statistic of interest beforehand. Nonparametric bootstrapping allows us to remove any a priori assumptions with regard to the sampling distribution of the statistic.

In this section we present an introduction to bootstrap methods in mathematical biology. There are two types of bootstrapping, parametric and nonparametric. We begin with the nonparametric case, and we illustrate the method with three examples. In Example 6.20, we develop a bootstrap interval estimate for the standard deviation. This example illustrates how bootstrapping can be used when there is no formula for the standard error, or the formula isn't well known, to create an interval estimate. In Example 6.23 we use bootstrapping to make inferences about the population mean. In this case, formulas do exist for the standard error, provided that we make assumptions about the distribution of the population; but with bootstrapping we make inferences about μ without any prior distribution assumptions. In Example 6.24 we illustrate the flexibility and simplicity of the bootstrap method by drawing inferences about the difference in medians between two populations.

6.6.1 Nonparametric Bootstrapping

Nonparametric bootstrap methods make no assumptions about the probability distribution of the parent population, and thus the sampling distribution. The basic idea is that we determine the sampling distribution for a statistic W, such as the mean, median, and variance, by assuming that the data set is a good representation of the population; then we "resample" the data set (our new "population") repeatedly to construct simulated data sets. These simulated data sets allow us to construct a relative frequency distribution for the simulated statistic W^* (Fig. 6.11). The underlying assumption behind bootstrapping is that the relative frequency distribution for the simulated statistic W^* provides an estimation of the actual sampling distribution for the statistic of interest W. Once the sampling distribution for W is estimated, we can use

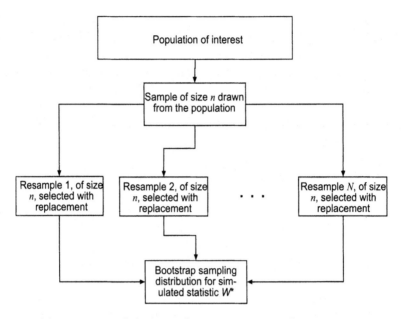

Figure 6.11 Schematic for nonparametric bootstraping.

it to make statistical inferences with respect to the parameter of interest.

Bootstrapping is actually a form of the *plug-in principle*, an intuitively pleasing principle which states that to estimate a parameter, simply use (plug-in) its corresponding sample statistic. For example, when performing statistical inference, we often need a parameter value for a probability distribution function F. The plug-in principle says that we can estimate the parameter for the distribution function F simply by using the corresponding value generated from the empirical distribution function. For example, when estimating the population mean μ, we should use the sample mean $\bar{x}$; to estimate the population median we should use the sample median, to estimate a population regression line, we should use the least squares regression line; and so on. Keeping with the theme of the plug-in principle, it seems reasonable that when the population is needed, we should simply use a random sample to provide an estimate for the population. This is the motivation behind bootstrapping.

Often, we are interested in finding an interval estimate for an unknown population parameter. With bootstrapping we use the interval between the $(100\alpha/2)$th percentile and the $100(1 - \alpha/2)$th percentile of the bootstrap distribution as the $100(1 - \alpha)\%$ interval estimate.

Definition 6.19

The $100(1 - \alpha)\%$ **bootstrap percentile confidence interval** for the parameter of interest is the interval between the $(100\alpha/2)$th and $(100(1 - \alpha/2))$th percentiles of the bootstrap distribution. □

Example 6.20

Consider the problem of estimating the standard error of the precipitation level (cm) for moisture in a particular growing region. The data available for a 12-year period in this region are the rain totals:

$$X = \{22, \ 8.6, \ 8.7, \ 15.2, \ 12.9, \ 10.2, \ 8.4, \ 5.9, \ 14.8, \ 17.6, \ 22.5, \ 11.3\}.$$

We use bootstrapping to determine a 95% percentile confidence interval for the population standard deviation, σ, of the annual precipitation level for the region of interest. We begin by sampling X with replacement to create 1000 simulated data sets, each containing 12 elements. The MATLAB m-file included at the end of this example produces a matrix A containing 1000 simulated data sets, a vector S containing the standard deviation of each of the bootstrap samples, and a relative frequency histogram of the standard deviations contained in the vector S, and it returns the upper and lower limits for a 95% bootstrap confidence interval for σ. The first row of the matrix A generated by the MATLAB function is

$$\{11.3 \ \ 8.7 \ \ 5.9 \ \ 10.2 \ \ 22.5 \ \ 17.6 \ \ 10.2 \ \ 22 \ \ 17.6 \ \ 10.2 \ \ 5.9 \ \ 17.6\}.$$

Note that the sampling is done with replacement, and therefore it is common to have duplicated data points in the simulated data sets. To draw inferences about the standard deviation of the rainfall totals, we estimate the sampling distribution for the standard deviation by examining a MATLAB-generated relative frequency histogram (Fig. 6.12). In the spirit of bootstrapping, we construct a 95% confidence interval for the median using a bootstrap percentile confidence interval. This consists of the interval between the 2.5% and 97.5% percentiles of the bootstrapped standard deviations. A short tutorial to finding percentiles is: (1) Sort the data to be used in ascending order; (2) find the location of the $k\%$ percentile by computing $d = N \times (k/100)$, where N is the number of elements in our data set; (3) if d is a whole number, add 0.5 to it (i.e., the location of the $k\%$ percentile is $d = d + 0.5$, and the $k\%$ percentile is the average of the two numbers on each side of this position); and (4) Otherwise, d is not a whole number and we round d upward to the next whole number. This rounded number is the correct location of the $k\%$ percentile [i.e., $d = \text{round}(d)$],

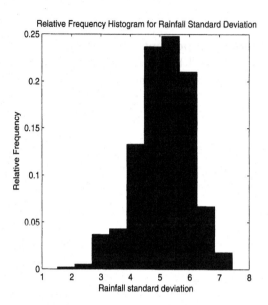

Figure 6.12 Relative frequency histogram for the distribution of the bootstrapped standard deviation. Note that the histogram appears to be mound shaped (approximately normal, with a slight negative skew).

and the value in this position is the $k\%$ percentile. For this example, the MATLAB function finds [3.045, 6.546] as a 95% percentile confidence interval for the standard deviation σ of the total rainfall for the region. □

```
function bootstrapmedian
X = [22 8.6 8.7 15.2 12.9 10.2 8.4 5.9 14.8 17.6 22.5 11.3];
A = [ ];
N = 1000;
for i = 1:N
    A(i,:) = [resample(data)];
    S(i) = std(u(i,:));
end
relfreq(S,N);
[Pk1, Pk2] = coninterval(S)

function relfreq(S,N)
[n,x] = hist(S);
bar(x,n/N,1)
axis square
```

```
title('Relative Frequency Histogram for Rainfall Standard Deviation')
xlabel('Rainfall standard deviation')
ylabel('Relative Frequency') function res = resample(vdata)
vs = length(vdata);
inds = ceil(vs*rand(vs,1));
res = vdata(inds);

function [Pk1, Pk2] = coninterval(M)
k1 = .025;
st = sort(M);
Nx = size(M,1);
result = k1*Nx;
[n,d] = rat(result);
if d == 1
      result = result + .5
else
      result = round(result);
end
if isequal(d,1)
      Pk1 = mean(st(result-0.5:result+0.5));
      Pk2 = mean(st(Nx - (result+.5):Nx - (result-.5)));
elseif result == 0
      result = result+1;
      Pk1 = st(result);
      Pk2 = st(Nx);
else
      Pk1 = st(result);
      Pk2 = st(Nx - result);
end
```

The nonparametric bootstrap process begins with collecting a sample of size n from a population with an unknown probability distribution function F. The goal is to draw inferences about a parameter ω (e.g., μ or σ). The steps for the nonparametric bootstrap method can be summarized as follows:

1. Create an empirical distribution function for the sample collected by assigning a probability of $1/n$ to each point x_1, x_2, $x_2, \ldots, x_n$ in the collected sample X. This ensures that each sample point is equally likely to be selected when resampling.

2. Using the empirical distribution function created in step 1, draw a random sample of size n using replacement. This is a "resample" of the original data.

3. For the resampled data sets, calculate the statistic of interest W corresponding to the parameter of interest ω. For example, if $\omega = \mu$, then $W = \bar{x}$. The statistic calculated from the resample data set is denoted by W^*, which is consistent with the notation for test statistics.

4. Repeat steps 2 and 3 N times, creating N resamples and N statistics $W_1^*, W_2^*, \ldots, W_N^*$. The size of N varies, depending on what we plan to do with the data. When finding confidence intervals, it is standard to choose N to be at least 1000.

5. Construct a relative frequency histogram for the statistics $W_1^*, W_2^*, \ldots, W_N^*$. The resulting distribution is the bootstrap sampling distribution that is used to estimate the sampling distribution for W. This distribution is then used to draw inferences about the population parameter ω.

When the bootstrap relative frequency histogram indicates a normal sampling distribution, we can often use the standard error to compute a *bootstrap t confidence interval*. This confidence interval is a good approximation provided that the bootstrap distribution is approximately normal and the the *bootstrap estimate of bias* is small.

Definition 6.21

The **bootstrap standard error**, SE_b, of a statistic is the standard deviation of the statistics generated by the resampling process. That is,

$$SE_b = \sqrt{\frac{\sum (W_i^* - \overline{W}_b)^2}{N-1}}, \tag{6.4}$$

where N is the size of the sample and $\overline{W}_b$ is the mean of the statistics generated, $W_1^*, W_2^*, \ldots, W_N^*$. The **bootstrap estimate of bias** is the difference between the mean of the bootstrap statistics, $\overline{W}_b$, generated and the value of the statistic computed from the original sample, W. $\square$

Definition 6.22

If the estimated sampling distribution for the statistic W is approximately normal with a relatively small bootstrap estimate of bias, an approximate $(1 - \alpha)100\%$ confidence interval for the parameter ω is the $(1 - \alpha)100\%$ **bootstrap t-confidence interval**, which is given by

$$W \pm t_{\alpha/2} SE_b,$$

where $t_{\alpha/2}$ is the value from the t-distribution with $n - 1$ degrees of freedom (n is the size of the original empirical sample) and $\alpha/2$ area to its right, SE_b is the bootstrap standard error, and W is the statistic generated from the original data set. $\square$

In the next example we carefully match up the notation used in the previous definitions in an application of bootstrap methodology to inferences about the population mean μ.

Example 6.23

For the rain data given in Example 6.20, we now use bootstrapping to construct a confidence interval for the mean precipitation level (cm) μ in a particular growing region. In this case the statistic of interest is $\omega = \mu$. We proceed much as in Example 6.20. We first resample the original data set,

$$X = \{22,\ 8.6,\ 8.7,\ 15.2,\ 12.9,\ 10.2,\ 8.4,\ 5.9,\ 14.8,\ 17.6,\ 22.5,\ 11.3\},$$

to simulate $N = 1000$ new data sets. We then compute the mean, $W^* = \overline{x}^*$, of the simulated data sets. Then we create a relative frequency histogram for the computed means to investigate the approximate sampling distribution of $\overline{x}$. We modify the previous MATLAB function simply by replacing the median vector by the mean vector. That is, we replace the line M = median(A,2) by M = mean(A,2). The resulting relative frequency histogram is shown in Fig. 6.13. In this case, the resulting frequency histogram seems to indicate an approximately normal sampling distribution. Thus, we can use a bootstrap t-confidence interval. The additional MATLAB, command

$$s = \mathsf{std}(\mathsf{M}),$$

inserted at the end of the primary function, computes the standard error using formula (6.4). The bootstrap t-confidence interval is $\overline{x} \pm t_{\alpha/2}\mathrm{SE_b} = 13.175 \pm (2.201)(1.4826)$, or $[9.9118, 16.4382]$. We note that the 95% percentile confidence interval is $[10.4625, 16.0375]$. $\square$

The similarity of the bootstrap t-confidence interval and the percentile confidence interval is not surprising since the bootstrap estimate of bias, $\overline{x}_b - \overline{x} = 13.1787 - 13.175 = 0.0013$, is very small relative to $\overline{x} = 13.175$. In addition, the sampling distribution (Fig. 6.13) appears to be approximately normal. When this occurs we expect confidence intervals constructed using these two methods to be similar. In general, if the bootstrap measure of bias is small relative to the value of the statistic, and the bootstrap sampling distribution is approximately normal, it is valid to use the bootstrap t-confidence interval. If the bias is not small, or the sampling distribution doesn't appear to be normal, we should not use the t-confidence interval, as this method is

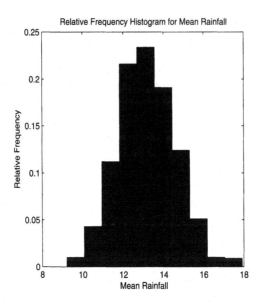

Figure 6.13 Relative frequency histogram for the distribution of the boot-strapped mean. Note that the histogram appears to follow an approximately normal distribution.

based on an assumption of normalcy. In the latter case it is still possible to use the bootstrap percentile confidence interval, but before doing so we recommend obtaining expert advice, as more advanced bootstrap techniques may be more appropriate.

Example 6.24

Suppose that we are interested in comparing a new method for increasing the fecundity of a particular endangered insect. We randomly select a small sample of 8 females out of a total of 18 females and subject them to the new treatment. The other 10 insects serve as the control group. The insects are monitored closely for a year and careful egg counts are kept. The table shows the recorded data:

Group	Data	N	Mean	SD
Treatment	46, 72, 8, 144, 65, 59, 21, 137	8	69	49.2
Control	51, 33, 48, 27, 65, 88, 24, 67,102, 54	10	55.9	25.4
		Difference	13.1	23.8

Suppose that we want to test for treatment differences and we know that the

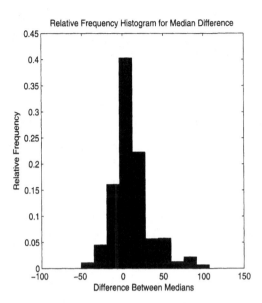

Figure 6.14 Relative frequency histogram for the distribution of the bootstrapped difference of medians. Note that the histogram appears to follow a normal distribution.

median is a better measure than the mean. We use bootstrapping to compute a confidence interval for the difference of medians. To do this we simply bootstrap both data sets. After each resampling, we compute the difference between each bootstrapped median. We use these differences to find a bootstrap percentile confidence interval. We modify the previous MATLAB program to find a 95% confidence interval for the median difference to be $[-31.5, 79]$. Because this confidence interval includes 0, we cannot conclude that the treatment had a beneficial effect on the median fecundity. The relative frequency histogram for the median differences is shown in Fig. 6.14. □

As the examples illustrate, nonparametric bootstrapping is extremely easy to apply and it allows us to make statistical inferences simply by considering a single data set. So why not always use nonparametric bootstrapping? The limitations of nonparametric bootstrapping can be broadly classified as sample size limitations and sensitivity to an extreme datum.

1. *Sample-size limitation.* In theory, bootstrapping is based on the idea that if the random sample provides a good estimate for the population, the bootstrapping method will provide a good estimate of the sampling distribution

for the statistic of interest. Generally, the bootstrap estimates improve as the sample size increases. The size of the samples needed for good bootstrap estimates depend greatly on the parameter of interest. For example, if it is the population mean that we want to estimate, sample sizes of 10 or more are usually sufficient (15 or more is better). On the other hand, if we apply nonparametric bootstrapping to estimate the population median, it can be shown that sample sizes of 600 or more are generally needed.

2. *Sensitivity to extreme datum.* For nonparametric bootstrapping to work correctly, the parameter we are trying to estimate can't be overly sensitive to extreme data values, since our empirical sample is unlikely to contain extreme data points. For example, consider trying to use bootstrapping to estimate the maximum value of a particular population. Then the largest value in our resampled data will simply be the largest value in the sample. Generally, each resample will contain one of three or four largest values in the sample. Thus, the resulting bootstrap sampling distribution will tend to be spiked at only three or four values, thus providing a meaningless bootstrap sampling distribution. Nonparametric bootstrapping will not produce useful results in this case, and we will have to resort to parametric bootstrapping, which we discuss in the next section.

6.6.2 Parametric Bootstrapping

Efron, in his seminal work in 1979, described two types of bootstrapping methods, nonparametric and parametric. In Section 6.6.1 we introduced nonparametric bootstrapping. We now focus on the parametric case. The two methods are actually quite similar, as they share the same key idea of using a single random sample, X_1, X_2, $X_3, \ldots, X_n$, taken from the population of interest, to generate statistical inferences with regard to various population parameters. Both require simulating data sets that are used to make statistical inferences. The difference between the two methods is in how the data sets are simulated. In nonparametric bootstrapping, we resampled, with replacement, the original data set. Recall that a primary strategy in this process was to assume that each element in the original sample had the same probability, $P(X_i) = 1/n$, of being selected. With parametric bootstrapping, this is not the case. Instead, we assume that the data come from a population with distribution function F_λ, which is well defined except for an undetermined parameter λ. We use our data set to obtain an estimate $\widehat{\lambda}$ for the parameter λ. Next, we simulate "pseudosamples" using a random number generator that selects values (simulated data) from the distribution $F_{\widehat{\lambda}}$. Finally, we use these pseudosamples to draw statistical inferences in the same manner as with nonparametric bootstrapping.

The parametric bootstrapping method can be summarized as follows.

1. Using the empirical sample $\{X_1, X_2, X_3, \ldots, X_n\}$, find estimates for the parameters in the population distribution function F. Let $\widehat{F}$ denote the resulting sample distribution function.

2. Simulate a pseudosample using the sample distribution function $\widehat{F}$. The simulated samples must be the same size as the empirical data set. These samples are created using a random number generator for $\widehat{F}$.

3. Calculate the statistic of interest for each pseudosample.

4. Repeat steps 2 and 3 a large number of times (2000 or more).

5. Draw inferences using the set of statistics generated in step 3.

Example 6.25

(**Distribution of the sample maximum**). Let $X_1, X_2, X_3, \ldots, X_n$ be iid random variables from a population with pdf f and cdf

$$F(x) = \Pr(X_i \leq x) = \int_{-\infty}^{x} f(x) \, dx.$$

Let $Y_n = \max(X_1, X_2, X_3, \ldots, X_n)$. Then Y_n is the maximum of the sample. To find the pdf and cdf for Y, we note that the cdf for Y is given by

$$H_n(y) = \Pr(Y_n \leq y) = \Pr(X_1 \leq y \text{ and } X_2 \leq \cdots \text{ and } X_n \leq y).$$

Because the random variable X is iid,

$$H_n(y) = \Pr(X_1 \leq y)\Pr(X_2 \leq y) \cdots \Pr(X_n \leq y) = [F(y)]^n.$$

We can find the pdf for Y_n by differentiating H_n to obtain

$$h_n(y) = n[F(y)]^{n-1} f(y).$$

As one can imagine, depending on the form of $F(y)$, both the cdf and the pdf for Y_n can be extremely complicated. In fact, if X is normally distributed, there is no closed form for the distribution of the sample maximum.   □

In situations where the distribution function of the variable of interest is intractable, yet we know the underlying distribution of the population, parametric bootstrapping is an invaluable tool. The next example illustrates how parametric bootstrapping can be used to draw inferences about the maximum value of a population.

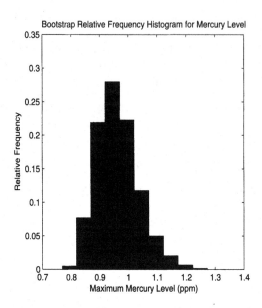

Figure 6.15 Relative frequency histogram for the distribution of the maximum value that results when employing parametric bootstrapping. Note the slight positive skew in the distribution.

Example 6.26

(**Parametric bootstrapping**) There is concern that fish in a particular region of North American have levels of methyl mercury that exceed the guideline, 1 part per million (ppm), for consumption. A sample of 120 fish are taken from this region and their levels of methyl mercury are recorded. The question of interest is: If 120 fish are taken from the region each year, what is the probability that the maximum concentration of methyl mercury will exceed 1 ppm? The summary of the data (in ppm) for the gathered sample is

$$\begin{pmatrix} n & \min & \max & \overline{x} & s \\ 120 & 0.07 & 1.13 & 0.52 & 0.172 \end{pmatrix}.$$

We perform a parametric bootstrap using the assumption that the underlying distribution of mercury levels in these fish follow's a normal distribution. The normal distribution includes the parameters μ and σ. We estimate these using the plug-in principle. That is, we use $\overline{x}$ and s in place of μ and σ. Next, we simulate 2000 pseudosamples from a normal distribution with mean 0.62 and standard deviation 0.172. The resulting relative frequency distribution for the maximum value for the bootstrapped samples is shown in Fig. 6.15. We find that there is an approximately a 24% chance that the sample maximum will exceed 1 ppm each year. In addition, we find a 95% percentile confidence interval for the population maximum for the given year to be [0.8452, 1.1269]. Because this interval includes 1 ppm, it would be prudent to recommend that people limit consumption of fish from this region. The MATLAB program for this example is very similar to that included in Section 6.6.1. The only changes occur in the first function, which is given following this example.   □

```
function bootstrapsmax
u = [ ];
count = 0;
N = 2000;
xbar = .52;
sd = .172;
u = xbar + sd*randn(2000,120);
Maxi = max(u,[ ],2);
relfreq(Maxi,N);
[Pk1, Pk2] = coninterval(Maxi)
for i =1:2000
    if Maxi(i) > 1
        count = count + 1;
    end
end
prob = count/N
```

6.6.3 Permutation Methods

Another type of resampling method, which uses sampling *without* replacement, allows us to perform significance testing (hypothesis testing) without any para-metric assumptions. Significance testing is the process of determining if an observation could be the result of selecting an "unusual" or "fluke" random sample, or if the observation is the result of some treatment. For example, we may be interested in exploring the effect of a certain pesticide treatment on crop yield, or if some environmental contaminant is affecting adversely the fe-cundity of a particular animal species. When performing such tests, we specify a value for the parameter of interest in the null hypothesis H_0. Working under the assumption that the value for the parameter specified in H_0 is true, we determine the appropriate sampling distribution for the statistic of interest. Next, we draw a random sample and compute a test statistic. As discussed previously, the test statistic can be used to compute a p-value, which we use to make a decision regarding the value of the parameter specified in the null hypothesis. In general, the idea behind significance testing can be summarized as follows:

1. Decide on the statistic of interest ($\overline{x}$, s, etc.). This is the statistic that measures the effect that is being measured.

2. Determine the sampling distribution, working under the assumption that the null hypothesis is true. That is, determine the sampling distribution under the assumption that the effect we are looking for is *not* present.

3. Use the random sample to compute the test statistic and then determine the p-value. If the test statistic falls in the tail of the distribution, it is unlikely that the sample that led to the test statistic is a not a true random sample (i.e., the p-value is considered significant).

Traditionally, an assumption is made with respect to the sampling distribution to which the statistic of interest belongs. This is the case in Section 6.5. Central to significance testing is the assumption that the value of the statistic is given in the null hypothesis. Because of this, normal bootstrapping will not work. Recall that for the bootstrap methods discussed so far, we worked under the assumption that the sample is a valid representation of the distribution. Thus, when we resampled, by sampling with replacement we were selecting samples that were centered at the *observed* value of the statistic. But with significance testing, the sampling distribution that we use is created under the assumption that the parameter value specified in the null hypothesis is true. Therefore, it is no longer appropriate to resample our data using replacement. Instead, we must now choose our resampling scheme in a way consistent with the idea that the null hypothesis determines the sampling distribution.

Example 6.27

It is believed that when a particular crop is treated with a pesticide, the yield is affected negatively. To test this hypothesis, researchers gather data from both treated and untreated crops. The data gathered are presented in the following table.

Yields from Treated Crops						Yields from Untreated Crops					
43	51	32	16	33	46	37	39	27	52	48	21
28	53	38	37	39	44	43	26	33	61	28	50
32	36	47	22	47	34	37	43	45	31	28	49
24						26	32	42			

The statistic of interest is the difference between the mean yield from treated crops and the mean yield from untreated crops. That is, we consider the statistic

$$x_d = \overline{x}_t - \overline{x}_u.$$

The sampling distribution for this statistic is the t-distribution. The null hypothesis is that there is no difference in yields between the treated and untreated crops. That is,

$$H_0 : x_d = \overline{x}_t - \overline{x}_u = 0.$$

If the null hypothesis is true, any difference observed between the means of the two groups is due simply to the random way in which the data are assigned to the two groups. In other words, any difference is due simply to the random way in which the data are permuted. $\square$

Example 6.27 gives us the insight required on how to resample the empirical data for significance testing. We just permute all the data from both groups and then break them into the two treatment methods. In Example 6.27 this means that we permute all 40 data pieces. Next, we use the first 19 values of the permuted set to simulate the treated crops and the remaining 21 to simulate the untreated crops. We repeat the process, each time recording the statistic that measures the difference of means between the two simulated data sets. Finally, we compare the simulated statistics with the statistic computed using the original empirical data. We count off the number of simulated statistics that are more extreme than the empirical statistic. The percentage of the simulated statistics that are more extreme than the empirical statistic represent the p–value that results from using permutations. The permutation method for comparing two samples can be summarized as follows:

1. Determine the statistic of interest W. For example, W may be the difference between the mean of sample 1 (the treatment group) and the mean of sample 2 (the control group). Let W_d represent the value of the statistic for the data gathered.

2. Take permutations of the entire data set. Suppose that the treatment group has n_1 data points and the control group has n_2 data points. Then $N = n_1 + n_2$ represents the total number of elements in the entire data set. Each permutation will contain N data points. Assign the first n_1 elements to the treatment group and the remaining n_2 to the control group.

3. Using the simulated data sets described in step 2, compute the statistic of interest. Call this W_p for the permuted data set, and record this statistic.

4. Repeat steps 2 and 3 a large number of times (more than 1000).

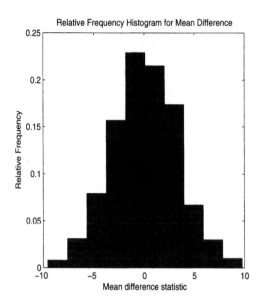

Figure 6.16 Relative frequency histogram for the difference of means. Theoretically, we expect the differences of means to follow a normal distribution. The results produced by the permutation method clearly agree with the theoretical prediction.

5. Determine the p-value by locating the value of the statistic for the original data, W_d, on the resulting permutated sampling distribution. This is carried out by counting the number of times that W_p is more extreme than W_d. The determination of what is "more extreme" is performed by examining the alternative hypothesis. If the statement for the alternative hypothesis contains the phrase "greater than," then by *more extreme* we mean those values of W_p that are greater than W_d. If the statement for the alternative hypothesis contains the phrase "less than," we count the number of times that W_p is less than W_d.

Example 6.28

(**Difference of Means**) For the data given in Example 6.27 we use the method outlined above to find the p-value. In this case, W represents the difference between the means of the two samples. Using MATLAB, we find that $\bar{x}_t = 36.94$ and $\bar{x}_u = 38$. Thus, $W_d = x_d = -1.06$. For the null and alternative hypotheses we have

$$H_0 : x_d = 0 \text{ vs. } H_1 : x_d < 0.$$

Using the MATLAB code following this example, we find that the permutation p-value is 0.3630, which is not significant. Thus, we fail to reject the null hypothesis. This implies that there is not enough evidence to conclude that the pesticide application is affecting crop yield adversely. The relative frequency histogram for the difference of means is given in Fig. 6.16. As we would expect from theory, the relative frequency distribution for the difference of the permuted means appears to be normally distributed. Although the permutation p-value does not depend on any parametric assumption, it is comforting to know that the permutation method produces results that agree with parametric results. □

```
function permutationtwosets
d1 = [43 51 32 16 33 46 28 53 38 37 39 44 32 36 47 22 47 34 24];
d2 = [37 39 27 52 48 21 43 26 33 61 28 50 37 43 45 31 28 49 26 32 42];
pstat = [ ];
Ndata = [d1 d2];
Ndata1 = length(d1);
N = length(Ndata);
NPerms = 1000;
datastat = mean(d1) - mean(d2);
extreme = 0;
for i = 1:NPerms
    rp = randperm(N);
    for j = 1:N
        pdata(j) = Ndata(rp(j));
    end
    pdata1 = pdata(1:Ndata1);
    pdata2 = pdata(Ndata1 + 1:N);
    xd1 = mean(pdata1);
    xd2 = mean(pdata2);
    pstat = [pstat xd1 - xd2];
    if pstat(i) > datastat
        extreme = extreme + 1;
    end
end
[n,x] = hist(pstat);
bar(x,n/NPerms,1)
```

```
axis square
title('Relative Frequency Histogram for Mean Difference')
xlabel('Mean Difference Statistic')
ylabel('Relative Frequency')
p_value = extreme/NPerms
```

In the case of significance testing for paired data, we can still use permutation methods, but now we must take care to maintain the paired structure. If the null hypothesis states that there is no difference between the paired data, then we should be able to permute the two values of paired data without observing any statistical difference. We illustrate the method in the following example.

Example 6.29

(**Difference of means, paired data**) An ecologist conjectures that an insect will change its food intake rate in the presence of predators. To test this hypothesis, the insect is monitored in an environment with no predator, and its food intake is recorded. The same insect is then placed in the presence of a predator, and its intake is again monitored. The predator is disarmed in such a way (e.g., putting wax on its fangs) that it is unable to harm the insect. The following table presents the results for 12 insects.

Insect	No Predator	Predator	Insect	No Predator	Predator
1	12.1	11.3	7	10.6	10.5
2	9.9	9.4	8	11.7	9.4
3	13.2	11.6	9	10.5	12.2
4	16.7	13.4	10	8.7	8.0
5	9.8	10.3	11	9.5	8.6
6	10.4	10.1	12	10.3	10.1

The permutations consist of permuting each of the pairs of data. That is, if we think of the results for each insect as an ordered pair, we randomly permute the elements in each ordered pair. Next, we compute the difference between the simulated means for the no predator and predator cases. The p-value is calculated using the same process as described previously. Using the MATLAB code following the end of this example, we find a p-value of 0.0384, which is considered significant. We conclude that there is evidence to support the claim that insects alter their food intake in the presence of a predator.   □

```
function permutationpaired
data = [12.1 11.3; 9.9 9.4; 13.2 11.6; 16.7 13.4;...
       9.8 10.3; 10.4 10.1; 10.6 10.5; 11.7 9.4; 10.5 12.2;...
       8.7 8.0; 9.5 8.6; 10.3 10.1];
simdata = [ ];
temp = [ ];
pstat = [ ];
N = size(data,1);
NPerms = 5000;
extreme = 0;
datastat = mean(data(:,1)) - mean(data(:,2));
for i=1:NPerms
    for j=1:N
        rp = randperm(2);
        temp = data(j,:);
        simdata(j,:) = [temp(rp(1)) temp(rp(2))];
    end
    colimean = mean(simdata,1);
    pstat(i) = colimean(1) - colimean(2);
    if pstat(i) > datastat
        extreme = extreme + 1;
    end
end
pval = extreme/NPerms
```

Caveat. Permutation methods provide a straightforward way to perform significance testing when comparing two data sets, paired data sets, or examining the relationship (correlation) between two quantitative variables (see the Exercises). Underpinning the method for all of these cases is the assumption in the null hypothesis that there is no difference (or no correlation) in the parameter of interest for the two data sets. In fact, this is a necessary condition for the use of permutation methods. Permutation methods can be used only under the assumption of no difference in the null hypothesis. In addition, permutation methods cannot be used for performing significance testing on a single data set. In this case, permuting the data clearly serves no purpose, as it remains the same data set.

When we are unable to use permutation methods, we may still be able to use bootstrapping to perform hypothesis testing. If the bootstrap confidence interval contains the value of the parameter specified in the null hypothesis H_0, we are unable to reject H_0. If the confidence interval does not contain the parameter value stated in H_0, we are able to conclude that we can reject H_0 in favor of the alternative hypothesis. Using the bootstrap confidence interval is not as accurate as using permutation methods for significance testing, but the beauty of bootstrapping is that its flexibility provides options for statistical inference when other methods fail.

EXERCISES

1. In treating the midgut of a grasshopper as a plug-flow chemical reactor, researchers found it necessary to measure the length of the midgut. The following data were collected from 12 adult grasshoppers. Data were gathered on the length of the midgut in adult grasshoppers. The following data were collected

$$X = \{8.8, \ 9.2, \ 9.5, \ 8.7, \ 9.2, \ 9.7, \ 8.8, \ 9.1, \ 9.4, \ 8.9, \ 9.3, \ 9.6\}.$$

   Employ bootstrapping techniques to create a 95% confidence interval for the mean gut length of all adult grasshoppers of this particular species.

2. Let X_1, X_2, ..., X_n be a random sample of size n from a Poisson distribution with mean τ. Let $k = \sum_{i=1}^{n} X_i$. Then an estimate for τ is given by $\hat{\tau} = k/n$. It can be shown that exact $(1 - \alpha)100\%$ confidence interval for τ is given by

$$[t_l, \ \tau_u] = \left[\frac{1}{2n} \chi^2_{2k}(\alpha/2), \ \frac{1}{2n} \chi^2_{2k+2}(1 - \alpha/2) \right],$$

   where χ^2_{2k} is the value from the chi-squared distribution with $2k$ degrees of freedom. For the following questions, assume a sample size of 10 and that $k = 32$.

   (a) Find an exact 95% confidence interval for the population mean τ.

   (b) Construct a bootstrap relative frequency distribution using parametric bootstrapping. To do this, use a Poisson random number generator, with $\hat{\tau} = k/n$ as the plug-in estimator for τ, to construct 3000 simulated samples.

   (c) Compute a 95% percentile bootstrap confidence interval for the population mean τ by using parametric bootstrapping. Use the 3000 simulated samples from part (b).

   (d) Compare the results of part's (a) and (c). Does it appear that the bootstrap confidence interval agrees with the exact confidence interval?

3. It has been hypothesized that for a particular species of grasshoppers, the mean length of the adult females is greater than that of their male counterparts. To test this hypothesis, grasshoppers were captured and their sex and length (cm) were recorded in the following table.

Male Body Length (cm)				Female Body Length (cm)			
1.83	1.82	1.85	1.91	1.84	1.92	1.88	1.78
1.79	1.9	1.77	1.84	1.93	1.76	1.81	1.75
1.74	1.92	1.83	2.00	2.1	1.87	2.15	1.82
1.85	1.92	1.84	1.79	1.86	1.83	1.97	

For the data above, use a permutation test to test the hypotheses

$$H_0: \ \mu_F - \mu_M = 0,$$
$$H_1: \ \mu_F - \mu_M > 0.$$

4. Assume we know that the underlying distribution of our sample observations X is the Bernoulli distribution with unknown parameter p, the population proportion. We know that for large n, the distribution of the statistic $\widehat{p}$ is approximately normally distributed with mean p and variance $np(1 - p)$. Suppose that we wish to construct a 95% confidence interval for p using the exact distribution of the sample from which we compute $\widehat{p}$. Describe the steps needed to do this if we are to use parametric bootstrapping.

5. Two methods are used to measure the concentration of a particular chemical. Test solution media of various concentrations are obtained and split into two portions. The concentration of the first portion is made using method 1, and the concentration of the second portion is made using method 2. It is thought that the second method tends to give a higher average than the first method. The results of the measurements are as follows

Sample Number	Method 1 (μ/mL)	Method 2 (μ/mL)
1	3.78	3.35
2	3.58	3.6
3	3.77	3.41
4	3.82	3.69
5	3.67	3.48
6	3.66	3.50
7	3.48	3.33
8	3.63	3.64
9	3.88	3.65
10	3.53	3.64

Use a permutation method to perform a significance test to determine if method 2 does indeed tend to give a higher average.

6. The following data set relates the body weight and lifespan for 11 mammals.

Species	Body Weight (lbs)	Maximum Life (years)
1	179	27
2	25	28
3	440	50
4	115	20
5	119	16
6	180	27
7	39	13
8	157	22
9	12	18
10	2	5
11	180	29

Of interest in many biological studies is the extent of linear correlation between two variables. A measure of this correlation is the correlation coefficient r. The correlation coefficient r is a normalized measure $(-1 \leq r \leq 1)$ of the strength of the linear relationship between two variables. A value of r near zero indicates that the data does not exhibit linear correlation, whereas a value near 1 or -1 indicates a strong linear correlation. The MATLAB command

$$\text{corrcoef}(\text{x, y})$$

computes r for the two column vectors $\mathbf{x}$ and $\mathbf{y}$. We can perform significance testing of r using permutation methods in much the same manner as we performed significance testing for two data sets. That is, in the null hypothesis we assume that $r = 0$.

(a) Explain why a permutation method would be applicable for performing a significance test of $r = 0$, and describe the steps that one would use to do the test.

(b) Make a scatter plot for the data given in the table.

(c) Calculate r for body weight vs. maximum lifespan. Does r appear to indicate a linear correlation exists between the variables?

(d) Is the value of r you computed in part (c) significantly different from $r = 0$? Use a permutation test to make your decision.

7. Consider the following data set:

$$
\begin{array}{cccccccccc}
1.3 & 5.3 & 7.6 & 2.1 & 3.7 & 4.8 & 1.6 & 1.4 & 3.1 & 2.3 \\
5.2 & 3.0 & 1.2 & 1.4 & 3.0 & 1.1 & 2.3 & 6.5 & 1.7 & 2.2
\end{array}
$$

Use bootstrapping to construct a 95% confidence interval for the mean. Use 1000 resamples. Is it valid to use the bootstrap t-confidence interval? Why or why not? What is the bootstrapping estimate of bias?

8. (Adapted from Roughgarden, 1998) Permutation and bootstrap techniques can be incorporated into exploring the effects of a particular variable in mathematical models. For example, Roughgarden (p. 146) proposes using the Beverton-Holt model to predict the population of wildebeest in the Serengeti Park. The model uses the fact that yearly rainfall totals play a major role in determining the amount of plant material available for food. The Beverton–Holt equation for resource-limited growth is

$$N_{t+1} = \frac{RN_t K}{K + (R - 1)N_t},$$

where R is the maximum geometric growth factor achieved as N goes to zero, and K is the carrying capacity. It has been shown that $K =$ (yearly rainfall) $\times$ 20,748 accurately approximates the dynamics of past wildebeest populations. What makes this model of particular interest is that rainfall amounts vary unpredictably from year to year. Also, harvesting of wildebeest is included in the model, allowing park officials to determine the maximum harvest rate that can be allowed without elevating the risk of the collapse of the herd. Let h represent the fraction of the herd that is harvested. The complete model is then given by

$$N_{t+1} = (1 - h)\frac{RN_t K}{K + (R - 1)N_t}.$$

Rainfall data for 34 consecutive years is given by 100, 38, 100, 104, 167, 167, 165, 79, 91, 77, 134, 192, 235,159, 211, 257, 204, 300, 187, 84, 99, 163, 97, 228, 208, 83, 44, 112, 191, 202, 137, 150, 158, 20. Suppose that five different harvest rates are being considered. The five rates are

$$h = [0.5 \ \ 0.075 \ \ 0.1 \ \ 0.125 \ \ 0.15].$$

(a) Explore the effect of the various harvesting rates on the wildebeest population by generating a computer program that performs 100 stochastic runs of 100-year simulations for the wildebeest population. Do this for each harvest rate. To determine the carrying capacity for each year, use $K =$ (yearly rainfall) $\times$ 20,748, where the yearly rainfall total is a random draw from the actual rainfall totals. Let $N_0 = 250{,}000$. What appears to be the best harvesting rate? Why?

(b) If a wildebeest population of 150,000 or less indicates a collapse of the herd, use the model to predict the percentage of years that a collapse will occur for each harvest rate.

6.7 Reference Notes

Statistics is one of the major areas of mathematics and other disciplines. As such, a very large number of outstanding statistics books are available, and it inadvisable to cite too many. The spectrum goes from popular books on statistics that attract general readership to advanced research works that are highly technical. Many texts are algebra-based and do not require calculus or much probability theory, whereas others are calculus-based; some require advanced mathematical analysis, probability theory, and linear algebra.

Two standard, classical references at the advanced level are Hogg et al. (2005) on mathematical statistics and Hogg & Tanis (2005) on statistical inference. There is more mathematical detail in these texts than is needed for most practitioners. Moving to a lower level, there are several excellent texts designed for undergraduate students in engineering, science, and mathematics: for example; Ross (1987), Walpole & Myers (1995), or Wackerly et al (2008). A good algebra-based text is one of the Triola-series texts (e.g., Triola, 2007).

Books on probability and stochastic processes are listed at the end of Chapters 5 and 7. Mangel's book (2006) gives an excellent overview of how advanced statistical ideas are useful in ecology.

References

Efron, B. & Tibshirani, R. J. 1993. *An Introduction to Bootstrap*, Chapman & Hall, New York.

Hogg, R. V. & Tanis, E. 2005. *Probability and Statistical Inference*, 6th ed., Prentice Hall, Upper Saddle River, NJ.

Hogg, R. V., McKean, J. W. & Craig, A. T. 2005. *Introduction to Mathematical Statistics*, 6th ed., Prentice Hall, Upper Saddle River, NJ.

Mangel, M. 2006. *The Theoretical Biologist's Toolbox*, Cambridge University Press, Cambridge, UK.

Milton, J. S. & Arnold, J.C., 2002. *Introduction to Probability and Statistics: Priciples and Applications for Engineering and the Computing Sciences*, 4th ed., McGraw-Hill, New York.

Ross, S. M. 1987. *Introduction to Probability and Statistics for Engineers and Scientists*, Wiley, New York.

Roughgarden, J. 1998. *Primer of Theoretical Ecology*, Prentice Hall, Upper Saddle River, NJ.

Triola, M. F. 2007. *Elementary Statistics*, 10th ed., Prentice Hall, Upper Saddle River, NJ.

Wackerly, D., Mendenhall, W. & Scheaffer, R.L. 2008. *Mathematical Statistics with Applications*, 7th ed., Thompson Publishing, Belmont, Ca.

Walpole, R. E. & Myers, R. H., 1995 *Probability and Statistics for Engineers and Scientists*, 5th ed. Prentice Hall, Upper Saddle River, NJ.

7

Stochastic Processes

Many time-dependent processes in science have elements of randomness. In fact, one could argue that most of the deterministic models we design probably have some random component that is being ignored. In this chapter we examine different ways to develop stochastic models of dynamic processes. One way is to add randomness in an ad hoc manner to an existing deterministic model, either a discrete model or a continuous model. The randomness comes in as demographic or environmental noise. Which choice to make and how to insert the stochasticity is a question of modeling. There are also processes that are random from the beginning and have no relation to existing difference or differential equations.

Sections 7.1 through 7.4 cover material that is accessible to most undergraduate students with good quantitative skills. Sections 7.5 through 7.8, which develop ideas about stochastic differential equations, require a higher-level knowledge of both ordinary and partial differential equations, as well as probability theory.

7.1 Introduction

We begin with an example. In the weather station at an outdoor ecological laboratory, a meter records the solar radiation falling on the site, measured in watts per square meter, every day at 2:00 P.M. This is a random process in time. On day 1 the meter measures $X(1)$, on day 2 it measures $X(2)$, and so

Mathematical Methods in Biology,
By J. David Logan and William R. Wolesensky
Copyright © 2009 John Wiley & Sons, Inc.

on, with $X(t)$ denoting the solar radiation on the site at 2:00 P.M. on day t. Note that $X(t)$ is a random variable, or random quantity, which is the result of a random experiment. We can represent the values we obtain over day 1, day 2, day 3, and so on, by a sequence of random variables $X(1)$, $X(2)$, $X(3)$,..., or just by $X(t)$, $t = 1, 2, 3, \ldots$.

As an aside, many authors use the notation X_t rather than $X(t)$, and we use them interchangeably in this book. The reader should be familiar with both notations.

A **stochastic process** (SP) is simply a collection of random variables. We denote this collection by $X(t)$, where t takes values in some well-defined set of times T, called the **index set**. In the solar radiation example, $T = \{1, 2, 3, \ldots\}$. Typically, the index set of times for a process is $T = \{1, 2, 3, \ldots\}$, $T = \{0, 1, 2, 3, \ldots\}$, or $T = [0, \infty)$. These index sets are the set of positive integers, the set of nonnegative integers, and the interval of real numbers $0 \le t < \infty$, respectively. The first two sets are discrete sets of time, and the interval is a continuum of times. In the latter case, consider a population of individuals; time varies continuously and we can regard the population as a collection of random variables $X(t)$, which represents the population an any time t that varies continuously. For notation, we often denote a stochastic process by $X(t)$, $t \in T$, where T is the index set; $X(t)$ takes values in some set of real numbers called the **state space**. The state space may be discrete, as in the population of individuals (population is integer-valued), or it may be a continuum (an interval), as in the example of solar radiation measurements. The word stochastic is a Greek word meaning "to aim at" or "to guess".

What characterizes a stochastic process? First, the time can be indexed as either discrete or a continuum, and the state space can be either a discrete set or a continuum. So there are four possibilities.

Example 7.1

The experiment is to roll a die sequentially. The state space is the set of possible outcomes, or $X(t) \in \{1, 2, 3, 4, 5, 6\}$, which is discrete; and the index set is $T = \{1, 2, 3, \ldots\}$, representing the first roll, second roll, and so on, which is also discrete. If we actually carry out an actual sequence of experiments, we might obtain the set of outcomes $\{2, 2, 3, 6, 5, 6, 3, 4, 1, \ldots\}$. Such a set is called a **realization** of the process, or a **sample path**. A realization is shown in Fig. 7.1. □

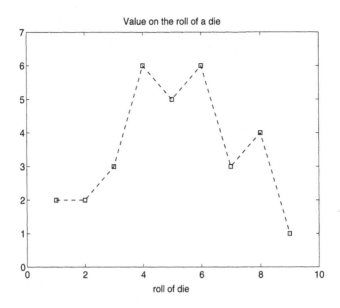

Figure 7.1 Sample path for nine rolls of a die. For visual clarity, the values are connected by a dashed line.

Example 7.2

The population of birds on an island changes because of births, deaths, immigration, and emigration. The state space is discrete (population) and the index set is the interval $[0, \infty)$, which is continuous. Here $t = 0$ is the time we begin counting. Of course, we do not count forever, but as a practical matter we usually consider no upper end to time even though our experiment has a maximum time limit. Also, in practice we can only take measurements at discrete times; but, in reality, the population varies over continuous time. □

Example 7.3

In the example of solar radiation measurements, state space is continuous (possible values in watts per square meter) and the index set is discrete $\{1, 2, 3, ...\}$.

Example 7.4

An example of both the state space and index set being continuous is the location x of an individual at any time t in a one-dimensional medium (e.g., the position of a fish in a stream at time t). Figure 7.2 shows a possible realization

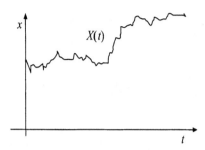

Figure 7.2 A sample path or realization for a continuous-time, continuous state SP.

of such a process. □

The main properties of a stochastic process are defining relations between the random variables $X(s)$ and $X(t)$ at different times s and t. For example, these could be dependence relationships such as conditional probabilities. (If you know the actual value obtained at time s, then what is implied about the probability that you will obtain a certain value at a later time t?). In particular, these relationships could be relationships between $X(t)$ and $X(t + dt)$, where dt is an infinitesimal increment of time. The defining relationships are myriad, and that is what the study of stochastic processes is about—modeling different types of random physical and biological processes by different types of defining relationships, and then studying the consequences. Some key questions are: Where does the process go, how long does it take to get there, and what is its variability? For example, what is the probability of a population going extinct? What is the average time of extinction? If a system can be in finitely many states and you know the probabilities of moving from one state to another in a given time interval, what is the long-term probability distribution of states? Overall, what is the long term behavior (e.g., the variance, the mean, or some other statistical property)?

In summary, there are three defining characteristics of a process:

1. The time index set (continuous or discrete)

2. The state space (continuous or discrete)

3. The dependencies between random variables at different times

We usually give names to stochastic processes. So there are, for example, random walks, Markov processes, Poisson processes, branching processes, Gaussian processes, Wiener processes, martingales, and Itô processes.

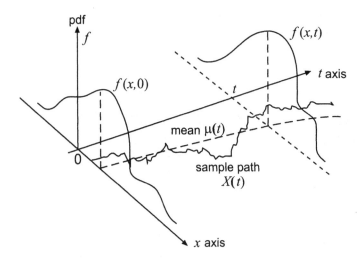

Figure 7.3 Time evolution of the pdf $f(x,t)$ and the mean $\mu(t)$ for a stochastic process; $X(t)$ is a realization, or sample path, lying in the xt plane.

In a stochastic process $X(t)$, for each fixed t the quantity $X(t)$ is an RV. To fix the idea, let $X(t)$ be a continuous process that is continuous in time. As such, we assume that there is a probability density function $f(x,t)$, which now depends on t. Therefore, by definition,

$$\Pr(a < X(t) \leq b) = \int_a^b f(x,t)\,dx,$$

which gives the probability that $X(t)$ takes a value between a and b at time t. As t varies, the mean and variance change as well. We define the **mean** and **variance** of a process by

$$\mu(t) = \mathrm{E}(X(t)) = \int_{-\infty}^{\infty} x f(x,t)\,dx,$$

$$\sigma^2(t) = \mathrm{Var}(X(t)) = \int_{-\infty}^{\infty} (x - \mu(t))^2 f(x,t)\,dx.$$

Figure 7.3 gives an overall generic view of a stochastic process showing the evolution of the pdf $f(x,t)$ in space-time, the mean $\mu(t)$, and a sample path, or realization, $X(t)$.

For a given SP, it is often desirable to determine the density function from the defining dependency relationships. This cannot always be accomplished. For some processes, information about the probabilities of how the system goes from one state to another, the transition probabilities, is enough to determine

the properties of the process. Also, stochastic processes can be generated by
recursion relations and differential equations by introducing stochasticity in a
deterministic law. We now consider examples of this type of a process.

7.2 Randomizing Discrete Dynamics

Perhaps the simplest way to create a random dynamical model is to take an
existing deterministic model and add stochasticity. In this section we look at
various ways to proceed for systems modeled by difference equations, or discrete
models.

From Chapters 1 and 2 we are familiar with the **geometric growth law**,
which states that some populations grow at a constant geometric rate. Specifically, let x_t denote the population of some group of organisms at time t, where
$t = 0, 1, 2, 3, ...$ Then the growth law is

$$x_{t+1} = rx_t, \tag{7.1}$$

where r is a constant growth rate (the offspring/parent ratio). If x_0 is a given
initial population size, iteration gives the population size at time t as

$$x_t = r^t x_0. \tag{7.2}$$

Equation (7.1) is a **deterministic law**. If the initial population is known, the
population size is determined exactly for all times and is given by (7.2).

Example 7.5

The deterministic dynamical model assumes that the growth rate is constant
for all times. The growth rate, however, is determined by many factors, some
of which may not even be well understood. Rather than constant, it seems
more reasonable that the growth rate is random, a stochastic process giving
a different value in each time interval. Therefore, we assume that the growth
rate is composed of a deterministic term r_0 and some random variations given
by Z_t. That is, the growth rate is a stochastic process $r_0 + Z_t$. We can think
of the random variations as "noise" around some average value. Then we can
replace the deterministic law (7.1) by

$$\begin{aligned} X_{t+1} &= (r_0 + Z_t)X_t \\ &= r_0 X_t + Z_t X_t. \end{aligned} \tag{7.3}$$

We have replaced the deterministic population size x_t by a stochastic process
X_t. Now, assuming that $X_0 = x_0$ is given, we can advance the population

size using the stochastic difference equation (7.3); at each time we sample the probability distribution of the random process Z_t to obtain a value at that step. One reasonable assumption on the variations Z_t are $E(Z_t) = 0$ (zero mean) and $var(Z_t) = \sigma^2 = $ const. For example, we could sample a normal probability distribution with mean 0 and variance σ^2 to obtain a value for Z_t for each time t.   □

Example 7.6

A different strategy to turn the deterministic growth law (7.1) into a stochastic law is to add a noise term directly to the equation itself to obtain, for example,

$$X_{t+1} = rX_t + Z_t. \tag{7.4}$$

The population at each time is again a random perturbation Z_t of a deterministic law. There is a large difference between (7.3) and (7.4). In (7.3) the perturbation about the deterministic part is $Z_t X_t$, which is proportional to the population itself. This means that if the population is growing, (7.3) will predict much larger changes in the realizations or sample paths than will (7.4).   □

Which of the two methods for adding stochasticity to a deterministic model presented above is correct? Both are correct, depending on how a researcher views the origins of random factors. Therefore, it is a question of modeling, or determining the best possible equations that describe the biological problem. Often this is what research is about—developing dynamic models by analyzing empirical data, and using intuition and experience.

We have shown only two ways to add stochasticity, both being ad hoc assumptions. There are other ways as well. A model can also be created, for example, using a Markov method and construct the probabilistic model directly by imposing conditional probabilities for the process. Yet another approach is to develop a stochastic population model from the beginning, a priori, ignoring any deterministic laws or conditional probabilities. In the sequel we illustrate these constructions. To repeat, which approach is best is a modeling issue.

There are many reasons for random variations in population birth and death rates. Ecologists often categorize these random effects as **environmental stochasticity** and **demographic stochasticity**. The former includes random weather patterns and other external sources where all individuals are affected equally and there is no individual variation; the latter involves natural variability in behavior, growth, vital rates, and other genetic factors that occur, even though the environment may be constant. Demographic stochasticity is often modeled by a Gaussian (normal) process, but environmental variations can take many forms. For example, an environment may suddenly have a catastrophic event such as a flood, or a bonanza year where the conditions for reproduction are ideal. In engineering one can identify similar processes, those that have external noise, akin to environmental stochasticity, and those that have internal or system noise, which is like demographic stochasticity. In Section 7.2.1 we give yet a different description of these two types of stochasticity.

Example 7.7

(**Sandhill cranes: environmental stochasticity**) Florida sandhill cranes nest in wetland areas, which at times are subject to extreme environmental conditions, such as floods, that destroy their nests. A deterministic model for the yearly population of the cranes is

$$x_{t+1} = (1 + b - d)x_t,$$

where $b = 5$ is the birth rate and d is the death rate. On average, a catastrophic flood occurs once every 25 years, lowering the birth rate and raising the death rate. In normal years $r = 1 + b - d = 1 + 0.5 - 0.1 = 1.4$, but in catastrophic years, $r = 1 + 0.6(0.5) - 0.1(1.25) = 0.575$. This is a stochastic population growth model where $r = 1.4$ ninety-six times out of a hundred, and $r = 0.575$ four times out of a hundred. In a simulation we can choose r at each time step from a uniform random variable on $[0, 1]$, as shown in the following MATLAB m-file:

```
xlist=[ ];
x=x0; xlist=x; N=20;
for n=1:N
test=rand;
if test< 0.04
r=0.575;
else
r=1.4;
end
x=r*x;
xlist=[xlist,x];
end
plot(0:N,xlist)     □
```

In the Exercises you are asked to carry out this simulation to obtain sample paths.

Example 7.8

(**Demographic stochasticity**) In the sandhill crane example, variability can occur even in non-catastrophic years just through natural variations in the animals themselves; this is demographic stochasticity. In this case we can assume, for example, that b and d are normally distributed with

$$b \sim N(0.5, 0.03), \quad d \sim N(0.1, 0.08).$$

The following MATLAB code will simulate this process (with no catastrophic years).

```
xlist=[ ];
x=x0; xlist=x; N=20;
for n=1:N
b=0.5+0.03*randn;  d=0.1+0.08*randn;
x=(1+b-d)*x;
xlist=[xlist,x];
end
plot(0:N,xlist)      □
```

Example 7.9

(**Ricker law**) Consider adding environmental stochasticity to the Ricker population model,

$$x_{t+1} = bx_t e^{-cx_t},$$

where b is the average yearly birth rate. Suppose that on average, there is a weather event once every 25 years that lowers the birth rate that year by 30%. We create a fixed, uniformly distributed random process Z_t having range $[0, 1]$. At each time step we modify the birth rate according to the rule: If $Z_t < 0.04$, *then* adjust the birth rate to $0.7b$; *else* make no adjustment. In this manner, the birth rate becomes a random variable B_t and we can write the dynamics as

$$X_{t+1} = B_t X_t e^{-cX_t}, \tag{7.5}$$

where X_t, the population, is now a random process. It is interesting to simulate this process and the Exercises will guide the reader through this activity.   □

One word about notation—on the right side of (7.5) we mean the specific value $X_t = x_t$ of the random variable, or population, calculated at the time step t. Some authors prefer to write the stochastic model (7.5) as $X_{t+1} = B_t x_t e^{-cx_t}$, where this caveat is made explicit.

Example 7.10

To model demographic changes we might assume that the birth rate is normally distributed. This assumption follows from the observation that birth rates do not vary much from the average, with large deviations rare. Then we have

$$X_{t+1} = G_t X_t e^{-cX_t},$$

where for each time t we choose the birth rate from the distribution $G_t \sim N(b, \sigma)$, a normal distribution with mean b and variance σ^2. □

Generally, a discrete model has the form

$$x_{t+1} = f(x_t, r), \tag{7.6}$$

where r is a parameter. (The analysis can clearly be extended to vector functions and sets of parameters.) In the examples above, stochasticity was added to the model by redefining the parameter r to be random process R_t, giving a stochastic process X_t satisfying

$$X_{t+1} = f(X_t, R_t).$$

The values of R_t are usually chosen from a fixed distribution with mean μ and variance σ^2. For example, we may choose $R_t = \mu + \sigma W_t$, where $W_t \sim N(0,1)$. Alternatively, the values of R_t may be chosen themselves to satisfy a linear autoregressive process [e.g., $R_t = aR_{t-1} + b + W_t$, where $W_t \sim N(0,1)$, and a and b are fixed constants]. To repeat an important point: How stochasticity is put into a model depends on what one seeks and is therefore strictly an issue of modeling. Different stochastic mechanisms produce different patterns of variability.

Sometimes researchers introduce stochasticity in a special way to simplify the analysis. Models of the form

$$Y_{t+1} = g(Y_t) + E_t,$$

where the noise E_t is additive, are called **nonlinear autoregressive** (NLAR) processes; such processes, where the variation has constant variance, are simpler to analyze and to which to apply data-fitting techniques. For example, environmental stochasticity can be introduced in (7.6) by adding noise on a logarithm scale, creating an NLAR process. Taking the logarithm of both sides of (7.6) and then adding randomness gives

$$\ln X_{t+1} = \ln f(X_t, r) + E_t, \tag{7.7}$$

where $E_t \sim N(\mu, \sigma)$. This is an NLAR process for the transformed variable $Y_t = \ln X_t$. Therefore, the model

$$X_{t+1} = f(X_t, r)e^{E_t}$$

is often studied as an environmental stochasticity model. Alternatively, demographic stochasticity is introduced in the discrete model (7.6) by adding stochasticity on a square-root scale, producing

$$X_{t+1} = (\sqrt{f(X_t, r)} + E_t)^2.$$

In some models the stochasticity may enter only through initial conditions. These situations are modeled by the process

$$X_{t+1} = f(X_t, r), \quad X_0 = C,$$

where C is a fixed random variable. The solution to this random equation can be obtained by solving the deterministic problem, if possible; in this case, the random initial condition is evolved deterministically. The solution has the form

$$X_t = G(t, C, r),$$

which is a transformation of a random variable C (treating t and r as parameters) and therefore the statistical properties of X_t can, in principle, be determined exactly.

Example 7.11

(**Random initial data**) Consider the random difference equation

$$\begin{aligned} X_{t+1} &= aX_t, \quad t = 0, 1, 2, ..., \\ X(0) &= X_0, \end{aligned}$$

where a is a positive constant and X_0 is a random variable. Treating the equation deterministically, we obtain

$$X_t = X_0 a^t.$$

Then

$$\begin{aligned} E(X_t) &= E(X_0 a^t) = a^t E(X_0), \\ \text{var}(X_t) &= \text{var}(X_0 a^t) = a^{2t} \text{var}(X_0). \end{aligned}$$

For example, if $0 < a < 1$, the expected value of X_t goes to zero as $t \to \infty$, and the variance goes to zero as well; this means that the spread of the distribution gets narrower and narrower. In fact, we can determine the pdf for X_t using methods from Chapter 5. Letting $f_{X_0}(x)$ denote the pdf of X_0, we have

$$
\begin{aligned}
\Pr(X_t \le x) &= \Pr(X_0 a^t \le x) = \Pr(X_0 \le x a^{-t}) \\
&= \int_{-\infty}^{x a^{-t}} f_{X_0}(y)\, dy = \int_{-\infty}^{x} f_{X_0}(z a^{-t}) a^{-t}\, dz.
\end{aligned}
$$

Therefore, the pdf of the process is

$$
f_{X_t}(x,t) = f_{X_0}(x a^{-t}) a^{-t}.
$$

For a specific case let X_0 be uniformly distributed on the interval $[0, b]$. That is,

$$
f(x) = \frac{1}{b} \mathbf{1}_{[0,b]}(x).
$$

Then

$$
f_{X_t}(x,t) = \frac{a^{-t}}{b} \mathbf{1}_{[0,b]}(x a^{-t}) = \frac{a^{-t}}{b} \mathbf{1}_{[0,ba^t]}(x),
$$

which is uniform. As an exercise, the reader should plot this density for $t = 1, 2, 3$ for $a > 1$ and $0 < a < 1$. This method is successful because this simple discrete model can be solved exactly.   □

The Exercises take you through different ways to develop random models.

7.2.1 Environmental and Demographic Stochasticity

In this subsection we take a different view of demographic and environmental stochasticity from that in our preceding discussion. As noted, random effects can enter a population in essentially two ways: through always-present *individual* variations caused by genetic mutations or natural differences in the members that affect changes in birth and death rates, and through environmental variations that affect each member of the population in the same way. For example, the growth rate of a population could be affected by environmental stochasticity if changes in food availability, weather, or other abiotic factors affect the average growth rate of a population. A rule of thumb is that environmental stochasticity parameter should be independent of the population size, while for demographic stochasticity the variance should be inversely proportional to population size, causing it to decrease as the population grows larger.

A simple models illustrates the idea of demographic stochasticity, where random events in *individual* reproduction, mortality, and genotype occur. Let N_t denote a population at time t, where $t = 1, 2, 3, \dots$. For simplicity, in this population we assume no deaths, and we assume that individual i gives birth to either a single offspring or no offspring in one generation. We let X_i be a Bernoulli random variable for the number of offspring for the ith individual in a generation, with $X_i = 1$ with probability p and $X_i = 0$ with probability $1 - p$. Then the total number of offspring ΔN_t added from the t to the $t + 1$st generation is

$$\Delta N_t = \sum_{i=1}^{N_t} X_i,$$

which is a binomial random variable with mean $N_t p$ and variance $N_t p(1 - p)$. We are assuming that individuals give birth independently. Therefore, if the population is large, ΔN_t is approximately normal (by the central limit theorem), or, precisely, in terms of a z-score,

$$\frac{\Delta N_t - N_t p}{\sqrt{N_t p(1 - p)}} \simeq Z_t,$$

where $Z_t \sim N(0, 1)$. Therefore,

$$\Delta N_t = p N_t + \sqrt{N_t p(1 - p)}\, Z_t.$$

Then, the total number of individuals in the next generation is

$$
\begin{aligned}
N_{t+1} &= N_t + \Delta N_t \\
&= N_t \left(1 + p + \frac{\sqrt{p(1 - p)}}{\sqrt{N_t}} Z_t \right).
\end{aligned}
$$

The term in parentheses is the growth rate; $1 + p$ is the deterministic part of the rate and the term

$$\sigma_d^2 = \frac{\sqrt{p(1 - p)}}{\sqrt{N_t}} Z_t$$

is called the demographic noise, or **demographic stochasticity**. The fact that the demographic stochasticity decreases as the population gets larger makes sense because individual random variations have a smaller effect in a large population.

Environmental stochasticity describes random events that affect all individuals in a similar manner. Now, instead of the probability p of each individual giving birth, we assume that at each fixed time t every individual has a probability P_t of reproducing, where P_t is a random variable with mean p and variance σ_e^2. As in the preceding discussion,

$$N_{t+1} = \lambda(t) N(t),$$

where

$$\lambda(t) = 1 + P_t + \frac{\sqrt{P_t(1 - P_t)}}{\sqrt{N_t}} Z_t$$

is a random variable representing the growth rate. We want to compute the mean and variance of $\lambda(t)$, We emphasize that N_t is fixed at the previous time and is not a random quantity; the random quantity N_{t+1}, the population at the next time, is computed using values of the RVs Z_t and P_t. Now let

$$D_t = \frac{\sqrt{P_t(1-P_t)}}{\sqrt{N_t}} Z_t.$$

Because P_t and Z_t are independent and $EZ_t = 0$, we have $ED_t = 0$. Therefore,

$$var\lambda(t) = \text{var } P_t + \text{var } D_t + 2 \text{ cov } (P_t, D_t).$$

But

$$
\begin{aligned}
\text{cov}(P_t, D_t) &= E(P_t - EP_t)(D_t - ED_t)) \\
&= E\left((P_t - p)\frac{\sqrt{P_t(1-P_t)}}{\sqrt{N_t}} Z_t\right) \\
&= 0,
\end{aligned}
$$

again by independence. Therefore,

$$
\begin{aligned}
\text{var } \lambda(t) &= \text{var } P_t + \text{var } D_t \\
&= \sigma_e^2 + \sigma_d^2.
\end{aligned}
$$

This equation describes the variation in the growth rate in terms of both environmental and demographic contributions. σ_e^2 describes how the random variable P_t changes year to year, and σ_d^2 describes the amount that births vary due to chance. If the population is large, $\sigma_d^2 \approx 0$ and the variation in growth is caused primarily by environmental stochasticity.

EXERCISES

1. This problem (adapted from Mooney & Swift, 1999) refers to Example 7.7. A deterministic model for the yearly population of Florida sandhill cranes is

$$x_{t+1} = (1 + b - d)x_t,$$

where $b = 0.5$ is the birth rate and $d = 0.1$ is the death rate. Initially, the population is 100.

   (a) Find a formula for the population after t years and plot the population for the first five years.

(b) On average, a flood occurs once every 25 years, lowering the birth rate 40% and raising the death rate 25%. Set up a stochastic model and perform 20 simulations over a five year period. Plot the simulations on the same set of axes and compare to the exact solution. Draw a frequency histogram for the number of ending populations for the ranges (bins) 200–250, 250–300, and so on, continuing in steps of 50.

(c) Assume that the birth and death rates are normal random variables,

$$N(0.5, 0.03^2) \quad N(0.1, 0.08^2),$$

respectively. Perform 20 simulations of the population over a five year period and plot them on the same set of axes. On a separate plot, for each year draw side-by-side box plots indicating the population quartiles. Would you say that the population in year 5 is normally distributed?

2. A "patch" has area a, perimeter s, and a strip (band) of width w inside the boundary of a from which animals disperse. Only those in the strip disperse. Let u_t be the number of animals in a at any time t. The growth rate of all the animals in a is r. The rate at which animals disperse from the strip is proportional to the fraction of the animals in the strip, with proportionality constant ϵ, which is the emigration rate for those in the strip.

(a) Argue that the deterministic dynamics is ruled by

$$u_{t+1} = ru_t - \epsilon \left(\frac{ws}{a} u_t \right).$$

(b) Determine conditions on the parameters r, w, s, ϵ, and a under which the population is growing.

(c) Why do you expect $s = k\sqrt{a}$ for some constant k? Write down the model in the case that the region is a circle.

(d) Suppose that at each time step the emigration rate is chosen from a normal distribution with mean 0.35 and standard deviation 0.5. With a growth rate of 1.06 in a circle of radius 100 m, with width $w = 10$ m, simulate the dynamics of the process. Take $u_0 = 100$. Sketch three simulations (realizations) on the same set of axes. Does the population grow or become extinct?

(e) In a given time step, suppose that the animals disperse with $\epsilon = 0.25$ or $\epsilon = 0.45$, each with equal probability. Simulate the dynamics of the process and sketch three realizations.

3. Consider the population growth law $x_{t+1} = r_t x_t$, where the growth rate r_t is a fixed positive sequence and x_0 is given.

   (a) Let x_T be the population at time T. Which quantity is a better measure of an *average* growth rate over the time interval $t = 0$ to $t = T$, the arithmetic mean $r_A = 1/T(r_0 + r_1 + \cdots + r_{T-1})$ or the geometric mean $r_G = (r_0 r_1 \cdots r_{T-1})^{1/T}$?

   (b) Consider the stochastic model process $X_{t+1} = R_t X_t$, where $R_t = 0.86$ or $R_t = 1.16$, each with probability $1/2$. Does the population eventually grow, die out, or remain the same? If $X_0 = 100$ is fixed, what is the expected value of the population after 500 generations? Run some simulations to confirm your answer.

4. Suppose you are playing a casino game where on each play you win or lose 1 dollar with equal probability. Beginning with $100, plot several realizations of a process where you play 200 times. Perform the same task if the casino has the probability 0.52 of winning. In the latter case, perform several simulations and estimate the average length of a game. The game ends when you go broke.

5. In a generation, suppose thaat each animal in a population produces two offspring with probability $\frac{1}{4}$, one offspring with probability $\frac{1}{2}$, and no offspring with probability $\frac{1}{4}$. Assume that an animal itself does not survive over the generation. Illustrate five realizations of the population history over 200 generations when the initial population is 8, 16, 32, 64, and 128. Do your results say anything about extinction of populations?

6. Suppose μ is the probability that a single individual in a population will die during a given year. If X_t is the population at year t, the probability that there will be n individuals alive the next year can be modeled by a binomial random variable,

$$\Pr(X_{t+1} = n | X_t = x_t) = \binom{x_t}{n}(1-\mu)^n \mu^{x_t - n}.$$

In other words, the population at the next time step is the probability of having n successes out of x_t trials, where $1 - \mu$ is the probability of success (living).

   (a) At time $t = 0$ assume that a population has 50 individuals, each having mortality probability $\mu = 0.7$. Use a computer algebra system to perform 10 simulations of the population dynamics over seven years and plot the results.

   (b) For the general model, find the expected value $E(X_{t+1})$ and the variance $\text{var}(X_{t+1})$.

7. Consider a random process X_t governed by the discrete model

$$X_{t+1} = aX_t + b,$$

where the initial condition X_0 is a random variable with density $f_{X_0}(x)$. Find a formula for the density $f(x, t)$ of X_t.

8. Experimental data for the yearly population of Serengeti wildebeest can be fit by the Beverton–Holt model

$$X_{t+1} = \frac{(1-h)rKX_t}{K + (r-1)X_t},$$

where K is the carrying capacity, r is a growth factor, and h is the annual harvesting rate The carrying capacity is a function of rainfall and is modeled by $K = 20748R$, where R is the annual rainfall. Actual measured rainfalls over a 25-year period are given in the vector $\mathbf{R} = (100, 36, 100, 104, 167, 107, 165, 71, 91, 77, 134, 192, 235, 159, 211, 257, 204, 300, 187, 84, 99, 163, 97, 228, 208)$. Using $X_0 = 250,000$ and $r = 1.1323$, simulate the population dynamics over 50 years for different harvesting rates, $h = 0.05, 0.075, 0.1, 0.125, 0.15$, each year randomly drawing the annual rainfall from the vector $\mathbf{R}$. Run enough simulations to make some conclusions about how the fraction of populations that collapsed depends on the harvesting rate. Assume that the threshold value of 150,000 wildebeest represents a collapsed herd.

7.3 Random Walk

One of the simplest stochastic processes is a random walk on the set of integers from 0 to N. So the state space is the set $\{0, 1, 2, 3, ..., N\}$. We assume that time progresses in unit amounts, so that the index set is $T = \{0, 1, 2, 3, ..\}$. First, in a biology context, suppose that at $t = 0$ an animal is located at position n. At each time step the animal moves one step to the right (up) with probability p or one step to the left (down) with probability $q = 1 - p$. Let X_k be the animal's position at time k. This is a discrete-time, discrete-space process. The time series ends when the animal reaches one of the boundaries, 0 or N. One can also think of this process as a gambling game at a casino where at each play, or turn, a player earns 1 dollar with probability p or loses 1 dollar with probability $q = 1 - p$. In this case X_k is the total earnings, on play k, and $X_0 = n$, which is the amount the player has at the beginning of the game. Thus, when $X_k = 0$ the player goes broke, and when $X_k = N$, the player quits the game.

The conditional probabilities that define the rules of the game are:

$$\Pr(X_{k+1} = n + 1 | X_k = n) = p,$$
$$\Pr(X_{k+1} = n - 1 | X_k = n) = q.$$

All two step moves, and beyond, have zero probability. So, for example, $\Pr(X_{k+1} = n + 2 | X_k = n) = 0$.

Here is a typical question about this process. What is the probability that if it starts at $X_0 = n$, the animal will reach N before it reaches 0? In gambling terms, what is the probability of winning N dollars before going broke? We can answer this question in the following way, using the total probability rule and conditioning on a single turn. This is a very useful technique that applies to a lot of problems. Consider the event A_n defined by "the animal is at position n and reaches N before it reaches 0" and let R be the event that it moves to the right on the first turn, and let $L = R^c$ be the event that it moves to the left. For simplicity, introduce the notation $u_n = \Pr(A_n)$, which is what we want to know. Then, using total probability,

$$
\begin{aligned}
u_n &= \Pr(A_n) \\
&= \Pr(A_n | R) \Pr(R) + \Pr(A_n | L) \Pr(L) \\
&= \Pr(A_n | R) p + \Pr(A_n | L) q.
\end{aligned}
$$

But if the animal moved to the right, it is now at position $n + 1$ and the event $A_n | R$ is the same as the event of reaching N before 0 if the animal is at $n+1$, or the event A_{n+1}. In other words, $\Pr(A_n | R) = u_{n+1}$. Similarly, $\Pr(A_n | L) = u_{n-1}$. Substituting these quantities into the last equation gives

$$u_n = p u_{n+1} + q u_{n-1}. \tag{7.8}$$

This equation relates three adjacent values of the sequence $u_0, u_1, u_2,..., u_N$. So (7.8) must hold for $n = 1, 2, ..., N - 1$. If $X_0 = 0$, it is clear that $u_0 = 0$ (if the animal begins at position 0 at time 0, it is on a boundary and the game ends; it cannot get to N) and if $X_0 = N$, then $u_N = 1$ (if the animal starts at N, it gets to N with probability 1). The conditions

$$u_0 = 0, \quad u_N = 1, \tag{7.9}$$

are called **boundary conditions** on the recursive sequence defined by (7.8). Equation (7.8) is a **difference equation** or **recurrence relation**.

If there is a relation between three adjacent values of a sequence, can we determine a formula for the sequence u_n (in terms of n) itself? That is, can we solve the difference equation subject to the given boundary conditions? The answer is yes, and in the next paragraph, using some algebra, we show how.

Consider the difference equation (7.8) with boundary conditions (7.9). The technique is to assume a solution of the form of a power function $u_n = \lambda^n$, where λ is some number to be determined. Substituting into (7.8) gives

$$\lambda^n = p\lambda^{n+1} + q\lambda^{n-1}.$$

Dividing every term by λ^{n-1} gives

$$\lambda = p\lambda^2 + q,$$

or

$$p\lambda^2 - \lambda + q = 0,$$

which is a quadratic equation (called the **characteristic equation**) for the unknown number λ. Using the quadratic formula, we obtain the two roots $\lambda_\pm$ given by

$$\lambda_\pm = \frac{1}{2p}(1 \pm \sqrt{1 - 4pq}).$$

Notice that the maximum value of $4pq$ is $\frac{1}{4}$, so the discriminant is nonnegative and the roots are real. But because $p + q = 1$, we can rewrite the square-root term as

$$\sqrt{1 - 4pq} = \sqrt{1 - 4p(1 - p)} = \sqrt{4p^2 - 4p + 1} = \sqrt{(2p - 1)^2} = |2p - 1|.$$

Therefore, the roots are

$$\lambda_\pm = \frac{1}{2p}(1 \pm |2p - 1|) = \left(\frac{1}{2p} \pm |1 - \frac{1}{2p}|\right) = \left\{\begin{array}{l} 1 \\ p/q \end{array}\right. .$$

We can conclude that both $1^n (= 1)$ and $(q/p)^n$ are solutions to the difference equation (7.8). In the special case $p = q = \frac{1}{2}$, the number 1 is a double root. We analyze these two cases separately.

If $p = q = \frac{1}{2}$, the general solution of (7.8) is

$$u_n = a + bn,$$

where a and b are arbitrary constants. Because $u_0 = 0$, we get $a = 0$; then, because $u_N = 1$, we get $b = 1/N$. Therefore, the solution of (7.8) subject to the boundary conditions is

$$u_n = \frac{n}{N}.$$

This gives the probability of reaching N before reaching zero if you start at n and move to the right or left with equal probability.

In the case $p \neq q$ the roots are 1 and q/p, giving two solutions 1^n and $(q/p)^n$. An arbitrary combination of these two solutions, namely,

$$u_n = a \cdot 1 + b\left(\frac{q}{p}\right)^n,$$

is the general solution of (7.8). Specific values of the constants a and b can be found from the boundary conditions. We have $u_0 = a + b = 0$, so $b = -a$. Then

$$u_n = c(\lambda_+^n - \lambda_-^n).$$

But then $u_N = 1$ forces $u_N = a - a(q/p)^N = 1$, which gives $a = 1 / \left(1 - (q/p)^N\right)$. Therefore, the formula for the probability u_n is given by

$$u_n = \frac{1 - (q/p)^n}{1 - (q/p)^N}.$$

Another question we can ask is about the expected number of time steps before the animal reaches a boundary 0 or N. Let T_n be the number of time steps before reaching a boundary if the animal begins at position n. Then T_n is a RV and it makes sense to find its expected value $E(T_n)$. As above, we condition on the first step and calculate the conditional expectation. Let R be the event "moves to the right" and L be the event "moves to the left" and denote $v_n = E(T_n)$. Then

$$
\begin{aligned}
v_n &= E(T_n) \\
&= E(T_n|R)\Pr(R) + E(T_n|L)\Pr(L) \\
&= E(T_n|R)p + E(T_n|L)q.
\end{aligned}
$$

But $E(T_n|R) = E(T_{n+1}) + 1 = v_{n+1} + 1$ because the animal moved from n to $n+1$ and it took one step; hence, the number of steps to the boundary from n is one more than the number of steps from $n+1$. Similarly, $E(T_n|L) = E(T_{n-1}) + 1 = v_{n-1} + 1$. Consequently, we have the difference equation

$$v_n = (v_{n+1} + 1)p + (v_{n-1} + 1)q$$

for the expected number of steps v_n from position n. We can rearrange this equation to get

$$pv_{n+1} - v_n + qv_{n-1} = -1. \tag{7.10}$$

If $n = 0$ or $n = N$ at the beginning, the animal is at the boundary already, so the expected number of steps is zero. This means that the boundary conditions are

$$v_0 = 0, \quad v_N = 0. \tag{7.11}$$

Solving (7.10)–(7.11) gives the expected number of steps to reach the boundary, starting at n. We leave this calculation as an exercise.

EXERCISES

1. In the random walk problem, what is the probability that the animal will reach 0 before it reaches N? [Hint: The boundary conditions change.]

2. Interpret the random walk problem as the gambler's ruin problem assuming that there is no upper state N; rather, assume that the state space is $\{0, 1, 2, 3, 4, ...\}$ and that X_k denotes the gambler's fortune on play k. If the gambler begins with a fortune $X_0 = n$, what is the probability that he will go broke? Consider the case $p \leq q$.

3. Solve (7.10)–(7.11). [Hints: For the case $p = q = \frac{1}{2}$, try a general solution of the form $u_n = a + bn - n^2$, where a and b are arbitrary constants. In the case $p \neq q$, to find the general solution, first find the general solution to the homogeneous equation, when the right side is zero, and add it to a particular solution to (7.10). The latter can be found by guessing a solution of the form $A + Bn$ for some constants A and B. Finally, apply the boundary conditions.]

7.4 Birth Processes

A birth process is a continuous-time, discrete-state SP. Imagine a process where at time $t = 0$ there are n_0 individuals, and at random future times there is a birth. How does the population evolve? Clearly, this idea can be extended to death processes and to birth–death processes. Before continuing, however, we need to learn some basic facts about linear differential equations. To determine the pdf of the process, we will need to solve such equations.

7.4.1 Linear Differential Equations

We noted earlier that there are some differential equations (e.g., separable equations) that can be solved by a formula, but most cannot. In studying birth processes we need to be able to solve the initial value problem (IVP)

$$y'(t) + ay(t) = q(t), \quad t > 0, \tag{7.12}$$
$$y(0) = y_0, \tag{7.13}$$

where a and y_0 are given constants, and $q(t)$ is a given function. This equation is classified as first order, because the highest derivative is first order, and nonhomogeneous, because there is a nonzero function $q(t)$ on the right side. If $q(t) = 0$, we say that the equation is homogeneous. Fortunately, this is an equation that can be solved easily using the following device.

First we multiply both sides by e^{at} to obtain

$$y'(t)e^{at} + ay(t)e^{at} = q(t)e^{at}.$$

The left side is the derivative of the product $y(t)e^{at}$. Thus,

$$(y(t)e^{at})' = q(t)e^{at}.$$

We can undo the derivative on the left side by integrating from 0 to t. So that we don't overwork the variable t, using it as both an upper limit of integration and a dummy integration variable, let us change the variable t in the last equation to s, and then do the integration. We get

$$\int_0^t (y(s)e^{as})' \, ds = \int_0^t q(s)e^{as} \, ds.$$

Now the *prime* in the integrand on the left is a derivative with respect to s. By the fundamental theorem of calculus[1] the left side is

$$y(t)e^{at} - y(0)e^{a\cdot 0} = \int_0^t q(s)e^{as} ds,$$

or

$$y(t) = y_0 e^{-at} + e^{-at} \int_0^t q(s)e^{as} ds, \tag{7.14}$$

which is the solution to (7.12)–(7.13).

Example 7.12

Find the solution to the IVP

$$\begin{aligned} y' - 3y &= 4, \quad t > 0, \\ y(0) &= 2. \end{aligned}$$

Here $a = -3$, $q(t) = 4$, and $y_0 = 2$. Then the solution (7.14) is

$$\begin{aligned} y(t) &= 2e^{3t} + e^{3t} \int_0^t 4e^{-3s} ds \\ &= 2e^{3t} + e^{3t} \cdot 4 \left[\frac{1}{-3} e^{-3s} \right]_0^t \\ &= 2e^{3t} + 4e^{3t} \left(\frac{1}{-3} e^{-3t} - \frac{1}{-3} e^{-3\cdot 0} \right) \\ &= 2e^{3t} - \frac{4}{3} (1 - e^{3t}) \\ &= \frac{10}{3} e^{3t} - \frac{4}{3}. \quad \square \end{aligned}$$

---

[1] One version of the fundamental theorem of calculus states that the integral of a derivative is the integrand evaluated at the endpoints, or

$$\int_a^b f'(t) \, dt = f(b) - f(a).$$

EXERCISES

1. Solve the following initial value problems.

   (a) $y' + y = e^{-t}$, $y(0) = 1$.

   (b) $y' - y = e^{-t}$, $y(0) = 0$.

   (c) $y' + 2y = t$, $y(0) = 1$.

2. State why the equation $y' + 2ty = e^{-t}$ with initial condition $y(0) = 1$ cannot be solved by formula (7.14).

3. Show that the equation $y' + 2ty = e^{-t}$ with initial condition $y(0) = 1$ can be solved by multiplying both sides by e^{t^2} and then observing that the left side is the derivative of a product. Your answer will involve an integral that has no simple antiderivative.

7.4.2 Simple Birth Process

A simple birth process can be characterized as follows. First, it is continuous in time with $0 \le t < \infty$, and it is discrete in space, with the state space the populations $\{n_0, n_0+1, n_0+2, n_0+3, ...\}$, where n_0 is the beginning population. At certain random times, to be determined, there is a birth. Here are the basic rules: In a very small time increment, which we call h, there is a single birth with a probability roughly proportional to h; in the same time increment, the probability of no births is roughly 1 minus that; the probability of more than one birth is very much smaller than h.

To be more precise about what we mean by "much smaller than" we review the order relation **little oh** of h. To say that h is small means that it is arbitrarily close to zero. We say that a quantity is $o(h)$, read "little oh of h" if the quantity goes to zero faster than h; symbolically, $o(h)$ is defined by

$$\lim_{h \to 0} \frac{o(h)}{h} = 0.$$

Notice that h^2 is much smaller than h for small h (e.g., if $h = 0.1$, then $h^2 = 0.01$); in fact, the quantity h^2 is $o(h)$ because $h^2/h \to 0$ as $h \to 0$. The quantity $\sqrt{h}$ is small and goes to zero as $h \to 0$, but it is not $o(h)$ because $\sqrt{h}/h$ does not go to zero as $h \to 0$. Observe that the symbol $o(h)$ includes constant multiples; thus, for example, $5 \times o(h) = o(h)$.

Now we are in position to define a linear birth process. These were first studied by Yule in 1924 in the context of the evolution of new species, and by Furry in 1937 regarding particle creation in cosmic rays. The processes are called **Yule processes**, or **Yule–Furry** processes. Let

$$N(t) = \text{number of individuals at time } t,$$

and let

$$p_n(t) = \Pr(N(t) = n), \quad n = n_0, n_0 + 1, n_0 + 2, \dots,$$

be the pdf. Let β be a positive constant, called the **birth rate**. The assumptions are that each individual can give birth, and each individual acts independent of the others. For a *single* individual in the population, we assume that

$$
\begin{aligned}
\Pr(1 \text{ birth in } (t, t+h)|N(t) = 1) &= \beta h + o(h), \\
\Pr(0 \text{ births in } (t, t+h)|N(t) = 1) &= 1 - \beta h + o(h), \\
\Pr(\text{more than one birth in } (t, t+h)|N(t) = 1) &= o(h).
\end{aligned}
$$

If there are n individuals in the population, and each acts independently, the probability of one birth is the same as the probability of the first individual having a birth, the probability of the second individual having a birth,..., or the probability of the nth individual having a birth. Then the probability of the n individuals having one birth is the sum of the probabilities of the first, second,... having one birth, each being $\beta h + o(h)$. Therefore,

$$\Pr(1 \text{ birth in } (t, t+h)|N(t) = n) = n(\beta h + o(h)) = n\beta h + o(h).$$

Similarly,

$$\Pr(0 \text{ births in } (t, t+h)|N(t) = n) = 1 - n\beta h + o(h)$$

and

$$\Pr(\text{more than one birth in } (t, t+h)|N(t) = n) = o(h).$$

Our goal is to obtain an equation for the pdf $p_n(t)$. First we note that the event that $N(t+h) = n$ can occur if $N(t) = n-1$ *and* there is one birth, or $N(t) = n$ *and* there are no births. Thus,

$$
\begin{aligned}
\Pr(N(t+h) = n) &= \Pr(N(t) = n-1)((n-1)\beta h \\
&\quad + o(h)) + \Pr(N(t) = n)(1 - n\beta h + o(h)).
\end{aligned}
$$

This means that

$$
\begin{aligned}
p_n(t+h) &= p_{n-1}(t)((n-1)\beta h + o(h)) + p_n(t)(1 - n\beta h + o(h)) \\
&= p_n(t) + (n-1)\beta h p_{n-1}(t) - n\beta h p_n(t) + o(h).
\end{aligned}
$$

We can turn this into a differential equation. We divide by h and rearrange to get a difference quotient on the left:

$$\frac{p_n(t+h) - p_n(t)}{h} = (n-1)\beta p_{n-1}(t) - n\beta p_n(t) + \frac{o(h)}{h}.$$

Taking the limit as $h \to 0$ gives

$$p_n'(t) = (n-1)\beta p_{n-1}(t) - n\beta p_n(t). \tag{7.15}$$

This equation holds for a range of n values: namely, $n = n_0 + 1, n_0 + 2, \ldots$.

Equations (7.15) form a recursive set of differential equations known as the **Kolmogorov equations**. If $p_{n-1}(t)$ is known, (7.15) is an equation for $p_n(t)$. These equations also come with initial conditions, or conditions at $t = 0$. Because at $t = 0$ the population is exactly n_0, we have

$$p_{n_0}(0) = \Pr(N(0) = n_0) = 1, \tag{7.16}$$
$$p_n(0) = \Pr(N(0) = n) = 0, \quad \text{for} \quad n > n_0. \tag{7.17}$$

To begin, we solve the first of (7.15) when $n = n_0$. The equation is

$$p_{n_0}'(t) = -n_0\beta p_{n_0}(t). \tag{7.18}$$

Observe that the first term on the right side of (7.15) is not present when $n = n_0$. Equation (7.18) is a decay equation having solution

$$p_{n_0}(t) = Ce^{-n_0\beta t}.$$

To determine the constant we use (7.16) and obtain $C = 1$. Therefore, the pdf for the population density is

$$p_{n_0}(t) = e^{-n_0\beta t}.$$

This equation says that the probability that the population at time t is n_0, the initial population, is decreasing exponentially.

To obtain the solution of (7.15) for larger values of n, we write (7.15) as

$$p_n'(t) + n\beta p_n(t) = (n-1)\beta p_{n-1}(t).$$

This equation is precisely in the form of the first-order nonhomogeneous equation (7.12), and we solve it in the same way. Multiply through by $e^{n\beta t}$ to obtain

$$p_n'(t)e^{n\beta t} + n\beta p_n(t)e^{n\beta t} = (n-1)\beta p_{n-1}(t)e^{n\beta t}.$$

The left side of this equation is the derivative of $p_n(t)e^{n\beta t}$. (Check this using the product rule.) Then we can write

$$\frac{d}{dt}\left(p_n(t)e^{n\beta t}\right) = (n-1)\beta p_{n-1}(t)e^{n\beta t}.$$

Now, by the fundamental theorem of calculus, integration of both sides will undo the derivative on the left side. To this end,

$$\int_0^t \frac{d}{dt}\left(p_n(t)e^{n\beta t}\right) dt = (n-1)\beta \int_0^t p_{n-1}(t)e^{n\beta t}dt,$$

or

$$p_n(t)e^{n\beta t} - p_n(0) = (n-1)\beta \int_0^t p_{n-1}(t)e^{n\beta t}dt.$$

Finally, using (7.17) and solving for $p_n(t)$ gives

$$p_n(t) = (n-1)\beta e^{-n\beta t}\int_0^t p_{n-1}(t)e^{n\beta t}dt, \quad n = n_0+1, n_0+2, ... \quad (7.19)$$

Hence, we have found a formula for the density $p_n(t)$ in terms of the preceding one.

Example 7.13

Compute $p_{n_0+1}(t)$. From (7.19) we have

$$
\begin{aligned}
p_{n_0+1}(t) &= n_0\beta e^{-(n_0+1)\beta t}\int_0^t p_{n_0}(t)e^{(n_0+1)\beta t}dt \\
&= n_0\beta e^{-(n_0+1)\beta t}\int_0^t e^{-n_0\beta t}e^{(n_0+1)\beta t}dt \\
&= n_0\beta e^{-(n_0+1)\beta t}\int_0^t e^{\beta t}dt \\
&= n_0\beta e^{-(n_0+1)\beta t}\left[\frac{1}{\beta}e^{\beta t}\right]_0^t \\
&= n_0 e^{-(n_0+1)\beta t}\left(e^{\beta t} - 1\right) \\
&= n_0 e^{-n_0\beta t}\left(1 - e^{-\beta t}\right).
\end{aligned}
$$

One can show that, in general, $p_n(t)$ has the form of a negative binomial distribution for each fixed time t:

$$p_n(t) = \binom{n-1}{n_0-1}e^{-n_0\beta t}(1-e^{-\beta t})^{n-n_0}, \quad n = n_0, n_0+1, n_0+2, ... \quad \Box \quad (7.20)$$

EXERCISES

1. Verify that (7.20) solves (7.19).

2. In the same way that we developed the model of a birth process, find the necessary equations and perform the required calculations to define a simple *death* process and determine the Kolmogorov equations.

3. (Poisson process) A discrete, continuous-time process $N(t) \in \{0, 1, 2, \ldots\}$, $t \in [0, \infty)$ is a Poisson process if $N(0) = 0$, $N(s) \leq N(t)$ for $s \leq t$, the number of events in $(s, t]$ is independent of the number prior to s, and there exists a constant λ, called the arrival rate, such that

$$\Pr(N(t + h) = n + 1 | N(t) = n) = \lambda h + o(h), \tag{7.21}$$
$$\Pr(N(t + h) = n | N(t) = n) = 1 - \lambda h + o(h), \tag{7.22}$$

and all other transitions are $o(h)$.

(a) Using the notation $p_k(t) = \Pr(N(t) = k)$, show that

$$\frac{dp_k(t)}{dt} = -\lambda p_k(t) + \lambda p_{k-1}(t), \quad k = 1, 2, 3, \ldots.$$

[Hint: Begin with the law of total probability, $p_k(t+h) = \sum_{i=0}^{k} \Pr(N(t+h) = k | N(t) = i) \Pr(N(t) = i)$, and expand out, eventually taking a limit.]

(b) What is the differential equation for $p_0(t)$? Show that $p_0(0) = 1$ and $p_k(0) = 0$ for $k \geq 1$.

(c) Show that the probability density is given by

$$\Pr(N(t) = k) = \frac{1}{k!} (\lambda t)^k e^{-\lambda t}.$$

(d) By comparing to a Poisson random variable, show that $\mathrm{E}(N(t)) = \lambda t$ and $\mathrm{Var}(N(t)) = \lambda t$.

7.4.3 Interevent Times and Simulation

As the birth process proceeds, at what times do births occur? These times are random variables themselves, T_1, T_2, T_3,..., where T_n is the time that the nth birth occurs. The time between two births is called the **interevent time**, and these times are RVs as well. We denote the interevent times by $X_1 = T_1 - 0$, $X_2 = T_2 - T_1$, $X_3 = T_3 - T_2$, and so on. Being RVs, we can ask how the interevent times are distributed; that is, what is the pdf?

We can make a good guess at the pdf without doing much calculation. First, let T_1 be the time of the first birth. Then the event $X_1 = T_1 > t$ is equivalent to $N(t) = n_0$. Therefore, from our earlier calculation,

$$\Pr(X_1 > t) = \Pr(T_1 > t) = \Pr(N(t) = n_0) = p_{n_0}(t) = e^{-n_0 \beta t}.$$

Therefore, the cdf is $F(t) = 1 - e^{-n_0 \beta t}$, and the pdf of X_1 is the derivative of that, or
$$n_0 \beta e^{-n_0 \beta t}.$$
This is the pdf for the exponential random variable X_1 with parameter $\lambda = n_0 \beta$. Therefore, we conclude that the average time for the first birth when there are n_0 individuals is $1/n_0 \beta$.

If we argue that the interevent times are independent, the pdf for X_2 should also be exponential with parameter $\lambda = (n_0 + 1)\beta$, because there is one more individual in the interval X_2. Therefore, continuing the argument, the pdf for the nth interevent time X_n is exponential with parameter $\lambda = (n_0 + n - 1)\beta$, because there are $n_0 + n - 1$ individuals at X_n.

If we simulate a birth process using a computer, we need a way to determine the times of the jumps, or the interevent times. Because these times are exponentially distributed, it requires choosing a value from an exponential distribution. Most software packages have built-in routines that allow selection of a random number from a standard normal distribution or a uniform distribution on $[0, 1]$ (e.g., randn and rand in MATLAB). But they may not have routines that select random numbers from other distributions.

Fortunately, there is a theorem in probability that allows us to represent any RV T in terms of a uniform RV U. So to simulate T, we just have to simulate U.

Theorem 7.14

Let T be a continuous RV on $[0, \infty)$ with cdf $F(t)$, and let U be a uniform RV on $[0, 1]$. Then
$$T = F^{-1}(U), \tag{7.23}$$
where F^{-1} is the inverse function of F. $\square$

Example 7.15

Let T be exponentially distributed with parameter λ. Therefore, $F(t) = 1 - e^{-\lambda t}$. We can compute the inverse function by setting $u = 1 - e^{-\lambda t}$ and solving for t. We find easily that
$$t = -\frac{\ln(1 - u)}{\lambda}.$$
The right side is $F^{-1}(u)$. Therefore, in terms of RVs,
$$T = -\frac{\ln(1 - U)}{\lambda}.   \square$$

The following string of equalities shows that $F^{-1}(U)$ and T have the same cdf, from which (7.23) follows:

$$\begin{aligned}
\Pr(F^{-1}(U) \leq t) &= \Pr(FF^{-1}(U) \leq F(t)) \\
&= \Pr(U \leq F(t)) \\
&= F(u) \\
&= \Pr(T \leq t).
\end{aligned}$$

EXERCISE

1. For a Poisson process the interevent time is exponentially distributed with parameter λ. Using this information, simulate a sample path for a Poisson process when $\lambda = 1$.

7.5 Stochastic Differential Equations

7.5.1 Examples

In Section 7.2 we noted a multitude of ways to introduce stochasticity into a discrete, deterministic model (a difference equation). Now we take up this question for continuous models, or differential equations. Let's begin with the simplest exponential growth model,

$$\frac{dx}{dt} = rx, \quad x(0) = x_0, \quad t > 0,$$

where x_0 is the initial population and r is the positive per capita growth rate. We know that the solution is $x(t) = x_0 e^{rt}$, which models exponential growth. Generally, we may surmise that the process is not exact, but has inherent random variations. To proceed, let's write this equation alternatively in differential form as

$$dx = rx\,dt.$$

Trivially, we can model stochasticity by adding a random noise term to the equation and obtain an equation for a stochastic process $X(t)$, which we write tentatively as

$$dX(t) = rX(t)\,dt + \text{ noise } \cdot dt.$$

What we mean by $dX(t)$ and by "noise" remains to be decided, but a good intuitive guess is that $dX(t) = X(t + dt) - X(t)$, which is the difference of two random variables. Intuitively, noise represents random fluctuations about some average value. These deviations enter the system through underlying environmental or demographic factors over time. Such factors are present in every system.

Yet another way to add noise is to modify the growth rate r and replace it by an average value (say, the value r) plus a deviation term. So, there is noise, or small variations over time, in the growth rate. To this end, we obtain

$$\begin{aligned} dX(t) &= (r + \text{noise})X(t)\,dt \\ &= rX(t)\,dt + X(t) \cdot \text{noise} \cdot dt, \end{aligned}$$

which is a different form of an equation describing a stochastic process.

As was the case for discrete models, we can add noise in many ways. But for continuous-time processes, there is a more-than-subtle difference. At discrete times, which are at a finite distance from preceding and later times, we can just select from a random distribution at each time. How do we select randomness from a *continuum* of times where there is no preceding or adjacent time? What properties do we want to impose on these random variations? Continuity? Independence? What is the form of the noise? These are all deep questions that require technical reasoning. A guiding principle is that we should choose a representation of the noise to end up with a reasonable, model equation that we can analyze satisfactorily and that is amenable to standard mathematical techniques. In the next section we will see that there is a suitable process, called a *Wiener process*, that fills the bill, but for now let us just be intuitive and denote the "noise" by $N(t)$, without tying down what it exactly is. So we represent our last two stochastic models, which are modifications of the exponential growth law, by

$$dX(t) = rX(t)\,dt + N(t)\,dt,$$
$$dX(t) = rX(t)\,dt + X(t)N(t)\,dt.$$

7.5.2 The Wiener Process

Now we come to the issue of noise and the meaning of $N(t)$. In 1827 the Scottish botanist Robert Brown observed microscopically an unexplained random, zigzag, noisy motion of small pollen grains suspended in a fluid. The key observations in this type of motion are:

- The paths of particles are continuous.

- The particles move independently.

- The motion is more active the smaller the particles.

– The motion is more active the less the fluid density.

– The motion is more active the higher the temperature.

It has generally been accepted that the origin of the motion is the bombard-ment of particles by molecules of the surrounding fluid and random collisions of the particles.

It wasn't until 1905 that Albert Einstein, in his miraculous year, explained the motion quantitatively in terms of the kinetic theory. Letting $u = u(x, t)$ be the probability density for a particle's location x, starting at the origin at time $t = 0$, Einstein derived the **diffusion equation**,[2]

$$u_t = Du_{xx}, \tag{7.24}$$

which is a partial differential equation for u. Here, $u(x, t)dx = \Pr(x < X \le x + dx)$ at time t. The positive constant D, called the diffusion constant, has dimensions length squared per unit time. Intuitively, if L is a characteristic length, such as the average distance between particles, and τ is a characteristic time, such as the average time between collisions, L and τ are related approx-imately by $D = L^2/\tau$. More precisely, assuming the particles are spheres of radius a, Einstein derived the formula

$$D = \frac{kT}{6\pi\eta a}$$

for the diffusion constant. Here T is temperature, η is the coefficient of viscosity of the fluid, and k is Boltzmann's constant. The solution to (7.24), when the particle begins at $x = 0$ at time $t = 0$, is

$$u(x, t) = \frac{1}{\sqrt{4\pi Dt}}e^{-x^2/4Dt}, \tag{7.25}$$

which means that the probability that the particle is in an interval $\alpha < x \le \beta$ at time t is

$$\int_\alpha^\beta u(x, t)\, dx.$$

Observe that the probability density (7.25) is the formula for a normal proba-bility density with mean 0 and variance $Dt/2$; thus, the density is a bell-shaped curve centered at the origin that widens with time. The greater D is, the faster it widens; therefore, the magnitude of D measures how fast the particles diffuse, or spread out, through the medium as they are bombarded with other fluid par-ticles. Notice that when t approaches zero the density becomes narrower and higher at the origin. The limit as t approaches zero is called the *delta function*

---

[2] Here, for simplicity, we are giving the equation in one spatial dimension rather than in the three-dimensional case.

$\delta(x)$. The expression $u(x, 0) = \delta(x)$ is interpreted to mean $\Pr(X(0) = 0) = 1$, or the probability that the particle is at $x = 0$ at time $t = 0$ is 1. If the particle begins at $x = x_0$ rather than $x = 0$, we can shift the density to the point x_0 by translating the x variable to get

$$u(x, t) = \frac{1}{\sqrt{4\pi Dt}} e^{-(x-x_0)^2/4Dt}.$$

This is the bell-shaped curve centered at x_0.

The path that a particle follows, called a *Brownian path*, is continuous, but it is nowhere differentiable because of the large number of zigzags caused by bombardment. This nondifferentiability caused significant mathematical difficulties that were sorted out in a general theory in the 1920s by Norbert Wiener. Thus, this type of diffusion process bears his name, and it is just the process we use to model random noise in physical, biological, economic, and other systems.

A **Wiener process** (also known as standard Brownian motion) $W(t)$ is a continuous SP over the interval $[0, \infty)$. We can think of $W(t)$ as the displacement of a small particle from the origin at time t. Between any two fixed times s and t $(s < t)$ the increment of displacement $W(t) - W(s)$ can be regarded as the sum of a very large number of infinitesimal displacements caused by multiple impacts occurring on a very small time scale. Therefore, the central limit theorem (roughly, the sum of a large number of random variables is approximately a normal distribution) implies that the increment of displacement $W(t) - W(s)$ is normally distributed. Displacement increments over two disjoint time intervals are also independent, and observations of the displacement increments do not depend on when the observations are made, but rather on the length of time of the observation, or $t - s$. Formally, we summarize the key properties of a Wiener process $W(t)$ as follows:

1. $W(t)$ is continuous in t.

2. $W(0) = 0$.

3. The increments $W(t) - W(s)$ and $W(v) - W(u)$ are independent for all $0 \le s < t \le u < v < \infty$.

4. The increment $W(t) - W(s)$ has a normal probability distribution with mean 0 and variance $t - s$ for all $0 \le s \le t < \infty$.

Property 4 (with $s = 0$) shows directly that $W(t)$ itself is a normal RV with mean 0 and variance t. Therefore, the pdf of $W(t)$ is

$$f(x, t) = \frac{1}{\sqrt{2\pi t}} e^{-x^2/2t}, \quad -\infty < x < \infty, \quad t > 0.$$

This means that the cdf is

$$F(x,t) = \Pr(W(t) \le x) = \int_{-\infty}^{x} \frac{1}{\sqrt{2\pi t}} e^{-y^2/2t} \, dy,$$

and

$$\Pr(a < W(t) \le b) = \int_{a}^{b} \frac{1}{\sqrt{2\pi t}} e^{-y^2/2t} \, dy.$$

Observe that this is precisely the density in (7.25) with $D = \frac{1}{2}$. We can easily check by direct substitution that f solves the diffusion equation

$$f_t = \frac{1}{2} f_{xx}.$$

We are usually interested in the small **displacement increments**

$$dW(t) = W(t + dt) - W(t).$$

From the preceding discussion:

- $E(dW) = 0$.

- $\text{var}(dW) = dt$.

- $E((dW)^2) = \sqrt{dt}$.

Figure 7.4 shows three MATLAB simulations of a Wiener process. The m-file finds sample paths of the Wiener process $W(t)$ by solving $dx = dW$ using the Euler method:

```
for paths=1:3;
T=1; N=100; h=T/N; t=0; x=0; v=x;
for n=1:N;
x=x+sqrt(h)*randn;
v=[v x];
t=t+h;
end
int=0:h:T; plot(int,v), hold on
end
xlabel('time'), ylabel('W(t)')
title('Sample Paths of Wiener Process') hold off
```

Now, back to the noise issue. Discarding our inexact notation $N(t)$ for noise, we now define the noise to be $dW(t)$, so that $N(t) \, dt = dW(t)$. This would mean, of course, that

$$N(t) = \frac{dW}{dt},$$

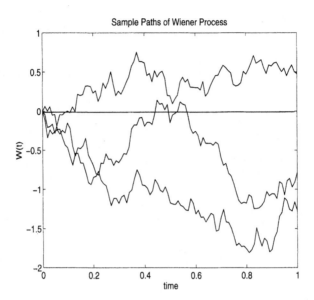

Figure 7.4 Three realizations of a Wiener process.

or that $N(t)$ is the derivative of the Wiener process. Because, as noted earlier, the Wiener path is continuous but not differentiable, this derivative does not exist in a classical sense. It can be shown, however, that it does exist in a generalized sense, and $N(t)$ is usually referred to as **white noise**. In a Fourier analysis sense, which is beyond our scope, white noise contains all frequencies of the spectrum—hence the name.

We next return to a few examples of adding noise, in an ad hoc manner, to the logistic differential equation

$$dx = rx \left(1 - \frac{x}{K}\right) dt.$$

First, replacing the growth rate r by $r + \sigma dW$, which contains noise, we obtain the stochastic model

$$dX = rX \left(1 - \frac{X}{K}\right) dt + \sigma X \left(1 - \frac{X}{K}\right) dW(t).$$

The change in X is the first term representing the average change, given by the deterministic model, plus the second term representing a deviation from the average; the coefficient of dW is the intensity of the noise. The deviation in this case is a demographic fluctuation because we perturbed the growth rate by a small random amount that is normally distributed. Differently, we can assume that the inverse carrying capacity $1/K$ has environmental fluctuations and we can replace $1/K$ by $1/K + \sigma dW$ to obtain the model

$$dX = rX\left(1 - \frac{X}{K}\right)dt + r\sigma X^2\, dW(t).$$

Finally, we could just add a noise term of intensity σX to the logistic equation directly to obtain

$$dX = rX\left(1 - \frac{X}{K}\right)dt + \sigma X dW(t).$$

How we add stochasticity is directly a modeling issue.

Example 7.16

The Langevin equation is $dV(t) = -aV(t)dt + \sigma dW(t)$. Here $V(t)$ is the velocity of a particle, and $-aV(t)$ is the usual force of air friction on the particle. The development of the Langevin equation was one of the first efforts to model friction by random forces on a moving particle. □

EXERCISES

1. Using substitution and differentiation, show directly that (7.25) solves (7.24).

2. Taking $D = \frac{1}{2}$, calculate the that probability that a Wiener path starting at the origin will pass through the gate $1 \le x \le 2$ at time $t = 4$. [Hint: Use a calculator or computer to calculate any necessary integrals.]

3. Define an SP $W_1(t)$ by $W_1(t) = cW(t/c^2)$, $c > 0$, where $W(t)$ is the standard Wiener process. Show that $W_1(t)$ has the same pdf as $W(t)$. Hint: Start with $\Pr(W_1(t) \le x)$.

4. Let $W(t)$ be a Wiener process. Find an integral expression for $\Pr(1 < W(t) \le 2)$, and write it in terms of the erf function, which is defined by

$$\operatorname{erf}(z) = \frac{2}{\sqrt{\pi}}\int_0^z e^{-s^2}\, ds.$$

7.5.3 Stochastic Differential Equations

An equation of the form

$$dX(t) = rX(t)dt + \sigma dW(t) \tag{7.26}$$

is an example of a **stochastic differential equation** (SDE). Along with an initial condition $X(0) = X_0$, it defines a stochastic process $X(t)$ which models a growth process with random noise of intensity σ.

The question is: What does (7.26) mean? A simple answer, which is really not an answer at all, is that (7.26) can be approximated by a discrete process. If we are interested in solving (7.26) on an interval $0 \leq t \leq T$, we can discretize the interval by choosing a partition $t_n = n\,dt$, $n = 0, 1, 2, ..., N$, where N is the number of subintervals in the partition, all of equal length dt. Then, using the notation X_n to denote the approximation for $X(t_n)$, (7.26) can be approximated by

$$X_{n+1} - X_n = rX_n dt + \sigma(W_{n+1} - W_n), \quad n = 0, 1, 2, ..., N - 1. \qquad (7.27)$$

As it turns out, it can be shown that the discrete solution to this equation converges in a special way to the solution of (7.26). Therefore, in some sense, we can think of (7.26) in terms of (7.27), or as a limiting form of (7.27). The numerical procedure defined by (7.27) is called the **Euler–Maruyama method**.

More generally, an **Itô SDE**[3] is an equation of the form

$$
\begin{aligned}
dX(t) &= a(X(t), t)dt + b(X(t), t)dW(t), \quad t > 0, & (7.28) \\
X(0) &= X_0, & (7.29)
\end{aligned}
$$

where a and b are given functions. To compute approximate solutions numerically at the equally spaced partition points $t_n = n\,dt$, $n = 0, 1, 2, ..., N$, we use the Euler–Maruyama method,

$$X_{n+1} = X_n + a(X_n, t_n)dt + b(X_n, t_n)(W_{n+1} - W_n), \quad , n = 0, 1, 2, ..., N - 1. \qquad (7.30)$$

Notice that $W_{n+1} - W_n = \sqrt{dt}\, Z_n$, where $Z_n \sim N(0, 1)$.

Example 7.17

(MATLAB program) The following MATLAB code numerically solves the stochastic logistic equation

$$dX(t) = rX(t)\left(1 - \frac{X(t)}{K}\right)dt + \sigma X(t)\left(1 - \frac{X(t)}{K}\right)dW(t), \quad X(0) = X_0 \qquad (7.31)$$

on the interval $[0, 8]$ with values of r, K, σ, and X_0 given in the program.

```
function EulerMaruyama
   r=0.8; K=100; sigma=0.2; T=8; N=500; dt=T/N; x0=10;
```

---

[3] K. Itô, working in the 1940s, developed the theory of SDEs and stochastic integrals.

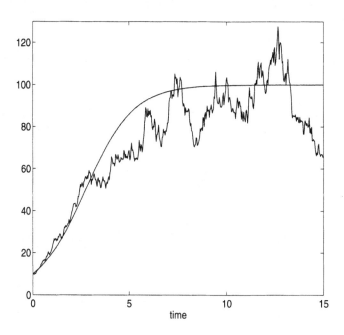

Figure 7.5 Solution of the SDE (7.31) by the Euler–Maruyama method and a plot of the expected value $E(X(t))$.

```
X=x0; Xlist=x0;
for t=1:N
    X=X+r*X*(1-X/K)*dt+sigma*X*(1-X/K)*sqrt(dt)*randn;
    Xlist=[Xlist,X];
end
Avg=x0; Avglist=x0;
for t=1:N
    Avg=Avg+r*Avg*(1-Avg/K)*dt;
    Avglist=[Avglist,Avg];
end
time=0:dt:T;
plot(time,Xlist,time,Avglist)
ylim([0,130]);        □
```

Without exposing all the difficulties, we now address the mathematical meaning of the Itô SDE (7.28). By way of motivation, recall that a first-order ODE can be reformulated as an integral equation, and vice versa. For example, using the fundamental theorem of calculus, the ODE

$$x'(t) = f(x(t), t), \quad 0 < t < T, \quad x(0) = x_0,$$

can be integrated from 0 to t to give the equivalent integral equation

$$x(t) - x_0 = \int_0^t f(x(s), s) \, ds.$$

We can formally integrate both sides of the Itô SDE (7.28) in the same way to obtain

$$X(t) - X(0) = \int_0^t a(X(s), s) \, ds + \int_0^t b(X(s), s) \, dW(s). \tag{7.32}$$

As it turns out, we can make sense of the two integrals in this equation, but not in the usual way. The second integral on the right, where the integration is with respect to a Wiener process, is called an **Itô stochastic integral**. A first guess might be to imitate the definition of a Riemann–Stieltjes integral and define

$$\int_0^T b(X(t), t) \, dW(t) = \lim_{\Delta t \to 0} \sum_{n=0}^{N-1} b(X(t_i), t_i)(W_{n+1} - W_n).$$

But the process $W(t)$ has too much variation (zigzag) in it (recall that its derivative does not exist), and the right side does not converge. So a less restrictive definition of convergence is required, called **mean-square convergence**. We say that

$$\int_0^T b(X(t), t) \, dW(t) = I$$

in a **mean-square sense** if

$$E\left(\sum_{n=0}^{N-1} b(X(t_i), t_i)(W_{n+1} - W_n) - I \right)^2 \to 0 \quad \text{as} \quad N \to \infty, \tag{7.33}$$

where $t_n = n \, \Delta t$, $n = 0, 1, 2, ..., N - 1$, is a partition of $[0, T]$. The first integral on the right of (7.32) is defined in a similar way, that is,

$$\int_0^T a(X(s), s) \, ds = J$$

in a **mean-square sense** if

$$E\left(\sum_{n=0}^{N-1} a(X(t_i), t_i) \Delta t - J \right)^2 \to 0 \quad \text{as} \quad N \to \infty.$$

Practically, it is difficult to calculate these stochastic integrals. This is really no different from the regular Riemann integral in elementary calculus—it is difficult to use the limit definition to calculate integrals. Overall, this topic is far beyond our scope, and we refer to Soong (1973) or L. J. S. Allen (2003) for a good elementary discussion of mean-square convergence. To sum up, we take (7.28) as merely a symbolic equation having the meaning (7.32).

Finally, it is always of interest to find the expected value $EX(t)$ of a process defined by an Itô SDE (7.28). It can be shown that

$$EX(t) = EX_0 + \int_0^t E(a(X(s), s))\, ds,$$

which, upon differentiation, can lead to a differential equation for $EX(t)$.

EXERCISES

1. Consider the Itô SDE

$$dX(t) = X(t)\, dt + \frac{1}{2}\sqrt{X(t)}\, dW(t), \quad X(0) = 1,$$

   on $0 \le t \le 1$. Find an exact formula for $E(X(t))$ Using the Euler–Maruyama method with $dt = 0.02$, plot 25 sample paths and $E(X(t))$ on the same set of axes.

2. Consider the logistic equation

$$\frac{dx}{dt} = rx\left(1 - \frac{x}{K}\right), \quad x(0) = x_0.$$

   Add stochasticity to the model by assuming that the inverse of the carrying capacity contains noise of the form $\sigma dW(t)$; that is, $1/K$ is replaced by $1/K + \sigma dW(t)$. Write the random model in the form of an Itô SDE.

3. A numerical method for solving (7.28)–(7.29) that is more accurate than the Euler–Maruyama method [see (7.30)] is a Runge–Kutta type method (called **Milstein's method**) (e.g., see E. Allen 2008, p. 102) defined by the difference equation

$$\begin{aligned}
X(t_{n+1}) &= X(t_n) + a(X(t_n), t_n)\Delta t + b(X(t_n), t_n)\Delta W(t_n) \\
&+ \frac{1}{2}b(X(t_n), t_n)\frac{\partial b}{\partial x}(X(t_n), t_n)\left[(\Delta W(t_n))^2 - \Delta t\right],
\end{aligned}$$

   where
$$\Delta W(t_n) = (W(t_{n+1}) - W(t_n)) = \sqrt{\Delta t}\, Z,$$

   with $n = 0, 1, 2, \ldots, N - 1$, and $Z \sim N(0, 1)$. Write a MATLAB program that solves the stochastic logistic equation given in Exercise 1 using Milstein's method, and plot several sample paths.

7.6 SDEs from Markov Models

We now present an alternative way to derive a SDE of Itô type using a discrete, Markov-type model. This method also gives a method to turn an ODE into a stochastic DE with a special form for the fluctuation term.

Suppose that a population is modeled by an ordinary differential equation of the form

$$\frac{dx}{dt} = \beta(x) - \alpha(x),$$

where $\beta = \beta(x)$ and $\alpha = \alpha(x)$ are population-dependent birth rates and death rates, respectively. Then, we can write the equation in differential form as

$$dx = \beta\, dt - \alpha\, dt$$

For example, the logistic model can be split into two terms as

$$dx = rx\, dt - \frac{rx^2}{K}\, dt \tag{7.34}$$

with $\beta = rx$ and $\alpha = rx^2/K$. This splitting of the right side of a DE is not unique; for example, the logistic equation can be written as

$$dx = \left(rx + \frac{rx^2}{2K}\right) dt - \frac{rx^2}{2K} dt \tag{7.35}$$

with $\beta = rx + rx^2)/2K$ and $\alpha = rx^2/2K$. Each splitting will lead to a different SDE.

To proceed, let $X(t)$ be a random process for the population $x(t)$. Assume that in a small time interval Δ the probability of a single birth is $\beta\Delta t$, and the probability of a single death is $\alpha\Delta t$, ignoring terms of o(Δt). Then the probability of no change in the population in time Δt is $1 - (\beta + \alpha)\Delta t$, and the probability of any other change is o(Δt). If $\Delta X(t) = X(t + \Delta t) - X(t)$ denotes the change in the population, we can conveniently write the previous assumptions in a table as follows:

Change	Probability
$\Delta X = 1$	$\beta\Delta t$
$\Delta X = -1$	$\alpha\Delta t$
$\Delta X = 0$	$1 - (\alpha + \beta)\Delta t$

We have ignored in the table o(Δt) terms. Then we can calculate the mean and variance of ΔX the usual way as

$$\mu = \mathrm{E}(\Delta X) = (+1)\beta\Delta t + (-1)\alpha\Delta t = (\beta - \alpha)\Delta t,$$
$$\mathrm{E}(\Delta X^2) = (+1)^2\beta\Delta t + (-1)^2\alpha\Delta t = (\beta + \alpha)\Delta t.$$

Therefore, to leading order,

$$\sigma^2 = \text{var}(\Delta X) = \text{E}(\Delta X^2) - \text{E}(\Delta X)^2 = (\beta + \alpha)\Delta t,$$

because $\text{E}(\Delta X) = o(\Delta t)$.

Now comes an important argument. Because ΔX can be regarded as the sum of many smaller changes, we can apply the central limit theorem and conclude that ΔX is a normal random variable. As such, we can write it as $\Delta X = \mu + \sigma Z$, where $Z \sim \text{N}(0,1)$. That is,

$$\Delta X(t) = (\beta - \alpha)\,\Delta t + \sqrt{\beta + \alpha}\sqrt{\Delta t}\,Z.$$

Observe that this equation represents a single iteration of the Euler–Maruyama method applied to the SDE

$$dX(t) = (\beta - \alpha)\,dt + \sqrt{\beta + \alpha}\,dW(t).$$

The numerical solution converges to the solution of the SDE as $\Delta t \to 0$ and $\sqrt{\Delta t}\,Z \to dW(t)$. Therefore, we have obtained an Itô SDE from the deterministic ODE (7.6). The first term, representing the mean change, is given by the deterministic model; the second term is the random fluctuation term, which enters in a special way.

Example 7.18

With the splitting (7.34), the logistic equation leads to the stochastic logistic equation

$$dX = rX\left(1 - \frac{X}{K}\right)dt + \sqrt{rX + \frac{rX^2}{K}}\,dW(t). \quad \square$$

In the next few paragraphs we show how a system of differential equations can be randomized to obtain a system of Itô SDEs. This method is similar to the method discussed for a single variable. We consider the simplest model from epidemiology: The evolution of an SIR disease where $s(t)$ and $i(t)$ denote the number of susceptible individuals and the number of infected individuals at time t, respectively. Both quantities are measured in individuals and they are considered dimensionless. Susceptible individuals move away from the susceptible class to the infected class at a rate bsi where b is the transmission rate, and individuals recover from the disease at rate ri, where r is the recovery rate. The deterministic dynamical equations are

$$\frac{ds}{dt} = -bsi, \quad \frac{di}{dt} = bsi - ri,$$

or

$$ds = -bsi\,dt, \quad di = bsi\,dt - ri\,dt.$$

The transmission rate b, measured in units of $(\text{days·infected})^{-1}$, is the fraction of total possible contacts si between s susceptibles and i infectives that result in a new infection. The recovery rate r, in units of $1/\text{days}$ measures how fast an infected individual recovers; for example, if an individual gets over the disease in an average of 6 days, $r = 1/6$. Recovered individuals do not become susceptible again.

We let $S(t)$ and $I(t)$ denote random variables for the susceptible class and infected class, respectively, and we hold these in the vector $\mathbf{X}(t) = (S(t), I(t))^{\mathrm{T}}$. During a small time period Δt the following table gives the assumed changes in $\Delta\mathbf{X}(t) = (\Delta S(t), \Delta I(t))^{\mathrm{T}}$ in both populations, where $\Delta\mathbf{X}(t) = \mathbf{X}(t + \Delta t) - \mathbf{X}(t)$:

Change	Probability
$\Delta\mathbf{X}(t) = (-1, 1)^{\mathrm{T}}$	$bSI\Delta t$
$\Delta\mathbf{X}(t) = (0, -1)^{\mathrm{T}}$	$rI\Delta t$
$\Delta\mathbf{X}(t) = (-1, 0)^{\mathrm{T}}$	$o(\Delta t)$
$\Delta\mathbf{X}(t) = (0, 0)^{\mathrm{T}}$	$1 - (bSI + rI)\Delta t - o(\Delta t)$

For example, the first entry is a move of one individual from the susceptible class to the infected class with no recoveries; the second entry is a recovery with no new infections. The third entry is a new infected *and* a recovery, which has the product probability $(bSI\Delta t)(rI\Delta t) = o(\Delta t)$. The last entry is no change at all, which is 1 minus the other probabilities. Any other possible change is assumed to be $o(\Delta t)$.

Now we compute the expected change $\mathrm{E}(\Delta\mathbf{X}(t))$ based on the four cases above (only the first two contribute a value):

$$\mathrm{E}(\Delta\mathbf{X}(t)) = \begin{pmatrix} -1 \\ 1 \end{pmatrix} bSI\Delta t + \begin{pmatrix} 0 \\ -1 \end{pmatrix} rI\Delta t + o(\Delta t)$$

$$= \begin{pmatrix} -bSI \\ bSI - rI \end{pmatrix} \Delta t + o(\Delta t).$$

Next, we recall from Section 5.6.2 that the covariance matrix is

$$C = \mathrm{cov}(\mathbf{X}(t)) = \mathrm{E}(\Delta\mathbf{X}(t)\Delta\mathbf{X}(t)^{\mathrm{T}}) - \mathrm{E}(\Delta\mathbf{X}(t))\mathrm{E}(\Delta\mathbf{X}(t))^{\mathrm{T}}.$$

But $E(\Delta \mathbf{X}(t))E(\Delta \mathbf{X}(t))^{\mathrm{T}} = o(\Delta t)$, so

$$
\begin{aligned}
C &= \operatorname{cov}(\mathbf{X}(t)) = E(\Delta \mathbf{X}(t)\Delta \mathbf{X}(t)^{\mathrm{T}}) + o(\Delta t) \\
&= \begin{pmatrix} -1 \\ 1 \end{pmatrix}(-1,1)bSI\Delta t + \begin{pmatrix} 0 \\ -1 \end{pmatrix}(0,-1)rI\Delta t + o(\Delta t) \\
&= \begin{pmatrix} bSI & -bSI \\ -bSI & bSI + rI \end{pmatrix}\Delta t + o(\Delta t).
\end{aligned}
$$

Now comes a key part of the argument. Because $\Delta S(t)$ and $\Delta I(t)$, and hence $\Delta \mathbf{X}(t)$, are the sums of many smaller random changes, the central limit theorem can be applied to conclude that $\Delta S(t)$ and $\Delta I(t)$ are normal random variables. By the representation result in Chapter 5, we can write

$$
\Delta \mathbf{X}(t) = E(\Delta \mathbf{X}(t)) + B\mathbf{Z},
$$

where $\mathbf{Z} = (Z_1, Z_2)^{\mathrm{T}}$, with $Z_1, Z_2 \sim N(0,1)$, and where $BB^{\mathrm{T}} = C$. Here, B is symmetric, so $B^2 = C$. The question is then about the existence of a square root of C.

From elementary linear algebra, we know that $\operatorname{tr} C = 2bSI+rI$, and $\det C = rbSI^2$, so $(\operatorname{tr} C)^2 - 4\det C = I^2(4b^2S^2 + r^2) > 0$; hence, C has real distinct eigenvalues and independent eigenvectors. Therefore C is diagonalizable by a matrix P, whose columns are the eigenvectors. This means that $C = PDP^{-1}$ where D is a diagonal matrix with the eigenvalues on the diagonal. It follows that $B = C^{1/2} = PD^{1/2}P^{-1}$, because

$$
B^2 = (PD^{1/2}P^{-1})(PD^{1/2}P^{-1}) = PDP^{-1} = C.
$$

Therefore, C has a square root.

For 2×2 matrices there is a simple formula for the square root of a symmetric matrix C given by

$$
C = \begin{pmatrix} \alpha & \beta \\ \beta & \gamma \end{pmatrix}.
$$

We have

$$
\sqrt{C} = \frac{1}{d}\begin{pmatrix} \alpha + \det C & \beta \\ \beta & \gamma + \det C \end{pmatrix}, \qquad d = \sqrt{\alpha + \gamma + 2\det C}.
$$

EXERCISES

1. Verify the formulas in (7.6) for the square root of a matrix.

2. Use the splitting (7.35) to obtain an Itô SDE from the logistic equation.

3. Let $R(t)$ denote the rainfall at a location at time t, which is a random process. In a small time Δt the following table gives, to first order, the probabilities of the change in rainfall from day t to day $t + \Delta t$:

change	probability
$\Delta R = \rho$	$\lambda \Delta t$
$\Delta R = 0$	$1 - \lambda \Delta t$

(a) Find $E(\Delta R)$ and $\mathrm{var}(\Delta R)$, and show that the Itô SDE associated with this process is
$$dR = \lambda \rho \, dt + \sqrt{\lambda \rho^2} \, dW(t).$$

(b) Use the Euler–Maruyama method to calculate a sample path when $R(0) = 0$ and $\lambda \rho = 18$, $\sqrt{\lambda \rho^2} = 6$.

4. An SIS model in epidemiology describes the case when susceptibles become infected and, after a recovery period, the infectives return to the susceptible class. If $s = s(t)$ and $i = i(t)$ denote the number of individuals in each class, the dynamics is given by
$$\frac{ds}{dt} = -bsi + ri, \qquad \frac{di}{dt} = bsi - ri,$$

where b and r are the transmision rate and the recovery rate, respectively, and $s + i = N$, where N is the total number of individuals. Letting $S(t)$ and $I(t)$ be random variables for the number of susceptibles and infectives, respectively. Set up a system of Itô SDEs that describes the stochastic dynamics of the disease. [It is clear that the deterministic model can be reduced to a single ODE for either $s(t)$ or $i(t)$, but in this exercise work with a system of two equations.]

5. The classical Lotka–Volterra predator–prey system is
$$\frac{dx}{dy} = x(a - by), \qquad \frac{dy}{dt} = y(cx - d).$$

(a) Set up an Itô stochastic system describing the random dynamics $X(t)$, $Y(t)$, of the prey and predator populations.

(b) Use the Euler–Maruyama method to calculate a sample path in both state space $[X(t)$ and $Y(t)$ vs. $t]$ and phase space $[Y(t)$ vs. $X(t)]$. Take $X(0) = 120$, $Y(0) = 40$, and $a = 1$, $b = 0.02$, $c = 0.01$, $d = 1$.

6. Consider a population $X(t)$ that satisfies the Itô SDE

$$dX(t) = -X(t)\,dt + \sqrt{X(t)+1}\,dW(t), \quad X(0) = 10.$$

Using the Euler–Maruyama method, perform 500 simulations and estimate the mean time to extinction (when the population becomes zero). What is $\mathrm{E}X(t)$?

7.7 Solving SDEs

For the Itô SDE

$$dX(t) = a(X,t)\,dt + b(X,t)\,dW(t), \quad X(0) = X_0,$$

we developed numerical algorithms to obtain an approximate solution. There are other questions that we may ask. Can we solve it and find a formula for $X(t)$? Can we determine the pdf, or at least find some equation that governs the pdf? Can we find $\mathrm{E}X(t)$ and $\mathrm{var}X(t)$?

To make some progress on obtaining answers to these questions it is necessary to delve into the calculus of stochastic processes. The rules of this calculus for computing derivatives and integrals are different from the rules of ordinary calculus. They come out of the definition of the Itô integral with respect to a Wiener process in (7.32). The key fact is the *Itô identity*, which gives a formula for the differential of a function of a Itô stochastic process; this identity is also called the *Itô chain rule*. Although we do not give a rigorous proof of this identity, we can give a convincing heuristic argument. This identity gives a method to compute Itô integrals, solve special SDEs, and determine an equation for the pdf of the process $X(t)$. As such, the following sections have a much higher level of mathematical sophistication than earlier sections, requiring knowledge of multivariable calculus and partial differential equations.

7.7.1 The Itô Identity

Let's take a moment to review the notion of a differential for an ordinary function $y = F(t, x)$. We assume that F possesses derivatives of all orders. If Δt and Δx are arbitrary increments of t and x, then

$$
\begin{aligned}
\Delta F &= F(t + \Delta t, x + \Delta x) - F(x, t) \\
&= F_t(t, x)\Delta t + F_x(t, x)\Delta x + \frac{1}{2}F_{tt}(t, x)\Delta t^2 + \frac{1}{2}F_{xx}(t, x)\Delta x^2 \\
&\quad + F_{tx}(t, x)\Delta t \Delta x + \text{ higher-order terms in } \Delta t \text{ and } \Delta x.
\end{aligned}
$$

The differential of F, denoted by dF, is the *linear,* or lowest-order, portion of ΔF, or
$$dF = F_t(t, x)\Delta t + F_x(t, x)\Delta x.$$

Thus, dF depends on t, x, Δt, and Δx. If x is a function of t, say, $x = x(t)$, then upon substitution of x into the function F gives a function of t, say, $y(t) = F(t, x(t))$. Because $\Delta x = x'(t)\Delta t + \cdots$, where the three dots denote higher-order terms in Δt, we can write the total change Δy in y over an interval $[t, t + \Delta t]$ as

$$\begin{aligned}\Delta y \quad = \quad & F_t(t, x(t))\Delta t + F_x(t, x)x'(t)\Delta t + \frac{1}{2}F_{tt}(t, x(t))\Delta t^2 \\ & + \frac{1}{2}F_{xx}(t, x(t))(x'(t)\Delta t)^2 + F_{tx}(t, x)x'(t)\Delta t^2 + \mathrm{O}(\Delta t^3).\end{aligned}$$

The differential dy, which is a function of t and Δt, is the linear part of the total change, so
$$dy = (F_t(t, x(t)) + F_x(t, x)x'(t))\Delta t.$$

What changes occur when $x(t)$ is replaced by an Itô stochastic process $X(t)$? Now, $Y(t) = F(t, X(t))$ is a stochastic process, where

$$\Delta X(t) \approx a(X(t), t)\Delta t + b(X(t), t)\Delta W(t),$$

where $\Delta W(t)$ is an increment of a Wiener process. For conciseness we omit writing the dependence of F and its partial derivatives, and a and b, on $X(t)$ and t, to get

$$\begin{aligned}\Delta Y \quad = \quad & F_t(t, X(t))\Delta t + F_x(t, X(t))\Delta X + \frac{1}{2}F_{tt}(t, X(t))\Delta t^2 + \frac{1}{2}F_{xx}(t, X(t))\Delta X^2 \\ & + F_{tx}(t, X(t))\Delta t\Delta X + \text{ higher-order terms in } \Delta t \text{ and } \Delta X \\ \approx \quad & F_t\Delta t + F_x(a\Delta t + b\Delta W(t)) + \frac{1}{2}F_{tt}\Delta t^2 + \\ & \frac{1}{2}F_{xx}(a^2\Delta t^2 + b^2\Delta W(t)^2 + 2ab\Delta t\Delta W(t)) + F_{tx}\Delta t(a\Delta t + b\Delta W(t)) \\ & + \text{ higher-order terms in } \Delta t \text{ and } \Delta W.\end{aligned}$$

In the latter expression we seek the lowest-order terms. Order, or *size,* of a random variable is measured by expected value rather than by magnitude. Because $\mathrm{E}(\Delta W(t)^2) = \Delta t$, the term $\frac{1}{2}F_{xx}b^2\Delta W(t)^2$ must be included with the order Δt terms. Moreover, because $\mathrm{E}(\Delta W) = \sqrt{\Delta t}$, the term $bF_x\Delta W(t)$ is order $\sqrt{\Delta t}$. Therefore, approximating $\Delta W(t)^2$ by Δt gives

$$\Delta Y \quad \approx \quad \left(F_t + aF_x + \frac{1}{2}F_{xx}b^2\right)\Delta t + bF_x\Delta W(t)$$
$$+ \text{higher-order terms in } \Delta t \text{ and } \Delta W.$$

Retaining the lowest-order terms in the preceding expression then gives the **Itô differential**,

$$dY(t) = \left(F_t + aF_x + \frac{1}{2}F_{xx}b^2\right) dt + bF_x dW(t). \qquad (7.36)$$

This is the **Itô identity** or Itô chain rule. It shows that if $X(t)$ is an Itô process, $Y(t) = F(t, X(t))$ satisfies an Itô SDE and is an Itô process as well.

7.7.2 Stochastic Integrals

We first list important rules for Itô integrals. The first three are common rules of integral calculus, and the last two are are useful results in calculating expected values of stochastic integrals. Accessible proofs of these facts, along with specific hypotheses, may be found in L. J. S. Allen (2003) and E. Allen (2007).

Proposition 7.19

Let $f(t)$ and $g(t)$ denote shorthand notation for the random functions $f(t, X(t))$ and $g(t, X(t))$, let G be any reasonable function, and let a, b, c, α, and β be constants. Then

1. $\int_a^b dG(t, W(t)) = G(b, W(b)) - G(a, W(a))$ for any reasonable function G.

2. $\int_a^b (\alpha f(t) + \beta g(t)) \, dW(t) = \alpha \int_a^b f(t) \, dW(t) + \beta \int_a^b g(t)) \, dW(t)$.

3. $\int_a^b f(t) \, dW(t) = \int_a^c f(t) \, dW(t) + \int_c^b f(t) \, dW(t)$.

4. $E\left[\int_a^b f(t) \, dW(t)\right] = 0$.

5. $E\left[\left(\int_a^b f(t) \, dW(t)\right)^2\right] = \int_a^b E(f^2(t)) \, dt$.   □

Example 7.20

This example shows how to apply a transformation and use the Itô identity to compute the integral

$$\int_0^t W(s) \, dW(s).$$

Let $dX(t) = dW(t)$, so that $a = 0$ and $b = 1$. Consider the transformation $Y = F(X) = X^2$. Then, by Ito's identity,

$$dY(t) = \frac{1}{2}(2) \, dt + 2X(t) \, dW(t) = dt + 2W(t) \, dW(t).$$

Then integration yields

$$Y(t) - Y(0) = t + 2 \int_0^t W(s) \, dW(s)$$

or

$$W^2(t) = t + 2 \int_0^t W(s) \, dW(s).$$

Therefore

$$\int_0^t W(s) \, dW(s) = \frac{1}{2} W^2(t) - \frac{1}{2} t. \quad \Box$$

7.7.3 Solving SDEs

Some simple Itô SDEs can be solved exactly. But as is the case with regular ODEs, it is more likely that SDEs cannot be solved. In these cases we usually resort to a numerical solution, using, for example, the Euler–Maruyama method, to obtain an approximate solution.

To find solutions to simple equations we use the Itô derivative and try to find a transformation $Y = F(t, X)$ that reduces the SDE in $X(t)$ to a simpler SDE in $Y(t)$.

Example 7.21

Consider the equation

$$dX(t) = -X(t)dt + \sigma \, dW(t).$$

Here $a = -X$ and $b = \sigma$. Letting $Y = F(t, X) = Xe^t$, by Ito's identity (7.36) we have

$$\begin{aligned} dY(t) &= (Xe^t - Xe^t + 0)dt + \sigma e^t \, dW(t) \\ &= \sigma e^t \, dW(t). \end{aligned}$$

Integrating both sides from 0 to t gives

$$Y(t) - Y(0) = \sigma \int_0^t e^s \, dW(s),$$

which gives

$$X(t) = X(0) + \sigma e^{-t} \int_0^t e^s \, dW(s). \quad \Box$$

Example 7.22

Solve the SDE

$$dX(t) = rX(t)dt + \sigma X(t)\,dW(t).$$

We can just substitute into the Itô identity (7.36) using the transformation $Y = F(X) = \sigma^{-1}\ln X$. This gives an integrable relation for $dY(t)$. However, here let us proceed differently by just using Taylor's theorem directly and retaining the second-order term in dX^2. To this end (dropping the t variable in X and W for simplicity),

$$
\begin{aligned}
dY(t) &= F_x\,dX + \frac{1}{2}F_{xx}\,dX^2 \\
&= \frac{1}{\sigma X}(rX\,dt + \sigma X\,dW) + \frac{1}{2}\left(\frac{-1}{\sigma X^2}\right)(rX\,dt + \sigma X\,dW)^2 \\
&= \frac{r}{\sigma}\,dt + dW - \frac{1}{2}\left(\frac{-1}{\sigma X^2}\right)\sigma^2 X^2\,dW^2,
\end{aligned}
$$

where we have retained only lower-order terms. But replacing dW^2 by dt, we have

$$dY(t) = \frac{r}{\sigma}dt + dW - \frac{1}{2}\left(\frac{-1}{\sigma X^2}\right)\sigma^2 X^2\,dt = \left(\frac{r}{\sigma} - \frac{\sigma}{2}\right)dt + dW.$$

Integrating from 0 to t gives

$$Y(t) - Y(0) = \left(\frac{r}{\sigma} - \frac{\sigma}{2}\right)t + W(t).$$

Finally, returning to the quantity X gives

$$\sigma^{-1}\ln X = \sigma^{-1}\ln X(0) + \left(\frac{r}{\sigma} - \frac{\sigma}{2}\right)t + W(t),$$

or, solving for X, we obtain

$$X(t) = X(0)e^{(r-\sigma^2/2)t+\sigma W(t)}. \qquad \square$$

EXERCISES

1. Use definition (7.33) of the stochastic integral to prove property (1) in Proposition 7.19.

2. Use Itô's identity with $Y = tX$, $dX(t) = dW(t)$, to show that

$$\int_0^t s\,dW(s) = tW(t) - \int_0^t W(s)\,ds.$$

3. Let $f(t)$ be a given function. Demonstrate the integration-by-parts formula

$$\int_0^t f(s)\, dW(s) = f(t)W(t) - \int_0^t f'(s)W(s)\, ds.$$

4. Use Itô's identity to verify that the solution to the SDE

$$dX(t) = X(t)dt + X(t)\, dW(t)$$

   is

$$X(t) = X(0)e^{t/2 + W(t)}.$$

5. Solve the Itô SDE
$$dX(t) = (rX(t) + \sigma)\, dW(t).$$

   [Hint: Let $Y = \sigma^{-1} \ln(rX + \sigma)$.]

6. Solve the SDE
$$dX(t) = a(t)X(t)\, dt, \quad X(0) = X_0.$$

7. Solve the SDE

$$dX(t) = a(t)X(t)\, dt + c(t)\, dW(t), \quad X(0) = X_0.$$

   [Hint: Let $Y(t) = X(t)\exp\left(-\int_0^t a(s)\, dx\right)$.]

8. Solve the SDE

$$dX(t) = b(t, X(t))\, dW(t), \quad X(0) = X_0.$$

   [Hint: Let $Y(t) = e^{ct}\int_0^t b(s)^{-1}\, ds$.]

7.8 The Fokker–Planck Equation

If $X(t)$ satisfies an Itô SDE

$$
\begin{aligned}
dX(t) &= a(X(t), t)\, dt + b(X(t), t)\, dW(t), \quad t > 0, \\
X(0) &= X_0,
\end{aligned}
$$

it is not hard to determine a partial differential equation for its pdf $f(x, t)$. Solving the PDE to find f is, of course, another story. One usually has to resort to numerical algorithms to obtain f approximately. Nevertheless, we want to show how to derive the equation for f, which is called the **Fokker–Planck equation**.

To start, let $\varphi(x)$ be a fixed but arbitrary function that is differentiable to all orders, and which is zero outside a closed, bounded interval. This implies, or course, that $\varphi(\infty) = \varphi(-\infty) = 0$. By Itô's identity, applied to φ,

$$d\varphi(X) = [a(X,t)\varphi'(X) + \frac{1}{2}b^2(X,t)\varphi''(X)]\,dt + a(X,t)\varphi'(X)\,dW(t).$$

Here, for notational simplicity, we stopped writing the dependence of X on t. Integrating from 0 to t yields

$$\varphi(X)-\varphi(x_0) = \int_0^t [a(X,s)\varphi'(X)+\frac{1}{2}b^2(X,s)\varphi''(X)]\,ds+ \int_0^t a(X,s)\varphi'(X)\,dW(s).$$

Now we take the expected value of both sides. By Proposition 7.19, the expected value of the rightmost integral is zero; and the expected value operation on the first integral on the right may be brought under the integral because expectation does not involve s. Thus,

$$E(\varphi(X) - \varphi(x_0)) = \int_0^t E[a(X,s)\varphi'(X) + \frac{1}{2}b^2(X,s)\varphi''(X)]\,ds.$$

Taking the derivative, we get

$$\frac{d}{dt}E(\varphi(X)) = E[a(X,t)\varphi'(X) + \frac{1}{2}b^2(X,t)\varphi''(X)].$$

Finally, using the definition of expected value,

$$\frac{d}{dt} \int_{-\infty}^{\infty} f(x,t)\varphi(x)\,dx = \int_{-\infty}^{\infty} f(x,t)[a(x,t)\varphi'(x) + \frac{1}{2}b^2(x,t)\varphi''(x)]\,dx,$$

then

$$\int_{-\infty}^{\infty} \frac{\partial}{\partial t} f(x,t)\varphi(x)dx = \int_{-\infty}^{\infty} f(x,t)a(x,t)\varphi'(x)\,dx+\frac{1}{2} \int_{-\infty}^{\infty} f(x,t)b^2(x,t)\varphi''(x)\,dx.$$

To get the integral on the left, the time derivative was brought under the integral sign. In the integrals on the right side we can now pull the derivatives off the φ using integration-by-parts; for example,

$$\begin{aligned}
\int_{-\infty}^{\infty} f(x,t)a(x,t)\varphi'(x)\,dx &= f(x,t)a(x,t)\varphi(x)|_{-\infty}^{\infty} \\
&\quad - \int_{-\infty}^{\infty} (f(x,t)a(x,t))_x\varphi(x)\,dx \\
&= -\int_{-\infty}^{\infty} \frac{\partial}{\partial x}(f(x,t)a(x,t))\varphi(x)\,dx.
\end{aligned}$$

Similarly, we can pull both the derivatives off φ in the second integral by integrating by parts twice. Then we get

$$\int_{-\infty}^{\infty} f(x,t)b^2(x,t)\varphi''(x)\,dx = \int_{-\infty}^{\infty} \frac{\partial^2}{\partial x^2}(f(x,t)b^2(x,t))\varphi(x)\,dx,$$

where we have set all the boundary terms generated by the integration by parts equal to zero. Therefore,

$$\int_{-\infty}^{\infty} \frac{\partial}{\partial t} f(x,t)\varphi(x)\,dx = -\int_{-\infty}^{\infty} \frac{\partial}{\partial x}(f(x,t)a(x,t))\varphi(x)\,dx$$
$$+ \frac{1}{2}\int_{-\infty}^{\infty} \frac{\partial^2}{\partial x^2}(f(x,t)b^2(x,t))\varphi(x)\,dx.$$

Rearranging gives

$$\int_{-\infty}^{\infty} \left(\frac{\partial}{\partial t} f(x,t) + \frac{\partial}{\partial x}(f(x,t)a(x,t)) - \frac{1}{2}\frac{\partial^2}{\partial x^2}(f(x,t)b^2(x,t)) \right) \varphi(x)\,dx = 0.$$

Because this equation is true for all $\varphi(x)$, the other factor in the integrand must be zero. Therefore, the pdf must satisfy the PDE

$$\frac{\partial}{\partial t} f(x,t) = -\frac{\partial}{\partial x}(f(x,t)a(x,t)) + \frac{1}{2}\frac{\partial^2}{\partial x^2}(f(x,t)b^2(x,t)),$$

which is the **Fokker–Planck equation**. It is accompanied by the initial condition

$$f(x,0) = f_0(x),$$

where f_0 is the pdf for the initial random variable variable X_0.

Example 7.23

Consider the SDE
$$dX(t) = r\,dt + \sigma\,dW(t).$$

Then the Fokker–Planck equation is

$$\frac{\partial f}{\partial t} = -r\frac{\partial f}{\partial x} + \frac{\sigma^2}{2}\frac{\partial^2 f}{\partial x^2}. \tag{7.37}$$

If $r = 0$, (7.37) reduces to the diffusion equation

$$\frac{\partial f}{\partial t} = \frac{\sigma^2}{2}\frac{\partial^2 f}{\partial x^2},$$

which we encountered in (7.24). When the initial condition is $u(x,0) = \delta(x)$, or the initial density is a point mass at zero, we found that the solution is $f(x,t) = \frac{1}{\sqrt{2\pi\sigma^2 t}}e^{-x^2/2\sigma^2 t}$. See (7.25). If $r \neq 0$, we can make a change of independent variables $\tau = t$, $z = x - rt$ to simplify (7.37). By the chain rule for several variables,

$$\frac{\partial f}{\partial t} = \frac{\partial f}{\partial \tau} - r\frac{\partial f}{\partial z}, \quad \frac{\partial^2 f}{\partial x^2} = \frac{\partial^2 f}{\partial z^2},$$

and (7.37) becomes

$$\frac{\partial f}{\partial \tau} = \frac{\sigma^2}{2}\frac{\partial^2 f}{\partial z^2},$$

which is again the diffusion equation. Therefore,

$$f(z,\tau) = \frac{1}{\sqrt{2\pi\sigma^2\tau}} e^{-z^2/2\sigma^2\tau},$$

or, in the original variables,

$$f(x,t) = \frac{1}{\sqrt{2\pi\sigma^2 t}} e^{-(x-r)^2/2\sigma^2 t}. \qquad \square$$

Example 7.24

In this example we add noise to a standard population model and calculate the long-time, or steady-state, probability distribution of the population from the Fokker–Planck equation. We assume that the deterministic dynamics is given by a Ricker growth law with a constant, per capita mortality rate μ:

$$\frac{x'}{x} = be^{-cx} - \mu.$$

We will assume, for simplicity and tractability, that x is near its stable, equilibrium population

$$x^* = \frac{1}{c}\ln\frac{b}{\mu}, \quad b > \mu.$$

Expanding the net per capita growth rate $f(x) = be^{-cx} - \mu$ in a Taylor series about x^* gives

$$\begin{aligned}
f(x) &= f(x^*) + f'(x^*)(x - x^*) + \cdots \\
&= \mu\ln\frac{b}{\mu} - c\mu x + \cdots.
\end{aligned}$$

We used the fact that $f(x^*) = 0$, and we are retaining only the lowest-order term. Denoting

$$\lambda = \mu\ln\frac{b}{\mu},$$

we have a deterministic model that holds near equilibrium:

$$\frac{x'}{x} = \lambda - c\mu x. \qquad \square$$

In an ad hoc way we add noise to the per capita rate and obtain the Itô SDE

$$dX(t) = (\lambda - c\mu X(t))X(t)\, dt + \sigma X(t)\, dW(t),$$

where σ is a positive constant representing the magnitude of the random fluctuation. The Fokker–Planck equation for the pdf $f(x, t)$ is

$$\frac{\partial f}{\partial t} = -\frac{\partial}{\partial x}\left((\lambda - c\mu x)xf\right) + \frac{1}{2}\sigma^2\frac{\partial^2}{\partial x^2}\left(x^2 f\right), \quad x, t > 0.$$

Instead of tackling this difficult, time-dependent PDE, let us ask what happens over a long time period. The deterministic dynamics approaches a constant, stable, steady state; therefore, we expect that over the long time term the pdf will approach a time-independent steady-state or time-independent solution, $f = f(x)$. This is interpreted as the **limiting distribution**. This steady-state distribution then satisfies the ordinary differential equation

$$-\frac{d}{dx}\left((\lambda - c\mu x)xf\right) + \frac{1}{2}\sigma^2\frac{d^2}{dx^2}\left(x^2 f\right) = 0 \quad x > 0.$$

We take $f(x) = 0$ for $x < 0$.

This steady-state Fokker–Planck equation can actually be solved to determine the long-term probability distribution $f(x)$ for the process $X(t)$. First, we can integrate both sides directly to obtain

$$-((\lambda - c\mu x)xf) + \frac{1}{2}\sigma^2\frac{d}{dx}\left(x^2 f\right) = 0,$$

where we have set the constant of integration equal to zero because f and its derivatives are assumed to vanish sufficiently fast at $\pm\infty$. Now we expand out the derivative and separate variables to obtain, after some algebra,

$$\frac{1}{f}\frac{df}{dx} = \frac{2}{\sigma^2}(\lambda - \sigma^2)\frac{1}{x} - \frac{2c\mu}{\sigma^2}.$$

Integrating, we get

$$\ln f(x) = \frac{2}{\sigma^2}(\lambda - \sigma^2)\ln x - \frac{2c\mu}{\sigma^2}x + D,$$

where D is a constant of integration. Exponentiating both sides gives

$$f(x) = Dx^{\alpha - 1}e^{-\beta x},$$

where

$$\alpha = \frac{2\lambda}{\sigma^2} - 1, \quad \beta = \frac{2c\mu}{\sigma^2}.$$

To determine D we use the fact that f is a pdf and therefore $\int f(x)\, dx = 1$. Therefore,

$$D = \frac{1}{\int_0^\infty x^{\alpha - 1}e^{-\beta x}\, dx}.$$

We can write the solution in a particularly familiar form using the gamma function. To review, the **gamma function** is a special function Γ used commonly in mathematics, engineering, and science, and it is defined by the improper integral,

$$\Gamma(\alpha) = \int_0^\infty x^{\alpha-1} e^{-x} \, dx, \quad \alpha > 1.$$

To use this function to rewrite D, we can change variables in the integral defining D by making the substitution $r = \beta x$, $dr = \beta \, dx$. Then, leaving the calculus to the reader, D can be written

$$D = \frac{1}{\int_0^\infty x^{\alpha-1} e^{-\beta x} dx} = \frac{\beta^\alpha}{\Gamma(\alpha)}.$$

Therefore, the limiting distribution is

$$f(x) = \frac{\beta^\alpha}{\Gamma(\alpha)} x^{\alpha-1} e^{-\beta x}, \quad x \geq 0,$$

which is the **gamma distribution**. The conclusion is that over a long period of time, the pdf of the population approaches a gamma distribution. It can be shown that

$$E(X) = \frac{\alpha}{\beta}, \quad \text{var}(X) = \frac{\alpha}{\beta^2}.$$

To review comments made in Chapters 5 and 6, the gamma distribution reduces to other common distributions for special values of the parameters. If $\alpha = 1$, the gamma distribution becomes the exponential distribution, and for $\alpha = \nu/2$, $\beta = 1/2$, it becomes a chi-squared distribution with ν degrees of freedom; as $\alpha \to \infty$, it approaches a Gaussian (normal) distribution.

EXERCISES

1. On the real line, in a small time interval Δt a particle located at an arbitrary point x moves randomly to $x + \Delta x$ (to the right) with probability p, or it remains where it is (no change) with probability $1 - p$. If $u(x, t) = \Pr(x < X(t) \leq x + dx)$, where $X(t)$ is its position at time t, why does

$$u(x, t + \Delta t) = (1 - p)u(x, t) + pu(x - \Delta x, t)?$$

In the limit as $\Delta t, \Delta x \to 0$, with

$$\lim \frac{\Delta x}{\Delta t} = c, \quad \lim \frac{\Delta x^2}{\Delta t} = 0,$$

show that u satisfies the PDE (the advection equation)

$$u_t = -pc u_x.$$

2. Consider the Itô SDE

$$dX(t) = \alpha X(t)\, dt + \beta X(t)\, dW(t), \quad X(0) = x_0,$$

where x_0 is a fixed constant.

(a) Write down the Fokker–Planck equation for the pdf $f(x,t)$.

(b) Let $\mu(t) = \mathrm{E}X(t)$. Show that $\mu'(t) = \alpha\mu(t)$, and therefore $\mu(t) = x_0 e^{\alpha t}$.

7.9 Reference Notes

Stochastic processes, or random occurrences that evolve in time, were part of program of probability theory from the 1800s. Now, stochastic processes are part of every scientific endeavor, and research in those areas is advancing to make it among the most active areas in mathematics.

We mentioned Brown's observations of random motion in 1827. Later, in his 1900 thesis, L. Bachelier modeled stock market deviations using Brownian motion, work apparently not known by Einstein in 1905 when he did his classic work on diffusion. A. A. Markov in 1906 defined processes that bear his name, and Langevin in 1908 modeled particles in a random force field. N. Wiener and Levy worked out much of the theory in the 1920s and 1930s, and Uhlenbeck and Ornstein did classic work in the 1930s. Kiyosi Itô, H. Kramers, and R. Feynman made important contributions in the 1940s. Stochastic processes also have great importance in finance and in engineering. One topic that we did not discussed is Markov chains; this important topic is covered in most books on stochastic processes and in some elementary-level texts on linear algebra.

The most accessible references for much of the material in this chapter are L. J. S. Allen (2003) and Mangel (2007). The latter takes an intuitive approach that includes many insights for ecologists. The other texts listed are more mathematical in nature, particularly Oksendal (2003), which requires measure theory.

Some of the ideas in Section 7.6 for constructing SDEs from ODEs work equally well for constructing stochastic partial differential equations. A good place to start is the E. Allen (2008) paper and the references therein.

We gratefully acknowledge the lectures of Ben Nolting, from which the section on environmental and demographic stochasticity was written. See also Case (2000).

References

Allen, E. 2007. *Modeling with Ito Stochastic Differential Equations*, Springer–Verlag, New York.

Allen, E. 2008. Derivation of stochastic partial differential equations, *Stoch. Anal. Appl.* **26**, 357–378.

Allen, L. J. S. 2003. *Introduction to Stochastic Processes with Applications to Biology*, Prentice Hall, Upper Saddle River, NJ.

Case, T. 2000. *An Illustrated Guide to Theoretical Ecology*, Oxford University Press, Oxford, UK.

Cyganowski, S., Kloeden, P., & J. Ombach, J. 2002. *From Elementary Probability to Stochastic Differential Equations with MAPLE*, Springer-Verlag, Berlin.

Gard, T. C. 1988. *Introduction to Stochastic Differential Equations*, Marcel Dekker, New York.

Lefebvre, M. 2007. *Applied Stochastic Processes*, Springer-Verlag, New York.

Mangel, M. 2007. *The Theoretical Biologist's Toolbox*, Cambridge University Press, Cambridge.

Oksendal, B. 2003. *Stochastic Differential Equations*, 6th ed., Springer-Verlag, Berlin.

Soong, T. T. 1973. *Random Differential Equations in Science and Engineering*, Academic Press, New York.

Hints and Solutions to Exercises

In this final section of the book, we provide hints and solutions to odd-numbered exercises. In many cases the solutions do not show every detail, and in some only the answer is given. Many of the plots required, which must be generated by software (e.g., MATLAB), are left to the reader.

Chapter 1

Section 1.1

1. $N_a = 4k_1\pi r^2$, $N_u = \frac{4}{3}k_2\pi r^3$. Then $N_a = N_u$ when $r = 3k_1/k_2$.

3. (a) The graph of S vs. dI/dt is a straight line with a vertical intercept at I_M and horizontal intercept at P. The graph of S vs. dE/dt is a straight line originating from the origin with slope E_M/P.

(b) The island with the maximum immigration rate.

Section 1.2

1. (a) Use a calculator to obtain the plots.

(b) Animals per unit time.

(c) Slope of tangent line is $f'(4) = \frac{1}{4}$ and the point on the curve is $(4, 2)$. The equation is $y = \frac{1}{4}t + 1$.

(d) $f_{\text{avg}} = \frac{f(4.2)-f(4)}{4.2-4} = \frac{0.049}{.2} = .245$.

Mathematical Methods in Biology,
By J. David Logan and William R. Wolesensky
Copyright © 2009 John Wiley & Sons, Inc.

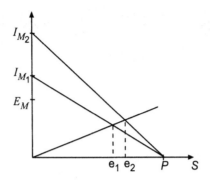

Figure A.1 Exercise 3(b), Sec. 1.1.

(e) Use $f'(4)(0.2) = (.25)(0.2) = 0.05$, or 5 animals per day.

(f) Growth rate $= f'(4) = 1/4$. Per capita growth rate $= f'(4)/f(4) = 0.25/2 = 0.125$ per time.

3. $\frac{dy}{dt} = r \implies y = rt + C.$ $y(0) = y_0 \implies C = y_0$ and $y = rt + y_0$.

5. $L'(10) = 0.6e^{-2}$.

7. The equilibrium is found by solving $g(x) = 0$ or $bxe^{-cx} - mx = 0 \implies x(be^{-cx} - m) = 0$ or $e^{-cx} = m/b$, so $x_e = -(1/c)\ln(m/b)$ is the nonzero equilibrium. The linearization is given by $g(x) \approx g(x_e) + g'(x_e)(x - x_e) = 0 + g'(x_e)(x - x_e)$. We find that $g'(x_e) = m\ln(m/b)$, which gives the linearization $g(x) \approx m/c\ln(m/b)(x + \ln(m/b))$.

9. The limiting growth rate is given by $\lim\limits_{N \to \infty} f(N) = a/b$. The quadratic approximation $P_2(N)$ for $f(N)$. is given by

$$P_2(N) = f(0) + f'(0)N + \frac{f''(0)}{2}N^2 = 0 + aN - abN^2.$$

11. We note that $(\sin at)' = a\cos at$. Then

$$\cos at = \frac{d}{dt}\left(\frac{1}{a}\sin at\right) = \frac{d}{dt}\left(\frac{1}{a}\sum_{k=0}^{\infty}\frac{(-1)^k}{(2k+1)!}(at)^{2k+1}\right) = \sum_{k=0}^{\infty}\frac{(-1)^k}{(2k)!}(at)^{2k}.$$

For $f(t) = \ln(1+t)$ we note that

$$f(0) = 0, \quad f'(0) = \frac{1}{1+0} = 1,$$

$$f''(0) = \frac{-1}{(1+0)^2} = -1, \quad f'''(0) = \frac{2}{(1+0)^2} = 2, \cdots.$$

Substituting into Taylor's theorem yields

$$\ln(1+t) = t - \frac{t^2}{2} + \frac{t^3}{3} - \frac{t^4}{4} + \cdots = \sum_{k=0}^{\infty} \frac{(-1)^k}{k+1} t^{k+1}.$$

For $g(t) = 1/(1+t)$ and note that

$$g(t) = \frac{1}{1+t} = \frac{d}{dt} \ln(1+t) = 1 - t + t^2 - t^3 + \cdots.$$

Then

$$\frac{1}{1+t^2} = g(t^2) = 1 - t^2 + t^4 - t^6 + \cdots = \sum_{k=0}^{\infty} (-1)^k t^{2k}.$$

13. The condition for optimal residence time requires that

$$G'(t) = \frac{G(t)}{t+3}.$$

Because $G'(t) = 3/\left((t+3)^2\right)$, the condition becomes

$$\frac{3}{(t+3)^2} = \frac{\frac{t}{t+3}}{t+3} = \frac{t}{(t+3)^2} \implies t = 3.$$

Thus, the optimal residence time is $t = 3$.

Section 1.3

1. Integration yields $y(t) = -2t + C$. Applying the initial condition $y(0) = 5$ gives $C = 5$, so $y(t) = -2t + 5$ is the solution. The graph is a straight line with slope -2 and y intercept 5..

3. Separating variables gives

$$\int \frac{dy}{10 - y} = \int 2 \, dt \implies \ln|10 - y| = -2t + C \implies y(t) = 10 - Ae^{-2t}.$$

Applying the initial condition $y(0) = 5$ gives $A = 5$, and the solution $y(t) = 10 - 5e^{-2t}$.

5. (a) Mass balance dictates that

$$\frac{dM}{dt} = \text{Rate } M \text{ comes in} - \text{Rate } M \text{ leaves.}$$

Applying this to M gives

$$\frac{dM}{dt} = q\gamma - \frac{M}{V} q, \quad M(0) = C_0 V.$$

Noting that $C = M/V_0$ or $M = V_0 C$ transforms the IVP for mass into the following IVP for C

$$\frac{dC}{dt} = \frac{q\gamma}{V} - \frac{q}{V}C, \quad C(0) = C_0.$$

(b) Separating variables and solving for $C(t)$ gives

$$C(t) = \gamma - (\gamma - C_0)e^{-qt/N}.$$

(c) The volume of the pond at time t is given by

$$V(t) = V_0 + (q_i - q_o)t.$$

Applying mass balance yields

$$\frac{dM}{dt} = q_i\gamma - \frac{M}{V(t)}q_o.$$

Noting that $C = M/V$ or $M = CV$, then $\frac{dM}{dt} = C\frac{dV}{dt} + M\frac{dC}{dt}$ and $\frac{dV}{dt} = q_i - q_o$. Substitution yields

$$\frac{dC}{dt} + C\frac{q_i - q_o}{V_0 + (q_i - q_o)t} = \frac{q_i\gamma - Cq_o}{V_0 + (q_i - q_o)t}, \quad C(0) = C_0.$$

Section 1.4

1. We have $\theta_e = T + q/k$ or $\theta_e = T + q/50 \Rightarrow T = \theta_e - q/50$. If $\theta_e = 22$, then $T = 22 - q/50$, and if $\theta_e = 38$, then $T = 38 - q/50$. The region between these two curves is the climate space.

3. Use $\theta(t) = q/k + T - e^{-kt/mc}(q + kT - k\theta_0)k^{-1}$ and then solve for k when $\theta = 35$. Approximately 9.7 minutes.

5. (a) Using the command dsolve, we find that

$$S(t) = \frac{IP}{I - E} + \frac{S_0(I - E) - IP}{(I - E)^2}e^{(E-I)t/P}.$$

(c) To find the equilibria, solve $dS/dt = 0$. This gives $S^* = PI/(I + E)$ as the equilibrium value. Now graph the horizontal line $S = S^*$.

7. $R(\theta) = \frac{1}{224}(\theta - 12)$. To find the number of days for development at a constant temperature of $\theta = 35$. we need to solve $\int_0^T \frac{1}{224}(35 - 12) \, ds = 1$ for T. This gives $T = 9.7$ days. If $\theta = 30 + 10\cos(2\pi t)$, we need to solve $1 = \int_0^T 18 + 10\cos(2\pi t) \, dt$ for T, or $18T + (5/\pi)\sin(2\pi T) = 224$. Using MATLAB, we find that T is approximately 12.3 days.

Section 1.5

1. Using MATLAB's ode solver, dsolve, we find:

    (a) $y = 6e^{-0.1t}$.

    (b) $y(t) = 20/(1 + 9e^{-1.5t})$.

    (c) $S(t) = 18 - 12e^{-t/6}$.

    (d) $y(t) = \frac{1}{9}(-50e^{-t} + 104e^{(-0.1t)})$.

3. Let $\hat{r} = rb/a$ and $\tau = tb/\rho$. The resulting scaled differential equation is $d\hat{r}/d\tau = 1 - \frac{1}{3}\hat{r}$, $\hat{r}(0) = 0$. Using MATLAB, we find that $\hat{r}(\tau) = 1 - e^{-(1/3)\tau}$.

Section 1.6

1. (a) $\bar{x} = 142.58$, sum $= 7129$, $s = 87.8017$, median $= 123$, CV $= 0.6158$.

    (b) If A is the matrix containing the data, the MATLAB command hist(A,8) creates a histogram with eight bins.

Section 1.7

1. Taking the partial derivatives of S with respect to a_1 and a_2 yields

$$\frac{\partial S}{\partial a_1} = \sum 2(a_1 + a_2 x_i - y_i)/; \text{ and } \frac{\partial S}{\partial a_1} = \sum 2a_2(a_1 + a_2 x_i - y_i).$$

Setting these equal to zero and solving simultaneously for a_1 and a_2 gives the desired result.

3. Rewrite the logistic expression $y = L/(1 + Ce^{ax})$ as $z = a_1 w + a_0$ where $w = x$, $z = \ln(L/y - 1)$, and $a_0 = \ln C$. This model still has the three parameters a_0, a_1, and L. To make this tractable, assume that $L = 100$ and then find a_1 and a_0 using least squares.

5. Define the vectors

$$Y = [0\ 1\ 2\ 3\ 4\ 5\ 6]$$

and

$$E = [0.78\ 0.86\ 0.96\ 0.82\ 0.98\ 1.12\ 0.82].$$

The MATLAB command polyfit(Y,E,1) yields the linear regression model $E = 0.0235Y + 0.8350$. For this model, $R^2 = 0.4229$.

Chapter 2

Section 2.1

1. Rewrite as
$$\frac{x_{t+1}}{x_t} = \frac{R}{1 + \frac{R-1}{K}x_t},$$

or
$$\ln(\frac{x_{t+1}}{x_t}) = \ln R - \ln(1 + \frac{R-1}{K}x_t).$$

Converting to a continuous model gives
$$\frac{1}{x}\frac{dx}{dt} = \ln R - \ln\left(1 + \frac{R-1}{K}x\right).$$

Expanding $\ln(1 + (R-1)x/K)$ in it Taylor's series and keeping only the first order term gives
$$\frac{1}{x}\frac{dx}{dt} = \ln R - (1 + \frac{R-1}{K}x) \Rightarrow \frac{dx}{dt} = x(\ln R - 1) - \frac{R-1}{K}x^2.$$

This is a continuous logistic growth population model.

3. Using separation of variables or MATLAB (use dsolve), we find that the solution to the given differential equation is $L = 1/(mt + c)$. Applying the boundary condition $L(0, n) = A_n$ we find that $c = 1/A_n$. This gives $L = A_n/(A_n mt + 1)$. Next we apply the boundary condition $sL(T, n) = A_{n+1}$ and we find that $L = A_n/(A_n mT + 1) = A_{n+1}$, which is the desired result.

Section 2.2

1. Use the MATLAB m-file given at the end of this section. To get two graphs on the same plot, add the additional MATLAB command hold on.

3. In MATLAB, create the m-file

```
function dxdt = fun1(t,x)
dxdt = 1.5*x*(1 - x/30);
```

In the command window enter the command [t,x]=ode23(@sec22number3, [0 10], 3). Finally, to see your plot, simply use plot(t,x).

5. The graph is linear with $d\theta/dt$-intercept $(q + kT)/mc$ and θ-intercept $(q + kT)/K$. The equilibrium value is the θ-intercept; that is, $\theta_e = (q + kT)/K$. On the interval $(0, \theta_e)$, we have that $d\theta/dt$ is positive, so the arrow on the phase line points to the right. On the interval (θ_e, ∞), we have that $d\theta/dt$ is negative; thus, the arrow on the phase line points to the left. Hence θ_e is a stable equilibrium.

Section 2.3

1. (a) $dx/dt = 2/x \Rightarrow \int x \, dx = \int 2 \, dt \Rightarrow (1/2)x^2 = 2t + C$.

    (b) $dx/dt = (2t)/x^2 \Rightarrow \int x^2 \, dx = \int 2t \, dt \Rightarrow x^3/3 - t^2 = C$ or $x = (3t^2 + C)^{1/3}$.

    (c) $dx/dt = (x+3)/(t+2) \Rightarrow \int 1/(x+3) \, dx = \int 1/(t+2) \, dt \Rightarrow \ln|x+3| = \ln|t+2| + C$ or $x = -3 + A(t+2)$.

    (d) $dx/dt = x(1+x) \Rightarrow \int 1/(x(1+x)) \, dx$. Using partial fractions, we get $\ln|x| - \ln|1+x| = t + C \Rightarrow x/(1+x) = Ae^t$ or $x = 1/(Be^{-t} - 1)$.

    (e) $dx/dt = (xe^{-t})/\sqrt{t+1} \Rightarrow \int 1/x \, dx = \int e^{-t}/\sqrt{t+1} \, dt$ or $\ln|x| = \int_0^t e^{-s}/\sqrt{s+1} \, ds \Rightarrow x = e^{\int_0^t e^{-s}/\sqrt{s+1} \, ds} + K$. Applying the initial condition gives $K = 1$.

3. To find the equilibria, we solve $dx/dt = 0$. We get $x_1^* = 0$ and $x_2^* = ((b-c)/c)^{1/n}$. Using the graph, we see that on the interval $(0, x_2^*)$ that $dx/dt > 0$. Thus, the arrow on the phase line point's to the right on that interval. On the interval (x_2^*, ∞) we have $dx/dt < 0$. Thus, the arrow on the phase line points to the left. Therefore, $x_1^* = 0$ is unstable and $x_2^* = ((b-c)/c)^{1/n}$ is stable.

5. We have $dN/dt = -m(t)N$. Separating variables and integrating gives $\ln N = -(p+1)/p_0 \int (t/t_0)^p \, dt = -(1/p_0 t_0^p)t^{p+1} + C$. Exponentiating gives $N(t) = Ae^{-(t^{p+1}/p_0 t_0^p)} = N_0 e^{-(t^{p+1}/p_0 t_0^p)}$. Thus, $S(t) = N(t)/N_0 = e^{-(t^{p+1}/p_0 t_0^p)}$. If $p = 0$, we have high survivorship, and $p = 10$ would model low survivorship. Thus, $p = 0$ may be appropriate for some human populations, whereas $p = 10$ would be more fitting for fish.

7. We find the equilibria by solving $(\lambda - b)x - ax^3 = 0$. This yields $x_1^* = 0$, $x_2^* = \sqrt{(\lambda - b)/a}$, and $x_3^* = -\sqrt{(\lambda - b)/a}$. We have two cases: (i) if $\lambda \leq b$, then $x^* = 0$ is the only real equilibrium, and phase line analysis shows that it is stable. (ii) If $\lambda > b$, we have three equilibria. In this case, phase line analysis shows that $x_1^* = 0$ is unstable, $x_2^* = \sqrt{(\lambda - b)/a}$ is stable, and $x_3^* = -\sqrt{(\lambda - b)/a}$ is also stable. This yields the following bifurcation diagram as shown in Fig. A.2.

9. Finding the equilibria gives $x_1^* = 0$, $x_2^* = \sqrt{-\lambda}$, $x_3^* = -\sqrt{-\lambda}$. The bifurcation diagram is shown in Fig. A.3.

Section 2.4

1. The cusp occurs when $x = \sqrt{3}$. Substituting into equations (2.17) and (2.18) gives $r = 3\sqrt{3}/8$ and $k = 3\sqrt{3}$.

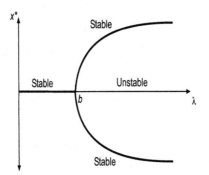

Figure A.2   Exercise 7, Sec. 2.3.

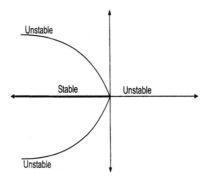

Figure A.3   Exercise 9, Sec. 2.3.

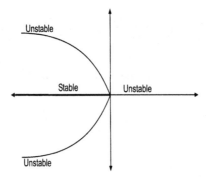

Figure A.4   Exercise 3, Sec. 2.4.

3. Using Figure 2.11, we get the bifurcation diagram in Fig. A.4.

5. (a) We solve $cp(1 - p) - \varepsilon p = 0$ and find equilibrium values $p_1^* = 0$ and $p_2^* = (c - \varepsilon)/c$. On the interval $(0, p_2^*)$ the patches increase (the arrow on the phase line points to the right) and on (p_2^*, ∞) the patches decrease (the arrow on the phase line points to the left). Thus, $p_1^* = 0$ is unstable and $p_2^* = (c - \varepsilon)/c$ is stable.

(b) You would expect the migration rate from the mainland to be proportional to the fraction of unoccupied patches. That is, the mainland migration rate is $m(1 - p)$. We now have only one nonnegative equilibrium, $p^* = \left((c - m - \varepsilon) + \sqrt{(m - c + \varepsilon)^2 + 4cm} \right)/2c$, and by plotting p vs. p', we find that it is a stable equilibrium.

(c) The quantity $D + p$ represents the total fraction of patches that are either destroyed or occupied. Thus, the model is now $p' = cp(1 - (D + p)) - cp$. We then have two equilibria, $p_1^* = 0$ and $p_2^* = (1/c)(c - cD - \varepsilon)$. Using phase line analysis [see the results for part (a)], we see that $p_1^* = 0$ is unstable and that $p_2^* = (1/c)(c - cD - \varepsilon)$ is stable.

Section 2.5

1. (a) $x_t = 100(1/2)^t$.

(b) $x_t = 25(-1.3)^t$.

(c) The equation is linear, so the solution is given by equation (2.24). That is, $x_t = 50(-0.2)^t + (0.7/1.2)(1 - (-0.2)^t)$ or $x_t = 49.4167(-2)^t + 0.5833$.

(d) Linear; thus, $x_t = \frac{1}{3}(4^{t+1} - 1)$.

3. The model for this is $x_{t+1} = Rx_t + (-p)$, where p is the monthly payment, R the monthly interest rate, and x_t is the amount remaining have the t payment. Solving gives $x_t = x_0 R^t - p(1 - R^t)/(1 - R)$. We want $x_{36} = 0$. Solving for p will give the desired monthly payment.

5. Substituting $x_t = r^t$ into the difference equation gives $r^{t+2} = -pr^{t+1} - qr^t$ or $r^t(r^2 + pr + q) = 0$, which implies that $r^2 + pr + q = 0$. To show that $x_t = Ar_1^t + Br_2^t$ is a solution, simply substitute into the original equation and then group terms, noting that $r_i^{t+2} - pr_i^{t+1} - qr_i^t = 0$, for $i = 1, 2$. For $x_0 = 1$ and $x_1 = \frac{1}{2}$ we find that $r_1 = 3$, $r_2 = -2$, $A = 2/5$, and $B = 3/5$.

7. Using Euler's representation, we see that if $r_1 = a + ib$ is a solution to the characteristic equation, $x_t = r_1^t = (\rho e^{i\theta})^t = \rho^t e^{i\theta t}$ is a solution to the difference equation. But $\rho^t e^{i\theta t} = \rho^t(\cos \theta t + i \sin \theta t)$ so we see that both $\rho^t \cos \theta t$ and $\rho^t \sin \theta t$ are solutions to the difference equation (note that the

real and imaginary parts must both satisfy the difference equation), and the general solution would be $x_t = A\rho^t \cos\theta t + B\rho^t \sin\theta t$. The characteristic equation for $x_{t+2} = x_{t+1} - x_t$ is $r^2 - r + 1 = 0$, which has roots $r = 1/2 \pm (\sqrt{3}/2)i$. Then $\rho = \sqrt{(1/2)^2 + (\sqrt{3}/2)^2} = 1$ and $\theta = \tan^{-1}(\sqrt{3}) = \pi/3$. Therefore, we have the general solution $x_t = A\cos(\pi/3)t + B\sin(\pi/3)t$. To find A and B, apply the initial states to get $A = 1$ and $B = 0$.

Section 2.6

1. To find the equilibria we solve $P^* = 2P^* - 3P^*/(1 + 2P^*)$. This yields $P_1^* = 0$ and $P_2^* = 1$ as the equilibria. Using a cobweb diagram, we see that $P_1^* = 0$ is stable and $P_2^* = 1$ is unstable.

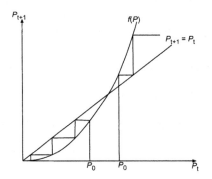

Figure A.5 Exercise 1, Sec. 2.6.

3. If we let $w_t = 1/x_t$ then $x_t = 1/w_t$ and upon substitution and simplifying we get $w_{t+1} = (b/a)w_t + (1/a)$. This is a linear difference equation and thus has a solution given by equation (2.24). If $R = b/a$ and $p = 1/a$, then $w_t = R^t w_0 + p((1 - R^t)/(1 - R))$. Substituting $w_t = 1/x_t$ we find that

$$x_t = \frac{(1 - R)x_0}{(1 - R)R^t + p(1 - R^t)x_0}.$$

Using algebra, we can rewrite x_t in a form analogous to the continuous logistic model.

5. We find the equilibria by solving $x = \ln x^2$. This has the solution $x^* = e^{1/2}$. We see that $f'(x) = 2\ln x + 2$ and that $f'(e^{1/2}) = 3 > 1$. Thus the equilibrium value $x^* = e^{1/2}$ is an unstable equilibrium.

7. We start by perturbing the equilibrium x^* by a small amount y_0 and then considering the time state of $x^* + y_t$. Substitution into the difference equation gives $x^* + y_{t+1} = a(x^* + y_t) + F(x^* + y_{t-1})$, which upon linearization becomes $y_{t+1} = ay_t + F'(x^*)y_{t-1}$. Denote $F'(x^*)$ by λ. We then have $y_{t+1} = ay_t + \lambda y_{t-1}$ We guess a solution of the form $y_t = m^t$. Substitution into the difference equation yields $m^{t-1}(m^2 - am - \lambda) = 0$. This has nonzero roots of $m = (a \pm \sqrt{a^2 + 4\lambda})/2$. For stability, it is required that $y_t = m^t \to 0$. This requires that $|m| < 1$. To determine conditions on λ [and thus $F(x^*)$], we consider the following cases.

(i) If $\lambda = -a^2/4$, then $m = a/2 < 1$; thus, we will have y_t converge to 0 and hence the equilibrium value is stable.

(ii) If $\lambda < -a^2/4$, then m is complex and using analysis similar to that given in Exercise 7 Section 2.5, we see that y_t will go to 0 (and thus the equilibrium is stable) provided that $|a^2/2 + \lambda| < 1$ or if $-1 - a^2/2 < \lambda < 1 - a^2/2$.

(iii) If $\lambda > -a^2/2$, we have two real roots, and we will have convergence provided that $\lambda < 1 - a$.

9. Solving $x^* = x^* e^{r - x^* - c(x^*)^2}$ yields $x^* = -1 + \sqrt{1 + 4rc}$ as the unique positive equilibrium. To find a condition for stability, let $f(x) = xe^{r-x-cx^2}$ and require that $|\lambda = f'(x^*)| < 1$.

11. (a) The equilibria occur when $\Delta P_t = 0$. Solving gives the equilibria $P_1^* = 0$ and $P_2^* = K$.

(b) The per capita growth rate is $\Delta P_t/P_t = r(1 - (P_t/K)^\theta)$.

(c) $P_{t+1} = P_t + rP_t \left(1 - \left(\frac{P_t}{K}\right)^\theta\right) = f(P_t)$. To determine stability, we consider $f'(P_1^*)$ and $f'(P_2^*)$. We find that $\lambda_1 = f'(0) = 1 + r > 0$; therefore, $P_1^* = 0$ is unstable. Next consider $\lambda_2 = f(K) = 1 - r\theta$. Then $|\lambda_2| < 1$ provided that $2 > r\theta > 0$. By using cobweb analysis, we see that for $r\theta = 2$ we also have stability.

(d) Graph $\theta = 2/r$. The region of stability is the region between the curve and the positive r and θ axes.

13. (a) For $N = 0$ we immediately see that $dN/dt = 0$. For $N = K$, use the fact that $\int_0^\infty Kk(\tau) \, d\tau = K$ and the result follows.

(b) Let $N = K + n(t)$. Substitution into the differential equation yields

$$\frac{dn}{dt} = -r \int_0^\infty n(t - \tau)k(\tau) \, d\tau - \frac{rn(t)}{K} \int_0^\infty n(t - \tau)k(\tau) \, d\tau.$$

Linearization yields the desired result.

(c) Using $n(t) = n_0 e^{\lambda t}$, we get

$$\lambda n_0 e^{\lambda t} = -r \int_0^\infty \lambda n_0 e^{\lambda(t-\tau)} k(\tau) \, d\tau.$$

Simplifying gives $\lambda = (-r-1)/T < 0$. Thus $n(t)$, goes to zero and stability follows.

Chapter 3

Section 3.2

1. The eigenpairs are 2, $(4,1)^T$ and -1, $(-2,1)^T$. Therefore, the general solution is

$$\begin{pmatrix} X_t \\ Y_t \end{pmatrix} = c_1 \begin{pmatrix} 4 \\ 1 \end{pmatrix} 2^t + c_2 \begin{pmatrix} -2 \\ 1 \end{pmatrix} (-1)^t.$$

The initial condition gives

$$\begin{pmatrix} 6 \\ 0 \end{pmatrix} = \begin{pmatrix} 4c_1 \\ c_1 \end{pmatrix} + \begin{pmatrix} -2c_2 \\ c_2 \end{pmatrix}.$$

Solving, we get $c_1 = 1$, $c_2 = -1$. So the solution is

$$X_t = 4 \cdot 2^t + 2(-1)^t,$$
$$Y_t = 2^t - (-1)^t.$$

3. (a) The eigenvalues are 1.0182, $-0.5091 \pm 0.6910i$. The magnitude of the complex eigenvalues are $\sqrt{(-0.5091)^2 + (0.6910)^2} = 0.858$, so $\lambda = 1.0182$ is the dominant eigenvalue; the corresponding eigenvector is

$$(0.9501, 0.28, 0.1375)^T.$$

Normalized so that the sum of the entries is 100, the stable age structure is $(69.5, 20.5, 10.1)^T$, which gives the long-term percentages of stages 1, 2, and 3, respectively. The entries 1 and 5 are the fecundities of stage 2 and stage 3 animals; 0.3 is the fraction that graduate from stage 1 to stage 2, and 0.5 is the fraction that graduate from stage 2 to stage 3.

5. (a) An equilibrium solution $\mathbf{x}$ must satisfy $\mathbf{x} = A\mathbf{x}$ or $(A-I)\mathbf{x} = \mathbf{0}$. If there are to be nontrivial solutions, then $\det(A - I) = 0$, or 1 is an eigenvalue of A. In the present case this condition becomes $(1 - ac) - bcs(1 - a) = 0$.

(b) If there is a single equilibrium, it must be $\mathbf{x} = \mathbf{0}$. By Exercise 6 we get asymptotic stability (convergence to zero) if, and only if,

$$ac < 1 - bcs(1 - a), \quad -bcs(1 - a) < 1.$$

The second inequality is always satisfied because the left side is negative. So we get stability when $ac < 2$.

7. (a) The population projection matrix is

$$A = \begin{pmatrix} 0 & f \\ g & s \end{pmatrix}.$$

(b) The characteristic equation is $\lambda^2 - s\lambda - gf = 0$, giving eigenvalues

$$\lambda_\pm = \frac{1}{2}\left(s \pm \sqrt{s^2 + 4gf}\right),$$

which are real and of opposite sign. The corresponding eigenvectors are

$$\mathbf{v}_+ = \begin{pmatrix} f \\ \lambda_+ \end{pmatrix}, \quad \mathbf{v}_- = \begin{pmatrix} f \\ \lambda_- \end{pmatrix}.$$

(c) The dominant eigenpair is λ_+, $\mathbf{v}_+$. The growth rate is λ_+. The long-term solution is

$$\mathbf{x}_t \sim c_1(\lambda_+)^t \mathbf{v}_+, \quad t \to \infty.$$

If $\lambda_+ < 1$ the solution decays to zero, and if $\lambda_+ > 1$, the solution grows without bound. The Jury conditions (Exercise 6) give decay to zero (stability of zero) when

$$s < 1 - gf < 2.$$

9. (a) The eigenvalues satisfy the characteristic equation $\lambda^2 - (2 - \frac{4}{3}a)\lambda + (1 - \frac{4}{3}a) = 0$. Clearly, $\lambda = 1$ is an eigenvalue. The other is $\lambda = 1 - \frac{4a}{3}$. (Use either factorization or the quadratic formula.)

(b) The dominant eigenvalue is $\lambda = 1$. The right eigenvector is $(1, 3)^\mathrm{T}$. A left eigenvector $(1, w)^\mathrm{T}$ satisfies $(1, w)^\mathrm{T} A = (1, w)^\mathrm{T}$. Writing out the two equations give $w = 3$.

(c) The eigenvector corresponding to $\lambda = 1 - 4a/3$ is $(1, -1)^\mathrm{T}$. Therefore the general solution has the form

$$x_t = c_1(1, 3)^\mathrm{T} + c_2(1, -1)^\mathrm{T}\left(1 - \frac{4a}{3}\right)^t.$$

11. The characteristic equation is $-\lambda^3 + a^3 = 0$, which gives eigenvalues $\lambda = a, ae^{4\pi i/6}, ae^{8\pi i/6}$. All the eigenvalues have modulus a, so there is not a unique dominant eigenvalue.

13. (a) The Leslie matrix is

$$L = \begin{pmatrix} 0 & 3.2 & 1.7 \\ 0.2 & 0 & 0 \\ 0 & 0.7 & 0 \end{pmatrix}.$$

(b) The age distribution after 3 years is

$$
L^3 \mathbf{x}_0 = \begin{pmatrix} 0 & 3.2 & 1.7 \\ 0.2 & 0 & 0 \\ 0 & 0.7 & 0 \end{pmatrix}^3 \begin{pmatrix} 2000 \\ 800 \\ 200 \end{pmatrix} = \begin{pmatrix} 2332 \\ 446 \\ 406 \end{pmatrix}.
$$

(c) The dominant eigenpair is $\lambda = 0.9455$, $\mathbf{v} = (0.9670, 0.2048, 0.1518)^{\mathrm{T}}$. So, the long-term behavior is

$$
\mathbf{x}_t \sim c_1 (0.9455)^t \begin{pmatrix} 0.9670 \\ 0.2048 \\ 0.1518 \end{pmatrix},
$$

where c_1 is a constant. (The eigenvector is not normalized.) The population is dying out.

15 If the equation has a single equilibrium $\mathbf{x}$, then $\mathbf{x} = A\mathbf{x}$ or $(A - I)\mathbf{x} = \mathbf{0}$. If this has a unique solution, $\det(A - I) \neq 0$. Therefore, 1 is not an eigenvalue of A.

17. (a) The eigenvalues are (rounded) $0, 0.031, 0.68$, with eigenvectors $(0, -1, 1)^{\mathrm{T}}$, $(0.07, 0.69, -0.73)^{\mathrm{T}}$, $(0.28, 0.13, 0.95)^{\mathrm{T}}$. The growth rate is 0.68 (-32%) and the population goes to zero as t gets large.

(b) The sensitivity matrix S shows that the yearly survivorship of adults (0.78) is the maximum element. We have

$$
S = \begin{pmatrix} 0.12 & 0.05 & 0.41 \\ 0.23 & 0.10 & 0.78 \\ 0.23 & 0.10 & 0.78 \end{pmatrix}.
$$

19. (a) The Leslie diagram. Graduation rates give arrows that go from J to Y and from Y to A; fecundity rates give arrows from Y to J and from A to J.

(b) The Leslie matrix is

$$
L = \begin{pmatrix} 0 & 0.8 & 2.2 \\ 0.5 & 0 & 0 \\ 0 & 0.6 & 0 \end{pmatrix}.
$$

(c) Develop a MATLAB m-file similar to that in Section 3.1 for three dimensions.

(d) The dominant eigenvalue is $\lambda = 1.02$ with eigenvector $(56, 28, 16)$, which gives the long-term percentage of each stage. The other two eigenvalues are $-0.51 \pm 0.62i$, which have modulus $\sqrt{0.55^2 + 0.62^2} = -0.802$.

(e) Hint: Consider a function $y = b\lambda^t$. Taking the logarithm, we get $\ln y = \ln b + (\ln \lambda)t$. Therefore, if y is an geometric function of t, $\ln t$ is a linear function of t.

(f) The sensitivity and elasticity matrices are

$$
S = \begin{pmatrix} 0.38 & 0.19 & 0.11 \\ 0.78 & 0.38 & 0.22 \\ 0.88 & 0.40 & 0.26 \end{pmatrix}, \quad E = \begin{pmatrix} 0 & 0.15 & 0.24 \\ 0.38 & 0 & 0 \\ 0 & 0.24 & 0 \end{pmatrix}.
$$

(g) The maximum element (corresponding to a nonzero entry in L) is 0.78, which is the survivorship from juveniles to youth.

(h) The elasticity of the dominant eigenvalue to adult fecundity is 0.25. This is the relative change of the dominant eigenvalue divided by the relative change in the adult fecundity. In other words, if the fecundity increases N%, the dominant eigenvalue will increase $0.25 \times$ N%.

(i) The elasticity of the dominant eigenvalue to the survivorship of juveniles to youth is 0.38 (from the elasticity matrix). Therefore, if survivorship increases 10%, the dominant eigenvalue (growth rate) increases $0.38 \times 10\%$, or 3.8%. So the new growth rate would be, approximately, $1.02 + 0.038 \times 1.02 = 1.059$. For comparison, if the survivorship goes from 0.5 to 0.55 (a 10% increase), the dominant eigenvalue is actually 1.061 (from MATLAB).

(j) We need to decrease the adult fecundity so that the growth rate is less than 1.0. So we must decrease the current growth rate by 2%, or -0.02. Therefore, from part (h), and the definition of elasticity,

$$
\frac{\Delta\lambda}{\lambda} = 0.25\frac{\Delta f}{f},
$$

or

$$
\frac{-0.02}{1.02} = 0.25\frac{\Delta f}{2.2},
$$

which gives the change in fecundity as $\Delta f = -0.173$. Thus, the fecundity must decrease, approximately, below 2.02.

Section 3.3

1. The equilibrium is at $P = Q = 0$. The nullclines are: $P = 0$, where $\Delta Q = 0$, and $Q = P/2$, where $\Delta P = 0$. In the domain $0 < P < 2Q$ we have $\Delta P < 0$ and $\Delta Q < 0$, so the direction of the orbits is down and to the left. In the domain $P > 2Q$ we have $\Delta P > 0$ and $\Delta Q < 0$, so the orbits go down and to the right. Along $P = 2Q$, the orbits go vertically downward. This

indicates (draw a PQ phase plane) that $(0,0)$ is unstable. The projection matrix is

$$\begin{pmatrix} 0 & -2 \\ -1 & 1 \end{pmatrix},$$

with eigenvalues $\lambda = 2, 1$, and corresponding eigenvectors $(1, -1)^{\mathrm{T}}$ and $(-2, 1)^{\mathrm{T}}$. Thus, the general solution is

$$\begin{pmatrix} P_t \\ Q_t \end{pmatrix} = c_1 \begin{pmatrix} 1 \\ -1 \end{pmatrix} 2^t + c_2 \begin{pmatrix} -2 \\ 1 \end{pmatrix} (-1)^t,$$

which can easily be plotted in MATLAB (or on a calculator) for different values of c_1 and c_2.

3. The equilibria (x, y) satisfy $x = x(r - y)$ and $y = y(x - s)$, which give equilibria $(0,0)$, $(s+1, r-1)$. Note that $r > 1$ for a viable equilibrium. The Jacobian matrix is

$$J(x, y) = \begin{pmatrix} r - y & -x \\ y & x - s \end{pmatrix}.$$

Therefore,

$$J(0,0) = \begin{pmatrix} r & 0 \\ 0 & -s \end{pmatrix},$$

which has eigenvalues $r, -s$. Because $r > 1$ the equilibrium $(0,0)$ is unstable. Next,

$$J(s+1, r-1) = \begin{pmatrix} 1 & -s-1 \\ r-1 & 1 \end{pmatrix}.$$

The characteristic equation is $\lambda^2 - 2\lambda + (1 + (r-1)(s+1)) = 0$. The Jury condition is

$$2 < 1 + \det J < 2,$$

which is never satisfied ($\det J > 0$), and therefore $(s+1, r-1)$ is not stable.

5. (a) The equilibria (x, y) satisfy the conditions

$$x = amxe^{-\beta x} + \frac{3}{2}mby, \quad y = axe^{-\beta x}.$$

Using the second equation to eliminate y from the first, we get

$$x = \frac{\ln R}{\beta}, \quad y = \frac{a \ln R}{\beta R}, \quad R = am\left(1 - \frac{3b}{2}\right).$$

The origin $(0,0)$ is also an equilibrium. Note that the condition on R is $R > 1$ for a viable (positive) equilibrium.

(b) The Jacobian matrix is

$$J(x, y) = \begin{pmatrix} am(1 - \beta x)e^{-\beta x} & \frac{3}{2}mb \\ a(1 - \beta x)e^{-\beta x} & 0 \end{pmatrix}.$$

Therefore,

$$J\left(\frac{\ln R}{\beta}, \frac{a \ln R}{\beta R}\right) = \begin{pmatrix} am(1 - \ln R)\frac{1}{R} & \frac{3}{2}mb \\ a(1 - \ln R)\frac{1}{R} & 0 \end{pmatrix} = \begin{pmatrix} \frac{1 - \ln R}{1 - 3b/2} & \frac{3}{2}mb \\ a\frac{1 - \ln R}{m(1 - 3b/2)} & 0 \end{pmatrix}.$$

The Jury conditions require, for asymptotic stability,

$$\left|\frac{1 - \ln R}{1 - 3b/2}\right| < 1 - \frac{3}{2}ba\frac{1 - \ln R}{(1 - 3b/2)} < 2.$$

Section 3.4

1. We have

$$A^{-1} = \begin{pmatrix} -2 & \frac{3}{2} \\ 1 & -\frac{1}{2} \end{pmatrix}.$$

So

$$\mathbf{x} = A^{-1}\begin{pmatrix} 2 \\ 1 \end{pmatrix} = \begin{pmatrix} -2 & \frac{3}{2} \\ 1 & -\frac{1}{2} \end{pmatrix}\begin{pmatrix} 2 \\ 1 \end{pmatrix} = \begin{pmatrix} -\frac{5}{2} \\ \frac{3}{2} \end{pmatrix}.$$

The augmented matrix is

$$\begin{pmatrix} 1 & 3 & 2 \\ 2 & 4 & 1 \end{pmatrix}.$$

Replacing the second row by the first row minus one-half the second row gives

$$\begin{pmatrix} 1 & 3 & 2 \\ 0 & 1 & \frac{3}{2} \end{pmatrix}.$$

The second row translates to $y = 3/2$, and the first row gives $x = 2 - 3(3/2) = -5/2$.

3. $\det(A - \lambda I) = 0$ is

$$\det\begin{pmatrix} 1 - \lambda & 3 \\ 2 & 4 - \lambda \end{pmatrix} = 0,$$

or $\lambda^2 - 5\lambda - 2 = 0$. Therefore,

$$\lambda = \frac{1}{2}\left(5 \pm \sqrt{33}\right).$$

5. The augmented matrix is

$$\begin{pmatrix} 2 & 3 & m \\ -6 & -9 & 5 \end{pmatrix}.$$

Replacing the second row by the first plus $1/3$ of the second row gives

$$\begin{pmatrix} 2 & 3 & m \\ 0 & 0 & m + \frac{5}{3} \end{pmatrix}.$$

Therefore, if $m \neq -\frac{5}{3}$, there is no solution. If $m = -\frac{5}{3}$, there are infinitely many solutions (x, y) that satisfy $2x + 3y = -\frac{5}{3}$.

7. Hint: think in terms writing the system in reduced row eschelon form.

9. The reduced row echelon form of the augmented matrix is

$$\begin{pmatrix} 1 & 0 & 3 & 1 \\ 0 & 1 & 10 & 2 \\ 0 & 0 & 1 & \frac{4}{9} \end{pmatrix},$$

giving $z = 4/9$, $y = -22/9$, $x = 1/3$.

11. Take a linear combination $c_1(2, -3)^{\mathrm{T}} + c_2(-4, 8)^{\mathrm{T}} = (0, 0)^{\mathrm{T}}$. This is the homogeneous system

$$\begin{pmatrix} 2 & -3 \\ -4 & -8 \end{pmatrix} \begin{pmatrix} c_1 \\ c_2 \end{pmatrix} = \begin{pmatrix} 0 \\ 0 \end{pmatrix}.$$

The coefficient matrix has nonzero determinant, so the system has only the trivial solution $c_1 = c_2 = 0$.

13. Taking a linear combination of the three vectors and setting it to the zero vector gives the linear system

$$\begin{pmatrix} 1 & 5 & -7 \\ 0 & -1 & 1 \\ 1 & 0 & -2 \end{pmatrix} \begin{pmatrix} c_1 \\ c_2 \\ c_3 \end{pmatrix} = \begin{pmatrix} 0 \\ 0 \\ 0 \end{pmatrix}.$$

The coefficient matrix has zero determinant, and therefore there will be nontrivial solutions. For example, take $c_1 = 2$, $c_2 = c_3 = 1$.

15. For matrix A, we seek the matrix V that diagonalizes A. That is, find V so that A factors into $A = VDV^{-1}$, where D is diagonal matrix. The matrix V is formed by the columns of eigenvectors, and D has the eigenvalues on the diagonal. Here,

$$V = \begin{pmatrix} 2 & 1 \\ 1 & 1 \end{pmatrix}, \quad D = \begin{pmatrix} 1 & 0 \\ 0 & 3 \end{pmatrix}.$$

One can check that $A = VDV^{-1}$.

$$V^{-1} = \begin{pmatrix} 1 & -1 \\ -1 & 2 \end{pmatrix}$$

Chapter 4

Section 4.2

1. Dividing the two equations we get the ratio $\frac{dy}{dx} = \frac{-2x}{3y^2}$; then separate variables to get $3y^2\,dy = -2x\,dx$. Integrating both sides gives

$$y^3 = -x^2 + C$$

or

$$y = \left(C - x^2\right)^{1/3}.$$

In the fourth quadrant, where $x > 0$ and $y < 0$, we have $x' > 0$ and $y' < 0$, and so the direction field points away from the origin, and thus the origin must be unstable.

3. Write the equations as

$$N' = N(r - cP - \rho E), \quad P' = P(bN - m - \sigma E).$$

Here, prime is a τ derivative. There are two equilibria, $(0,0)$ and (N^*, P^*), where

$$N^* = \frac{m + \sigma E}{b}, \quad P^* = \frac{r - \rho E}{c}.$$

To get a positive equilibrium, assume that $r > \rho E$. Now scale N and P by their equilibria, and scale time by $1/r$. That is,

$$x = \frac{N}{N^*}, \quad y = \frac{P}{P^*}, \quad t = r\tau.$$

Then the equations become

$$x' = (1 - a)x(1 - y), \quad y = -qy(1 - x),$$

where prime now denotes t, and

$$a = \frac{\rho E}{r}, \quad q = \frac{m + \sigma E}{r}$$

are dimensionless constants. These equations, the fishing model, have the same form as the nonfishing Lotka–Volterra model, and their behavior is therefore the same. The equilibria are now $(0,0)$ and $(1,1)$, in scaled x ,

y coordinates, in both models. In the nonfishing model $\rho = \sigma = 0$. The solution curves are periodic cycles around the equilibrium.

On a given cycle, define the average prey and predator densities by

$$\bar{x} = \frac{1}{T} \int_0^T x(t)\, dt, \quad \bar{y} = \frac{1}{T} \int_0^T y(t)\, dt,$$

where T is the period. We calculate this. Dividing the first differential equation by xT gives

$$\frac{x'}{Tx} = \frac{1-a}{T}(1-y),$$

and integrating from 0 to T gives

$$\frac{1}{T}(\ln x(t) - \ln x(0)) = \frac{1-a}{T} \int_0^T (1-y)\, dt = (1-a)(1-\bar{y}).$$

The left side is zero by periodicity, so $\bar{y} = 1$, the same as the equilibrium, as might have been expected. Similarly, $\bar{x} = 1$. Returning to dimensioned coordinates, we have

$$\bar{P} = P^* = \frac{r - \rho E}{c}.$$

But in the case of nonfishing, $\bar{P} = r/c$ (because $\rho = 0$). Therefore, when fishing is present, the average predator population increases. This is Volterra's principle.

5. When $R = 0$ (no refuge), the equilibria are $(0,0)$ and $(m/b, r/c)$. When there is refuge, the equilibria are $(0,0)$ and

$$\left(\frac{m}{b} + R, \frac{r}{c} + \frac{rb}{cm} R \right).$$

So both the predator and the prey have increased equilibria.

Section 4.3

1. (a) The eigenvalues are -2 and 4 with eigenvectors $(-1,-1)^{\mathrm{T}}$ and $(-1,1)^{\mathrm{T}}$. The origin is an unstable saddle point with separatrices in the directions of the eigenvectors.

(b) The eigenvalues are -1 and 1 with eigenvectors $(1,0)^{\mathrm{T}}$ and $(1,2)^{\mathrm{T}}$. The origin is an unstable saddle point with separatrices in the directions of the eigenvectors.

(c) The eigenvalues are $3i$ and $-3i$ with eigenvectors $(-i,3)^{\mathrm{T}}$ and $(i,3)^{\mathrm{T}}$. The origin is a stable center, but not asymptotically stable.

(d) The eigenvalues are 2 and -3 with eigenvectors $(1, 1)^{\mathrm{T}}$ and $(-1, 4)^{\mathrm{T}}$. The origin is an unstable saddle point with separatrices in the directions of the eigenvectors.

(e) The eigenvalues are 1 and 1 (1 is an eigenvalue of multiplicity 2) with a single eigenvector $(2, 1)^{\mathrm{T}}$. The origin is an unstable node.

(f) The eigenvalues are $-2 + 3i$ and $-2 - 3i$ with eigenvectors $(1, -i)^{\mathrm{T}}$ and $(1, i)^{\mathrm{T}}$. The origin is an asymptotically stable spiral.

3. The characteristic equation is

$$\det \begin{pmatrix} a - \lambda & b \\ c & d - \lambda \end{pmatrix} = \lambda^2 - (a + d)\lambda + ad - bc$$
$$= \lambda^2 - \mathrm{tr}(A)\lambda + \det A = 0.$$

The eigenvalues are

$$\lambda = \frac{1}{2}\left(\mathrm{tr}(A) \pm \sqrt{\mathrm{tr}(A)^2 - 4\det A}\right).$$

Assume that the trace is negative and the determinant is positive. If $\mathrm{tr}(A)^2 > 4\det A$, the roots are real and both negative (asymptotically stable node); if $\mathrm{tr}(A)^2 < 4\det A$, both roots are complex with negative real part (asymptotically stable spiral).

5. The eigenvalue are solutions of

$$\lambda^2 + 2a\lambda + (a^2 + 1) = 0$$

or

$$\lambda = \frac{1}{2}\left(-2a \pm \sqrt{4a^2 - 4(a^2 + 1)}\right) = -a \pm i.$$

If $a > 0$, then we have a stable spiral; if $a = 0$ we have a center, and if $a < 0$, we have an unstable spiral.

Section 4.5

1. (a) The x nullcline is the circle $x^2 + y^2 = 4$, and the y nullcline is the straight line $y = 2x$. These intersect at the critical points

$$\left(\frac{4}{\sqrt{5}}, \frac{8}{\sqrt{5}}\right), \quad \left(\frac{-4}{\sqrt{5}}, \frac{-8}{\sqrt{5}}\right).$$

The Jacobian is

$$J(x, y) = \begin{pmatrix} 2x & 2y \\ -2 & 1 \end{pmatrix}.$$

At the first critical point, in the first quadrant, the eigenvalues are complex, $\lambda = 1.39 \pm i2.65$, which gives an unstable (clockwise) spiral. The second critical point, in the fourth quadrant, is a saddle point with eigenpairs -3.41, $(0.91, 0.41)^{\mathrm{T}}$ and -2.62, $(0.63, 0.78)^{\mathrm{T}}$. The eigenvectors give the directions of the stable and unstable manifolds (separatrices) at the saddle point. This information is enough to plot the phase diagram.

(b) The only critical point is the origin $(0,0)$, where the Jacobian is the zero matrix with $\det J = 0$. Therefore, none of the linearity results apply. However, the x and y nullclines are the x axis and y axis, respectively. In the first quadrant the direction field is $(+, -)$, in the second quadrant, $(+, +)$, in the third quadrant, $(+, +)$, and in the fourth quadrant, $(+, -)$. The origin is clearly unstable. In this problem we can find the orbits by dividing the equations and separating variables to get $3y^2\, dy = -2x\, dx$. Integrating gives the orbits $y^3 = -x^2 + C$.

(c) The critical points are $(0,0)$ and $(2,4)$. The Jacobian is

$$J(x, y) = \begin{pmatrix} 8 & -2y \\ 2x & -1 \end{pmatrix}.$$

The eigenvalues of $J(0,0)$ are $8, -1$, and so the origin is a saddle point. The eigenvectors are $(1,0)^{\mathrm{T}}$ and $(0,1)^{\mathrm{T}}$, which define the directions of the unstable and stable manifolds (separatrices) at the origin, respectively; the eigenvalues of $J(2,4)$ are $7/2 \pm i\sqrt{47}/2$, which have a positive real part; so $(2,4)$ is an unstable spiral.

3. (a) The Bendixson–Dulac criterion holds. Note that

$$\frac{\partial f}{\partial x} + \frac{\partial g}{\partial y} = \frac{\partial}{\partial x}(y) + \frac{\partial}{\partial y}\left((x^2+1)y - x^5\right) = x^2 + 1 > 0$$

for all x, y in the plane. Therefore, there can be no periodic solutions.

(b) Note that

$$\frac{\partial f}{\partial x} + \frac{\partial g}{\partial y} = 1 + x^2 + 5y^2 > 0,$$

so by Bendixson–Dulac there are no periodic solutions.

(c) Notice that there are no critical points and $y' > 0$ everywhere in the plane. Therefore, no periodic orbits can exist.

(d) By the Bendixson–Dulac criterion,

$$\frac{\partial f}{\partial x} + \frac{\partial g}{\partial y} = -1 - 5y^4 < 0,$$

and therefore no periodic orbits can exist.

(e) There are no critical points, so there can be no periodic orbits.

5. Dividing the two equations and separating variables gives

$$ydy = -f(x)\,dx.$$

Integrating gives the orbits

$$\frac{1}{2}y^2 = -\int_0^x f(u)\,du + C$$

or

$$y = \pm\sqrt{2}\sqrt{C - V(x)},$$

where

$$V(x) = \int_0^x f(u)\,du.$$

By the given properties, $V(x) > 0$, $x \neq 0$, and $V(0) = 0$. Moreover, because $V'(x) = f(x)$, we have $V'(x) > 0$ if $x > 0$, and $V'(x) < 0$ if $x < 0$. Finally, $V''(x) > 0$, so V is concave up. Note that $C \geq 0$. [Sketching a generic plot of $V(x)$, we see, for example, a concave-up, parabolic-shaped graph.] Therefore, for any $C > 0$, there are exactly two roots, $a < 0$ and $b > 0$, to $V(x) = C$. We have $C - V(x) > 0$ in (a, b), and therefore $\sqrt{C - V(x)} > 0$ in (a, b), and undefined for $x < a$ and $x > b$. So the oribit lies within the interval $[a, b]$ and is symmetric above and below the x axis (because of the $\pm$ sign). The direction of the orbit is clockwise because $x' = y$. Note that there is a vertical tangent at the $x = a$ and $x = b$. Therefore, any orbit is periodic. In this problem, a picture is worth a thousand words!

7. The Jacobian is

$$J(x,y) = \begin{pmatrix} 0 & 1 \\ -p_x(x,y)y - xh'(x) - h(x) & -p(x,y) - yp_y(x,y) \end{pmatrix},$$

so

$$J(x,y) = \begin{pmatrix} 0 & 1 \\ -h(0) & -p(0,0) \end{pmatrix}.$$

The trace $-p(0,0)$ is negative and the determinant $h(0) > 0$, and therefore the origin is asymptotically stable.

9. (a) The equations are

$$P' = \phi - aP - bHP, \quad H' = \varepsilon bHP - cH.$$

Here aP is the per capita growth rate of the plant biomass, and bHP is the rate at which herbivores consume the plants (a type I response); cH is the per capita mortality rate of herbivore biomass, and εbHP is the rate that plant biomass consumption is coverted into herbivore biomass. Clearly, ϕ is plant biomass per time time, a and c are rates and have the units time^{-1},

b is (time · herbivore biomass)$^{-1}$, and ε is the yield, measured in herbivore biomass per unit plant biomass.

(b) The critical points are

$$\left(\frac{\phi}{a}, 0\right), \quad (P^*, H^*) = \left(\frac{c}{b\varepsilon}, \frac{\varepsilon\phi - ac/b}{c}\right).$$

The P nullcline is

$$H = \frac{\phi - aP}{bP},$$

and the H nullclines are $H = 0$ and $P = c/b\varepsilon$. There are two cases here, $\phi/a > c/b\varepsilon$, and $\phi/a < c/b\varepsilon$. In the latter case the second equilibrium is not in the first quadrant and thus is unphysical. So we examine the first case. The Jacobian is

$$J(P, H) = \begin{pmatrix} -a - bH & -bP \\ \varepsilon bH & \varepsilon bP - c \end{pmatrix}.$$

Then

$$J\left(\frac{\phi}{a}, 0\right) = \begin{pmatrix} -a & -b\frac{\phi}{a} \\ 0 & \varepsilon b\frac{\phi}{a} - c \end{pmatrix}.$$

By assumption, $\varepsilon b\phi/a - c > 0$, and hence $(\phi/a, 0)$ is a saddle point. Next,

$$J(P^*, H^*) = \begin{pmatrix} -a - bH^* & -b\frac{c}{b\varepsilon} \\ \varepsilon bH^* & \varepsilon b\frac{c}{b\varepsilon} - c \end{pmatrix} = \begin{pmatrix} -a - bH^* & -\frac{c}{\varepsilon} \\ \varepsilon bH^* & 0 \end{pmatrix}.$$

The trace is negative and the determinant is positive, and therefore (P^*, H^*) is an asymptotically stable coexistent state.

For $\phi/a < c/b\varepsilon$ the vertical nullcline lies to the right of ϕ/a and $(\phi/a, 0)$ is an asymptotically stable node. All orbits tend to this equilibrium and the herbivores die out. This occurs when the plant production ϕ is small.

Therefore, a bifurcation occurs when $\phi = ac/b\varepsilon$. For ϕ less than this critical value there is a single equilibrium (a stable node); at this critical value the node branches into an unstable saddle point and there appears a coexistent stable node or spiral.

11. The x nullcline is $y = x$ and the y nullcline is the form of a type II function response, $y = 5x^2/(4 + x^2)$. There are three intersections, and hence two critical points, $(0, 0)$, $(1, 1)$, and $(4, 4)$. The Jacobian is

$$J(x, y) = \begin{pmatrix} -1 & 1 \\ \frac{40x}{(4+x^2)^2} & -1 \end{pmatrix}.$$

Hence,

$$J(0,0) = \begin{pmatrix} -1 & 1 \\ 0 & -1 \end{pmatrix}, \quad J(1,1) = \begin{pmatrix} -1 & 1 \\ \frac{8}{5} & -1 \end{pmatrix}, \quad J(4,4) = \begin{pmatrix} -1 & 1 \\ \frac{2}{5} & -1 \end{pmatrix}.$$

The origin is therefore a stable node (eigenvalues $-1, -1$), and $(1,1)$ is a saddle (eigenvalues $-1 \pm \sqrt{8/5}$), and $(4,4)$ is an asymptotically stable node (eigenvalues $-1 \pm \sqrt{2/5}$). Putting in direction arrows along the nullclines and in the regions in between gives the phase plane diagram.

13. (a) F is the rate at which fruit falls on the ground, and aRP is the rate at which fruit, the resource, is eaten by predators (mass action). The per capita mortality rate of predator biomass is d, and $kaRP$ represents the rate at which fruit biomass is converted to predator biomass; k is the yield, or conversion factor. We scale R and P by their equilibria, and time τ by by d^{-1}, or

$$R = \frac{x}{d/ka}, \quad P = \frac{y}{kF/d}, \quad t = d\tau.$$

Then the model becomes

$$x' = c(1 - xy), \quad y' = y(x - 1), \quad c = \frac{akF}{d^2}.$$

(b) Now the critical point is $(1,1)$. The x nullcline is the hyperbola $xy = 1$, and the y nullclines are $x = 1$ and $y = 0$. The Jacobian is

$$J(x,y) = \begin{pmatrix} -cy & -cx \\ y & x - 1 \end{pmatrix}.$$

Then

$$J(1,1) = \begin{pmatrix} -c & -c \\ 1 & 0 \end{pmatrix},$$

with negative trace and positive determinant. So $(1,1)$ is asymptotically stable. The eigenvalues are $\lambda = \frac{1}{2}(-c \pm \sqrt{c^2 - 4c})$; if $c > 2$, then $(1,1)$ is a node, and if $c < 2$, then $(1,1)$ is a spiral.

15. We write the first equation as

$$x' = ax\left(1 - \frac{x}{a/b}\right) - cxy,$$

then the prey satisfy the logistic law of growth and are consumed by the predators with a type I response (mass action kinetics). The second equation can be written

$$y' = \delta y\left(1 - \frac{y}{\delta x/f}\right),$$

which shows that the predators satisfy a logistic growth law with a carrying capacity $\delta x/f$, which depends on the prey population. In the first equation there is a rate of consumption but there is no direct growth rate in the predator equation resulting from consumption; the effect of prey appears indirectly in the carrying capacity.

17. Susceptibles become infected at the rate bSI, and infectives get removed at rate γI. After a period of time in the removed class, individuals become susceptible again at the rate aR. The equations are

$$S' = -bSI + aR, \quad I' = bSI - \gamma I, \quad R' = \gamma I - aR.$$

We can reduce the analysis to two equations by using $R = N - I - S$, where N is the total population. Then

$$S' = -bSI + a(N - I - S), \quad I' = bSI - \gamma I.$$

The critical points are

$$(N, 0), \quad \left(\frac{\gamma}{b}, \frac{\gamma + a}{a(N - \gamma/b)} \right).$$

So, there are two cases, $N > \gamma/b$ and $N < \gamma/b$. In the first case there are two equilibria, and in the second case there is only a single equilibrium, $(N, 0)$. The basic reproductive number is $R_0 = \gamma N/b$, so these cases are $R_0 > 1$ and $R_0 < 1$. The I nullclines are $S = \gamma/b$ and $I = 0$. The horizontal nullcline is

$$I = \frac{a(N - S)}{bS + a}.$$

In the first case there is a positive equilibrium in the first quadrant, and in the second case the equilibrium lies in the fourth quadrant, which is nonphysical. The Jacobian is

$$J(S, I) = \begin{pmatrix} -bI - a & -bS - a \\ bI & bS - \gamma \end{pmatrix}.$$

For $R_0 > 1$ we have

$$J(N, 0) = \begin{pmatrix} -a & -bN - a \\ 0 & bN - \gamma \end{pmatrix},$$

and the eigenvalues are $-a$ and $bN - \gamma > 0$, so $(N, 0)$ is a saddle point. For the positive equilibrium

$$(S^*, I^*) = \left(\frac{\gamma}{b}, \frac{\gamma + a}{a(N - \gamma/b)} \right),$$

we have

$$J(S^*, I^*) = \begin{pmatrix} -bI^* - a & -\gamma - a \\ bI^* & 0 \end{pmatrix}.$$

The trace is negative and the determinant is positive, so (S^*, I^*) is an asymptotically stable equilibrium (node or spiral). Thus, for $R_0 > 1$ there is an endemic state. A phase plane diagram shows that any initial state tends to this endemic state. For $R_0 < 1$, the eigenvalues of $J(N, 0)$ are $-a$ and $bN - \gamma < 0$, and the equilibrium $(N, 0)$ is an asymptotically stable node. The vertical nullcline $S = \gamma/b$ lies to the right of N. All orbits tend to $(N, 0)$, and the infection dies out.

19. The positive, coexistent equilibrium is

$$u^* = \frac{a+1}{1-ab}, \quad v^* = \frac{b+1}{1-ab}.$$

Therefore, $ab < 1$ (so the nullclines cross). The Jacobian is

$$J(u, v) = \begin{pmatrix} 1 - 2u + av & au \\ bcv & c + cbu - 2cv \end{pmatrix}.$$

We have, after simplification,

$$J(u^*, v^*) = \begin{pmatrix} -u^* & au^* \\ bcv^* & -cv^* \end{pmatrix}.$$

Clearly, the trace is negative and the determinant is positive; thus (u^*, v^*) is asymptotically stable. The eigenvalues at $(0, 0)$ are 1 and c, so $(0, 1)$ is an unstable node. The eigenvalues at $(0, 0)$ are $1 + a$ and c, so $(1, 0)$ is a saddle. Finally, $(1, 0)$ is a saddle, with eigenvalues -1 and $c(b + 1)$. When there is no positive coexistent state, the nonzero nullclines do not cross and the orbits all tend to infinity $(+\infty, +\infty)$ in the first quadrant.

Chapter 5

Section 5.1

1. Because A and A' are disjoint sets, by the additivity condition $\Pr(S) = \Pr(A \cup A') = \Pr(A) + \Pr(A') = 1$.

3. $\Pr(A \cap B) = \Pr(A|B) \Pr(B) = \Pr(A) \Pr(B)$, where the last equality is by independence.

5. Proof of Bayes' theorem: For a partition B_i of A,

$$\Pr(B_i|A) = \frac{\Pr(B_i \cap A)}{\Pr A} = \frac{\Pr(A \cap B_i)}{\Pr A} = \frac{\Pr(A|B_i)PB_i}{\Pr A}$$
$$= \frac{\Pr(A|B_i)PB_i}{\sum_i \Pr(A|B_i)\Pr B_i},$$

where in the last line we used the law of total probability.

7. Let A be the event that bird A lives at least six months, and let B be the event that bird B lives at least six months. Then we are give $\Pr(A) = 0.5$ and $\Pr(B) = 0.7$. Notice that the events are independent, but not disjoint.

Section 5.2

1. We are given the frequencies x, y, z of the genotypes aa, ab, and bb, respectively:

$$x = \frac{1998}{2000}, \quad y = \frac{2}{2000}, \quad z = 0.$$

Therefore, in this initial generation, the frequencies p and q of a and b are

$$p = x + \frac{1}{2}y = \frac{1999}{2000}, \quad q = z + \frac{1}{2}y = \frac{1}{2000},$$

Now, in the next generation, we have (using independence)

$$x = \Pr(aa) = \Pr(\text{parent 1 gives } a \text{ and parent 2 gives } a)$$
$$= \Pr(\text{parent 1 gives } a) \cdot \Pr(\text{parent 2 gives } a) = \frac{1999}{2000} \cdot \frac{1999}{2000}.$$

Also,

$$y = \Pr(ab) = 2\Pr(\text{one parent gives } a \text{ and a second parent gives a } b)$$
$$= 2\Pr(\text{one parent gives } a) \cdot \Pr(\text{a second parent gives } b) = 2\frac{1999}{2000} \cdot \frac{1}{2000}.$$

Finally,

$$z = \Pr(bb) = \Pr(\text{parent 1 gives a } b \text{ and parent 2 gives a } b)$$
$$= \Pr(\text{parent 1 gives a } b) \cdot \Pr(\text{parent 2 gives a } b) = \frac{1}{2000} \cdot \frac{1}{2000}.$$

Section 5.3

1. (a) We have X = length of gestation, and $X \sim N(266, 16^2)$. Therefore,

$$\Pr(X \leq 275) = \Pr(266 + 16Z \leq 275)$$
$$= \Pr\left(Z \leq \frac{9}{16}\right) \approx 0.70.$$

Software or a table can be used to calculate the probability.

(b) We have

$$\Pr(275 < X < 290) = \Pr(275 < 266 + 16Z < 290)$$
$$= \Pr\left(\frac{9}{16} < Z < \frac{24}{16}\right)$$
$$= \Pr\left(Z < \frac{24}{16}\right) - \Pr\left(Z < \frac{9}{16}\right)$$
$$= 0.231.$$

(c) Let g be the number of gestation days to be in the top 10%, or $\Pr(X \geq g) = 0.1$. Then

$$0.1 = \Pr(X \geq g) = \Pr(266 + 16Z \geq g)$$
$$= \Pr\left(Z \geq \frac{g - 266}{16}\right) = 1 - \Pr\left(Z < \frac{g - 266}{16}\right).$$

Therefore,

$$\Pr\left(Z < \frac{g - 266}{16}\right) = 0.9,$$

and from a table,

$$\frac{g - 266}{16} = 1.29, \quad \text{or} \quad g \approx 287 \text{ days.}$$

3. (a) We must have

$$1 = \int_0^2 cx^2 \, dx = c\frac{x^3}{3}\Big|_0^2 = \frac{8}{3}c \implies c = \frac{3}{8}.$$

We have

$$EX = \int_0^2 x\frac{3}{8}x^2 \, dx = \frac{3}{8}\frac{x^4}{4}\Big|_0^2 = \frac{3}{2},$$
$$EX^2 = \int_0^2 x^2\frac{3}{8}x^2 \, dx = \frac{3}{8}\frac{x^5}{5}\Big|_0^2 = \frac{12}{5}.$$

(b) Therefore,

$$\text{var } X = EX^2 - (EX)^2 = \frac{12}{5} - \frac{9}{4} = \frac{3}{20}.$$

(c) The cdf is

$$F(x) = \int_0^x \frac{3}{8}s^2 \, ds = \frac{x^3}{8}, \quad 0 \leq x \leq 2.$$

(d) We also have

$$\Pr(X > 1) = 1 - \Pr(X \leq 1) = 1 - F(1) = 1 - \frac{1}{8} = \frac{7}{8}.$$

5. (a) $M_X(t) = \int_{-\infty}^{\infty} f(x)e^{tx} \, dx = E(e^{tX})$, so

$$M_X'(t) = \int_{-\infty}^{\infty} xf(x)e^{tx} \, dx$$

and

$$M_X'(0) = \int_{-\infty}^{\infty} xf(x) \, dx = E(X)$$

(b) Taking another t-derivative gives us

$$M_X''(t) = \int_{-\infty}^{\infty} x^2 f(x)e^{tx} \, dx,$$

and therefore

$$M_X''(0) = \int_{-\infty}^{\infty} x^2 f(x) \, dx = E(X^2).$$

Hence,

$$\text{var } X = E(X^2) - (EX)^2 = M_X''(0) - M_X'(0)^2.$$

(c) We have

$$M_X(t) = \int_0^{\infty} \lambda e^{-\lambda x} e^{tx} \, dx = \lambda \int_0^{\infty} e^{-(\lambda - t)x} \, dx$$

$$= \frac{\lambda}{-(\lambda - t)} e^{-(\lambda - t)x} \Big|_0^{\infty} = \frac{\lambda}{\lambda - t},$$

provided that $\lambda > t$ (for convergence of the integral).

(d) If $X \sim N(0,1)$ then

$$
\begin{aligned}
M_X(t) &= \frac{1}{\sqrt{2\pi}} \int_{-\infty}^{\infty} e^{-x^2/2+tx} dx \\
&= \frac{1}{\sqrt{2\pi}} \int_{-\infty}^{\infty} e^{-\frac{1}{2}(x^2-2tx+t^2)+t^2/2} dx \\
&= \frac{1}{\sqrt{2\pi}} e^{t^2/2} \int_{-\infty}^{\infty} e^{-\frac{1}{2}(x-t)^2} dx \\
&= \frac{\sqrt{2}}{\sqrt{2\pi}} e^{t^2/2} \int_{-\infty}^{\infty} e^{-z^2} dz = e^{t^2/2}.
\end{aligned}
$$

[In the second line we completed the square in the exponent in the integrand; in the next-to-last line we made the substitution $z = (x-t)/\sqrt{2}$, and for the final equality we used the standard result that the value of the integral is $\sqrt{\pi}$.]

7. We have

$$
\int_0^{\infty} (1 - F(x)) \, dx = \int_0^{\infty} \left(1 - \int_0^x f(y) \, dy \right) dx = \int_0^{\infty} \left(\int_x^{\infty} f(y) \, dy \right) dx.
$$

Now we do integration by parts with $u = \int_x^{\infty} f(y) dy$, and $dv = dx$. Then $du = -f(x)$ and $v = x$. Hence, the last integral is

$$
x \int_x^{\infty} f(y) \, dy \Big|_0^{\infty} + \int_0^{\infty} f(x) x \, dx = EX,
$$

because the first term vanishes.

9. Each equality follows from simple substitutions in the integrals and definitions of the cdf and pdf.

11. We have, by definition,

$$
\Pr(Y \le y) = \Pr(g(X) \le y) = \Pr(X \le g^{-1}(y)) = \int_{-\infty}^{g^{-1}(y)} f(x) \, dx.
$$

Now make the substitution $x = g^{-1}(z)$, $dx = (g^{-1}(z))' \, dz$ to get

$$
\int_{-\infty}^{f^{-1}(y)} f(x) \, dx = \int_{-\infty}^{y} f(g^{-1}(z))(g^{-1}(z))' \, dz.
$$

Therefore,

$$
f_Y(z) = f(g^{-1}(z))(g^{-1}(z))'.
$$

13. We have $\lambda = a + bt$. By Example 5.23, the hazard rate $F(t)$ satisfies

$$a + bt = \frac{F'(t)}{1 - F(t)}.$$

Integrating from 0 to t gives

$$at + bt^2/2 = \int_0^t \frac{F'(s)}{1 - F(s)} ds = -\ln(1 - F(t)).$$

Therefore,

$$F(t) = 1 - e^{-(at+bt^2/2)}.$$

Then

$$f(t) = F'(t) = (a + bt)e^{-(at+bt^2/2)}.$$

15. The expected value of the gamma distribution is

$$\begin{aligned}
\mathrm{E}X &= \frac{1}{\Gamma(a)} \int_0^\infty x\lambda^a e^{-\lambda x} x^{a-1} \, dx \\
&= \frac{1}{\lambda\Gamma(a)} \int_0^\infty \lambda e^{-\lambda x} (\lambda x)^a \, dx \\
&= \frac{\Gamma(a+1)}{\lambda\Gamma(a)} = \frac{a}{\lambda}.
\end{aligned}$$

We leave a similar calculation of $\mathrm{E}X^2$, and hence the variance, $\mathrm{var}\, X = a/\lambda^2$, to the reader.

Section 5.4

1. For a *Bernoulli RV* we have $\Pr(X = 0) = p$, $\Pr(X = 1) = 1 - p$. Therefore, $\mathrm{E}X = 0 \cdot p + 1 \cdot (1 - p) = 1 - p$, and $\mathrm{E}X^2 = 0^2 \cdot p + 1^2 \cdot (1 - p) = 1 - p$. Therefore, $\mathrm{var}\, X = (1 - p) - (1 - p)^2 = p(1 - p)$. For a *geometric RV* we have $\Pr(X = n) = p(1 - p)^{n-1}$. Therefore,

$$\mathrm{E}X = \sum_{n=1}^\infty npq^{n-1} = p\sum_{n=0}^\infty \frac{d}{dq}q^n = p\frac{d}{dq}\sum_{n=0}^\infty q^n = p\frac{d}{dq}\frac{1}{1-q} = \frac{1}{p}.$$

$$\mathrm{E}X^2 = \sum_{n=1}^\infty n^2 pq^{n-1} = p\sum_{n=1}^\infty n\frac{d}{dq}q^n = p\frac{d}{dq}\sum_{n=1}^\infty nq^n = p\frac{d}{dq}\left(\frac{q}{1-q}\mathrm{E}X\right)$$

$$= p\frac{d}{dq}\frac{q}{(1-q)^2} = \frac{2}{p^2} - \frac{1}{p}.$$

Therefore,

$$\mathrm{var}\, X = \frac{2}{p^2} - \frac{1}{p} - \left(\frac{1}{p}\right)^2 = \frac{1 - p}{p^2}.$$

3. The probability that a patient is a woman is 0.55. Assuming a large population of patients, we have a binomial situation. Therefore, letting X be the number of women patients,

$$\Pr(X \leq 2) = \sum_{k=0}^{2} \binom{12}{k} (0.55)^k (0.45)^{12-k}.$$

5. Let X be the number of visits until a success occurs for the first time. Success is mating, with probability $p = 0.25$, and failure is the female leaving, with probability $1 - p = 0.75$. Clearly, X is a geometric random variable. Then

$$\Pr(X = k) = p(1 - p)^{k-1} = (0.25)(0.75)^{k-1}.$$

The expected value and variance of X are

$$EX = 1/p = 4, \quad \text{var } X = \frac{1-p}{p^2} = 12.$$

The probability that mating will occur on visit 1 or visit 2 is

$$\Pr(X \leq 2) = (0.25)(0.75)^0 + (0.25)(0.75)^1 = 0.44.$$

7. This is a multinomial RV. With the notation in the text, the desired probability is

$$1 - \Pr(X_1 = x_1, X_2 = x_2, X_3 < 3)$$

$$= \sum_{\substack{x_1+x_2+x_3=30, \\ x_3<3}} \binom{30}{x_1 \; x_2 \; x_3} (0.6)^{x_1} (0.38)^{x_2} (0.02)^{x_3}.$$

9. This is a Poisson random variable and the mean is $\lambda = \frac{1}{13}t$, where t is in years. Thus,

$$\Pr(X = k) = \frac{(t/13)^k}{k!} e^{-t/13}, \quad k = 0, 1, 2, \ldots$$

(a) In 10 years

$$\Pr(X = 0) = \frac{(10/13)^0}{0!} e^{-10/13} = 0.463.$$

(b) The probability of no floods in four years is

$$\Pr(X = 0) = \frac{(4/13)^0}{0!} e^{-4/13} = 0.735.$$

(c) We have

$$\Pr(X = 1) = \frac{\left(\frac{13}{13}\right)^1}{1!} e^{-13/13} = e^{-1} = 0.368.$$

(d) The three consecutive decades are independent events, so the probability that there is exactly one flood in each decade is

$$\left(\frac{(10/13)^1}{1!} e^{-10/13}\right)^3 = \left(\frac{10}{13}\right)^3 e^{-30/13} = 0.45.$$

11. We have, by definition of Z_n,

$$|Z_n - 2|^2 = \begin{cases} 1, & \text{with probability } 1/n \\ 0, & \text{with probability } 1 - 1/n. \end{cases}$$

Therefore,

$$E|Z_n - 2|^2 = 1 \cdot \frac{1}{n} + 0 \cdot \left(1 - \frac{1}{n}\right) = \frac{1}{n} \to 0.$$

13. Write the likelihood function in the form

$$L(\mu, \sigma) = \left(\frac{1}{\sigma\sqrt{2\pi}}\right)^n \prod_{k=1}^n e^{-(x_k - \mu)^2/2\sigma^2}.$$

Taking the logarithm and using the logarithm rules,

$$\mathcal{L} = \log L(\mu, \sigma) = -n \ln\left(\sigma\sqrt{2\pi}\right) - \frac{1}{2\sigma^2} \sum_{k=1}^n (x_k - \mu)^2.$$

To maximize $\mathcal{L}$ we need to take the two partial derivatives and set them equal to zero. First, taking the partial derivative with respect to μ gives

$$\frac{\partial \mathcal{L}}{\partial \mu} = \frac{1}{\sigma^2} \sum_{k=1}^n (x_k - \mu) = \frac{1}{\sigma^2}\left(\sum_{k=1}^n x_k - n\mu\right).$$

Setting this to zero gives

$$\mu = \frac{1}{n} \sum_{k=1}^n x_k = \overline{x}.$$

Next,

$$\frac{\partial \mathcal{L}}{\partial \sigma} = -\frac{n}{\sigma} + \frac{1}{\sigma^3} \sum_{k=1}^n (x_k - \mu)^2 = 0$$

implies

$$\sigma^2 = \frac{1}{n} \sum_{k=1}^n (x_k - \mu)^2.$$

15. The X_k, $k = 1, 2, ..., 25$, are each Poisson random variables with mean $\mu = \lambda$ and variance $\sigma^2 = \lambda$. We know from the CLT that, approximately, for large n,

$$\frac{S_n - n\mu}{\sqrt{n\sigma^2}} = Z_n,$$

where Z_n is a standard normal random variable. Hence,

$$\overline{X} = \frac{S_n}{n} = \mu + \frac{\sqrt{\sigma^2}}{\sqrt{n}} Z_n = \lambda + \frac{\sqrt{\lambda}}{5} Z_n.$$

Therefore,

$$\Pr(\overline{X} < \lambda) = \Pr(Z_n < 0) = \frac{1}{2}$$

and

$$\Pr\left(\overline{X} < \lambda + \frac{\sqrt{\lambda}}{5}\right) = \Pr\left(\lambda + \frac{\sqrt{\lambda}}{5} Z_n < \lambda + \frac{\sqrt{\lambda}}{5}\right) = \Pr(Z_n < 1) = 0.84.$$

17. The likelihood function is

$$L(p|8) = \binom{10}{8} p^8 (1 - p)^2.$$

Taking the derivative with respect to p (using the product rule) and setting it to zero gives

$$\binom{10}{8} p^7 (1 - p)(8 - 10p) = 0.$$

Therefore $p = 0.8$, as might be expected. To plot the likelihood function note that $\binom{10}{8} = 45$ and use the following MATLAB commands: p=0:0.01:1; L=45*(p.^8).*(1-p).^2; plot(p,L).

Section 5.5

1. Mariginals do not determine the joint density uniquely. To give a general method for constructing an example, consider the joint density p_{ij} $(i, j = 1, 2)$ given in the nonnegative matrix

$$\begin{pmatrix} a & b \\ c & d \end{pmatrix}.$$

We require that $a + b + c + d = 1$. Then the marginals are given by the row and column sums. Take, for example,

$$a + b = \frac{1}{4}, \quad c + d = \frac{3}{4},$$
$$a + c = \frac{1}{2}, \quad b + d = \frac{1}{2}.$$

Therefore, the four unknowns a, b, c, and d satisfy the preceding five linear equations. We can form the augmented and perform row reduction to obtain the solution

$$a = -\frac{1}{4} + d, \quad b = \frac{1}{2} - d, \quad c = \frac{3}{4} - d,$$

where d is arbitrary. But d is constrained by the condition

$$\frac{1}{4} < d < \frac{1}{2},$$

so that all the entries are positive and less than 1. Take, for example, $d = 3/8$ or $d = 5/16$; both choices lead to the same joint density.

3. (a) We have $f(x, y) = 8xy$ for $0 \le x \le 1$, $0 \le y \le x$, and zero otherwise. Hence,

$$f_X(x) = \int_0^x 8xy \, dy = 4x^3, \quad 0 \le x \le 1,$$

and zero otherwise. Also,

$$f_Y(y) = \int_y^1 8xy \, dy = 4y(1 - y^2), \quad 0 \le y \le 1,$$

and zero otherwise.

(b) Note that $f(x, y) \ne f_X(x)f_Y(y)$, so X and Y are not independent.

(c) We have

$$\mathrm{E}X = \int_0^1 \left(\int_0^x x 8xy \, dy \right) dx = \int_0^1 8x^2 \cdot \frac{x^2}{2} \, dx = \frac{4}{5}.$$

Similarly, the integrals for $\mathrm{E}Y$ and $\mathrm{E}(XY)$ are

$$\mathrm{E}Y = \int_0^1 \left(\int_0^x y 8xy \, dy \right) dx, \quad \mathrm{E}(XY) = \int_0^1 \left(\int_0^x xy 8xy \, dy \right) dx.$$

You may also calculate these expectations using the marginal densities, for example,

$$\mathrm{E}X = \int_0^1 x \cdot 4x^3 \, dx = \frac{4}{5}.$$

5. We are given $f(x, y) = e^{-(x+y)}$ for $x, y \ge 0$, and zero otherwise. Let $Z = X/Y$. Then

$$\Pr(Z \le z) = \Pr(X \le zY).$$

This is an integration over the wedge W in the first quadrant of the XY plane bounded by the straight lines (rays) $X = 0$ and $Y = (1/z)X$. (Draw a picture.) Thus,

$$
\begin{aligned}
\Pr(X \le zY) &= \int \int_W e^{-(x+y)} \, dx \, dy \\
&= \int_0^\infty \left(\int_0^{zy} e^{-(x+y)} \, dx \right) dy = \int_0^\infty e^{-y} \left(\int_0^{zy} e^{-x} \, dx \right) dy \\
&= \int_0^\infty e^{-y}(1 - e^{-zy}) \, dy = \int_0^\infty \left(e^{-y} - e^{-(1+z)y} \right) dy \\
&= -e^{-y} + \frac{1}{1+z} e^{-(1+z)y} \Big|_0^\infty = 1 - \frac{1}{1+z} = \frac{z}{1+z}.
\end{aligned}
$$

7. By independence, the joint pdf is the product of the densities, or

$$
f(x_1, x_2) = \frac{1}{a^2}(h(x_1) - h(x_1 - a))(h(x_2) - h(x_2 - a)), \quad 0 \le x_2, x_2 \le a,
$$

and zero otherwise. Because X_1 and X_2 are both uniform on $[0, a]$, and thus each has expected value $a/2$, we have

$$
E(W) = E(X_1 + X_2) = E(X_1) + E(X_2) = \frac{a}{2} + \frac{a}{2} = a.
$$

By independence,

$$
E(X_1 X_2) = E(X_1)E(X_2) = \frac{a}{2} \cdot \frac{a}{2} = \frac{a^2}{4}.
$$

Otherwise, using the joint density, we have

$$
\begin{aligned}
E(X_1 X_2) &= \frac{1}{a^2} \int_0^a \int_0^a x_1 x_2 (h(x_1) - h(x_1 - a))(h(x_2) - h(x_2 - a)) \, dx_1 \, dx_2 \\
&= \frac{1}{a^2} \left(\int_0^a x_1(h(x_1) - h(x_1 - a)) \, dx_1 \right) \\
&\quad \times \left(\int_0^a x_2(h(x_2) - h(x_2 - a)) \, dx_2 \right) \\
&= \frac{1}{a^2} \int_0^a x_1 \, dx_1 \cdot \int_0^a x_2 \, dx_2 = \frac{1}{a^2} \frac{a^2}{2} \frac{a^2}{2} = \frac{a^2}{4}.
\end{aligned}
$$

Section 5.6

1. We have

$$EX = 1 \cdot \frac{1}{2} + 2 \cdot \frac{1}{2} = \frac{3}{2},$$

$$EY = 1 \cdot \frac{3}{16} + 2 \cdot \frac{2}{16} + 3 \cdot \frac{5}{16} + 4 \cdot \frac{3}{8} = \frac{23}{8},$$

$$EX^2 = 1^2 \cdot \frac{1}{2} + 2^2 \cdot \frac{1}{2} = \frac{5}{2},$$

$$EY^2 = 1^2 \cdot \frac{3}{16} + 2^2 \cdot \frac{2}{16} + 3^2 \cdot \frac{5}{16} + 4^2 \cdot \frac{3}{8} = \frac{152}{16},$$

$$\mathrm{var}X = EX^2 - (EX)^2 = \frac{1}{4}, \quad \mathrm{var}Y = EY^2 - (EY)^2 = 79/64.$$

and,

$$EXY = 1 \cdot 1 \cdot \frac{1}{8} + 1 \cdot 2 \cdot \frac{1}{16} + 1 \cdot 3 \cdot \frac{3}{16} + 1 \cdot 4 \cdot \frac{1}{8}$$
$$+ 2 \cdot 1 \cdot \frac{1}{16} + +2 \cdot 2 \cdot \frac{1}{16} + 2 \cdot 3 \cdot \frac{1}{8} + 2 \cdot 2 \cdot \frac{1}{4}$$
$$= \frac{71}{16}.$$

Therefore,

$$\mathrm{cov}(X.Y) = EXY - EXEY = \frac{1}{8},$$

$$\rho(X, Y) = \frac{1}{\sqrt{\mathrm{var}\, X}\sqrt{\mathrm{var}\, Y}}\mathrm{cov}(X.Y) = 0.225.$$

Conclusion: The random variables are essentially not correlated.

3. Note that the pdf does not factor, so the random variables are not independent.

(a) By definition,

$$EXY = \frac{1}{40} \int_{-3}^{3} \int_{-1}^{1} xy(x + y)^2 \, dx \, dy$$
$$= \frac{1}{40} \int_{-3}^{3} \left(\int_{-1}^{1} (yx^3 + 2x^2 y^2 + xy^3) \, dx \right) dy$$
$$= \frac{1}{40} \int_{-3}^{3} \frac{4}{3} y^2 \, dy = \frac{3}{5}.$$

Also, showing the appropriate integrals but leaving the calculation of the integrals to the reader,

$$EX = \frac{1}{40} \int_{-3}^{3} \int_{-1}^{1} x(x+y)^2 \, dx \, dy, \quad EY = \frac{1}{40} \int_{-3}^{3} \int_{-1}^{1} y(x+y)^2 \, dx \, dy,$$

$$EX^2 = \frac{1}{40} \int_{-3}^{3} \int_{-1}^{1} x^2(x+y)^2 \, dx \, dy, \quad EY^2 = \frac{1}{40} \int_{-3}^{3} \int_{-1}^{1} y^2(x+y)^2 \, dx \, dy,$$

$$\text{var } X = EX^2 - (EX)^2, \quad \text{var } Y = EY^2 - (EY)^2.$$

(b) By definition,

$$\rho(X,Y) = \frac{1}{\sqrt{\text{var } X}\sqrt{\text{var } Y}} \text{cov}(X.Y) = \frac{1}{\sqrt{\text{var } X}\sqrt{\text{var } Y}} (EXY - EXEY).$$

Chapter 6

Section 6.1

1. (a) $z_{\alpha/2} = 1.64$

   (b) $z = 1.15$

Section 6.2.1

1. The confidence interval for μ is given by $\bar{x} \pm (\sigma/\sqrt{n})z_{\alpha/2} = 70.2 \pm (5/\sqrt{15})(1.96)$.

3. The sample size is given by $n = z_{\alpha/2}^2 \sigma^2/(\text{SE})^2 = (1.96)^2(3.2)^2/(0.2)^2 = 983.4$, round up to $n = 984$ as the required sample size.

Section 6.2.2

1. For two degrees of freedom, the chi-squared distribution is

$$f(x) = \frac{1}{2\Gamma(1)} x^0 e^{-x/2} = \frac{1}{2} e^{x/2},$$

which is the exponential distribution.

3. $P(-\sqrt{y} < X < \sqrt{y}) = P(-\sqrt{y} < Z < \sqrt{y}) = (2/\sqrt{2\pi}) \int_0^{\sqrt{y}} e^{-x^2/2} \, dx$. Let $u = x^2$, then $du = 2x \, dx$ or $dx = \frac{1}{2} u^{-1/2}/, du$ substitution yields $P(-\sqrt{y} < Z < \sqrt{y}) = (1/\sqrt{2\pi}) \int_0^y u^{-1/2} e^{-u/2} \, du = P(X^2 < y)$. Note that $f(u) = (1/\sqrt{2\pi})u^{-1/2}e^{-u/2}$ is the chi-squared density function with 1 degree of freedom [note that $\Gamma(1/2) = \sqrt{\pi}$].

Section 6.2.4

1. (a) $t = 2.571$. (b) $P(-1.5 < t < 1.5) = 2(0.097) = 0.194$.

3. The 90% confidence interval is given by $((n-1)s^2/b, (n-1)s^2/a)$ where $P(\chi^2 \leq a) = 0.05$ and $P(\chi^2 \geq b) = 0.05$. For this problem we have $n - 1 = 12$, $s^2 = 10.7$, $a = 5.266$, and $b = 21.02$. Substitution yields the desired interval.

5. Use the interval given in the solution to Exercise 3. For this problem, calculate s^2 using the data $n = 7$ and $\alpha/2 = 0.025$.

Section 6.3

1. We need to find values L_1 and L_2 such that $P(L_1 < p_1 - p_2 < L_2) = \alpha$. For a standard normal random variable Z, we know that $L_1 = -z_{\alpha/2}$ and $L_2 = z_{\alpha/2}$. This gives

$$P\left(-z_{\alpha/2} < \frac{(\widehat{p}_1 - \widehat{p}_2) - (p_1 - p_2)}{\sqrt{p_1(1-p_1)/n_1 + p_2(1-p_2)/n_2}} < z_{\alpha/2}\right) = \alpha.$$

Solving for $p_1 - p_2$ gives L_1 and L_2 as

$$(\widehat{p}_1 - \widehat{p}_2) - z_{\alpha/2}\sqrt{p_1(1-p_1)/n_1 + p_2(1-p_2)/n_2}.$$

Now replace p_1 and p_2 under the radical by their estimates $\widehat{p}_1$ and $\widehat{p}_2$ and you have the desired limits for a $1 - \alpha$ confidence interval for $p_1 - p_2$.

3. (a) Use the confidence interval given in equation (6.3) with $n = 1000$, $\widehat{p} = 0.25$, $z_{.025} = 1.96$, and the result will follow.

   (b) Same as part (a), but now with $\widehat{p} = 0.275$.

   (c) Use the result in Exercise 1.

5. Let $\widehat{p}_1 = \frac{10}{20}$ and $\widehat{p}_2 = \frac{15}{20}$. Now compute a 95% confidence interval for $p_1 - p_2$ using the result given in Exercise 1. If this interval *does not* include 0, one can conclude that *there is* a significant difference between germination rates. If the interval *does* include 0, we *cannot* conclude that there is a significance difference (at least at the 0.05 level of significance).

Section 6.4

1. $a = 15.09$.

3. $P(X^2 \geq 10) = 0.0752$. This is not significant at the $\alpha = 0.05$ level of significance, but it is significant at the $\alpha = 0.1$ level of significance.

5. Proceeding as in the coin-toss example, we have $X^2 = 3^2/50 + 2^2/25 + (2.5)^2/12.5 + (2.75)^2/6.25 = 1.445$. Then $\Pr(X^2 \geq 1.445) = .2293$. Thus, we *cannot reject* her hypothesis that the weekly mortality rate is 0.5.

Section 6.5

1. (a) $H_0: p \geq 0.5$, vs. $H_1: p < 0.5$.

   (b) $\widehat{p} = \frac{230}{500} = 0.46$ and

   $$z = \frac{\widehat{p} - p}{\sigma} = \frac{.46 - .5}{\sqrt{.25/500}} = -1.7889.$$

   The p-value is given by

   $$p\text{-value} = \Pr(Z < -1.7889) = 0.0368$$

   and is classified as significant. We conclude that there is evidence to support the claim that the proportion who favor drilling is less than 50%.

   (c) We are subject to a type I error. We have rejected the null hypothesis, and it is possible that it is true.

3. (a) $H_0 : \mu \geq 22.6$ vs. $H_1 : \mu < 22.6$

   (b) We have $\bar{x} = 19.75$ and $s = 4.5751$. Then the p-value $= \Pr(x < 10.75) = \Pr\left(t < (19.75 - 22.6)/(4.5751/\sqrt{12}) = -2.1579\right)$. Using 11 degrees of freedom, we find that the p-value $= 0.0269$. This is classified as significant; thus, we reject H_0 and conclude that there is evidence to support the claim that the average number of crabs in a pod is less than 22.6.

5. We find that $\bar{x} = 12.9$ and $s = 14.1063$ and compute the p-value $= 2P(t > 12.9/(14.1062/\sqrt{10}) = 2.8919)$. This gives a p-value of 0.018. We conclude that there is evidence that the medication lowers blood pressure.

Section 6.6.3

1. Proceed as in Example 6.23.

3. Proceed as in Example 6.28.

5. Proceed as in Example 6.29.

7. Proceed as in Example 6.23.

Chapter 7

Section 7.2

1. (a) $X_{t+1} = (1 + 0.5 - 0.1)X_t = 1.4X_t$. Therefore the solution is $X_t = 100(1.4)^t$. Your graph will show the population geometrically growing 40% per year.

(b) In catastrophic years the birth rate is $0.60 \times 0.5 = 0.3$, and the death rate is $1.25 \times 0.1 = 0.125$. At these new rates, $1 + b - d = 1.175$, or 17.5% per year. A catastrophic year occurs once every 25 years, or with probability 0.04. To set up a model of the stochatic process we pick a uniform random number in $[0, 1]$ and choose the growth rate to be 1.175 when the random number is less than 0.04, and choose it to be 1.4 otherwise. The following MATLAB m-file performs this task.

```
function sandhill
clear all
X=100; Xhistory=X; T=5;
t=1:T
c=rand;
if c<=0.04, r=1.175;
else r=1.4;
end
X=r*X; Xhistory=[Xhistory,X];
end
plot(0:T,Xhistory)
```

(c) To simulate a normal random variable $b \sim N(\mu, \sigma^2)$, we use $b = \mu + \sigma z$, where $z \sim N(0, 1)$. In MATLAB, the command randn selects a normal RV in $[0, 1]$. To obtain a simulation as requested, in the previous program replace the time loop by

```
for t=1:T
b=0.5+0.03*randn; d=0.1+0.08*randn;
X=(1+b-d)*X;
Xhistory=[Xhistory,X];
end
```

3. (a) By iteration, the population at time T is $x_T = r_0 r_1 r_2 \cdots r_{T-1} x_0$. If we assume the average growth rate is r, then $x_T = r^T x_0$, so $r^T = r_0 r_1 r_2 \cdots r_{T-1}$, or

$$r = (r_0 r_1 r_2 \cdots r_{T-1})^{1/T} = r_G.$$

(b) Take T even. Notice that $X_T = R_0 R_1 R_2 \cdots R_{T-1} x_0$, where half the R's are 0.86 and the other half are 1.16. The order is immaterial and therefore

$$X_T = (0.86)^{T/2}(1.16)^{T/2} x_0 = (0.9989)^T x_0.$$

The growth rate is less than 1. Thus,

$$X_{500} = (0.9989)^{500} 100 = 54.84 \approx 55.$$

5. Below is a MATLAB program that produces a time series for the population. The smaller the initial population, the more likely that an extinction will occur at an earlier time.

```
clear all
x=128; xlist=x; sum=0; N=200;
for t=1:N
sum=0;
for j=1:x
r=rand;
if r < 0.25, birth=0;
elseif r > =0.25 & r< 0.75, birth=1;
else r > =0.75, birth=2;
end
sum=sum+birth;
end
xlist=[xlist,sum];
end
T=0:N;
plot(T,xlist)
```

7. The deterministic solution to $X_{t+1} = aX_t + b$ is $X_t = Ca^t + \frac{b}{1-a}$, where c is an arbitrary constant. Applying the initial condition, we get $C = X_0 - \frac{b}{1-a}$. Therefore,

$$X_t = \left(X_0 - \frac{b}{1-a} \right) a^t + \frac{b}{1-a} = X_0 a^t + \frac{b}{1-a}(1 - a^4).$$

Therefore,

$$
\begin{aligned}
\Pr(X_t \le x) &= \Pr\left(X_0 a^t + \frac{b}{1-a}(1-a^t) \le x\right) \\
&= \Pr\left(X_0 \le x a^{-t} - \frac{b}{1-a}(1-a^t)a^{-t}\right) \\
&= \int_{-\infty}^{x a^{-t} - \frac{b}{1-a}(1-a^t)a^{-t}} f_{X_0}(y)\,dy \\
&= \int_{-\infty}^{x} f_{X_0}\left(z a^{-t} - \frac{b}{1-a}(1-a^t)a^{-t}\right) a^{-t}\,dz,
\end{aligned}
$$

where we made the substitution

$$
y = z a^{-t} - \frac{b}{1-a}(1-a^t)a^{-t}, \quad dy = a^{-t}dz
$$

in the next-to-last step. Therefore

$$
f_{X_t}(x,t) = f_{X_0}\left(x a^{-t} - \frac{b}{1-a}(1-a^t)a^{-t}\right) a^{-t}.
$$

Section 7.3

1. We have the same equation,

$$
p u_{n+1} - u_n + q u_{n-1} = 0,
$$

but the boundary conditions are now given by

$$
u_0 = 1, \quad u_N = 0.
$$

The general solution to the difference equation is given in the text. For $p = q$ the general solution is $u_n = a + bn$. Applying the boundary conditions, we get $a = 0$ and $b = -1/N$. Therefore,

$$
u_n = 1 - \frac{n}{N}, \qquad p = q.
$$

For the case $p \ne q$ the general solution is

$$
u_n = a + b\left(\frac{q}{p}\right)^n.
$$

Applying the two boundary conditions gives a and b and we get

$$
u_n = \frac{(q/p)^n - (q/p)^N}{1 - (q/p)^N}, \qquad p \ne q.
$$

3. The equations are

$$pv_{n+1} - v_n + qv_{n-1} = -1,$$
$$v_0 = v_N = 0.$$

When $p = q$ we have

$$\frac{1}{2}v_{n+1} - v_n + \frac{1}{2}v_{n-1} = -1.$$

By the hint we let

$$v_n = a + bn - n^2,$$

where a and b are to be determined. From the boundary conditions we find that $a = 0$ and $b = N$. Thus,

$$v_n = n(N - n), \qquad p = q.$$

When $p \neq q$ the solution to the homogeneous equation, as before, is

$$v_n = a + b\left(\frac{q}{p}\right)^n.$$

Assume a solution to the nonhomogeneous equation of the form $w_n = Bn$. Substituting into the full nonhomogeneous equation we find $B = -1/(p-q)$. The general solution to the nonhomogeneous equation is the sum of the solution to the homogenous equation and a solution to the nonhomogeneous equation (this is true for all linear, nonhomogeneous equations). Therefore

$$v_n = a + b\left(\frac{q}{p}\right)^n - \frac{n}{p - q}.$$

Because $v_0 = v_N = 0$, we get $a + b = 0$ and $a + b\,(q/p)^N = N/(p-q)$. Solve these two equations to obtain a and b and therefore the solution.

Section 7.4.1

1. (a) An integrating factor is e^t, so multiply the equation by e^t and note that the equation becomes $(ye^t)' = 1$. Taking antiderivatives, we get

$$ye^t = t + C.$$

At $t = 0$ we have $y = 1$, so $1 = C$. Therefore, the solution is $ye^t = t + 1$.

$$y = (t + 1)e^{-t}.$$

(b) Now the integrating factor is e^{-t}. Multiply by this and turn the equation into $(ye^{-t})' = e^{-2t}$. Integrating yields

$$ye^{-t} = -\frac{1}{2}e^{-2t} + C.$$

At $t = 0$ we have $y = 0$, so $C = 1/2$. Therefore,

$$y = \frac{1}{2}\left(1 - e^{-2t}\right)e^t.$$

(c) The integrating factor is e^{2t}. Multiplying by the factor and writing the left side as a total derivative we get $(ye^{2t})' = te^{2t}$. Integrating gives us

$$ye^{2t} = \int te^{2t}\,dt + C.$$

To find the integral we can use integration by parts or look up the integral in a table of integrals. We find that

$$ye^{2t} = \frac{1}{4}e^{2t}(2t - 1) + C.$$

When $t = 0$, $y = 1$, we get $C = 5/4$. Then

$$ye^{2t} = \left(\frac{1}{4}e^{2t}(2t - 1) + \frac{5}{4}\right)e^{-2t} = \frac{1}{4}\left((2t - 1) + 5e^{-2t}\right).$$

2. The coefficient of y on the left side, namely, $2t$, is not a constant, but a function of t. This invalidates the formula.

3. Multiplying the equation by e^{t^2} gives

$$\left(ye^{t^2}\right)' = e^{-t}e^{t^2}.$$

Integrating, we find that

$$ye^{t^2} = \int e^{-t}e^{t^2}\,dt + C.$$

There is no formula for the integral (not all functions have simple antiderivatives). So, rather, integrate both sides of the equation from 0 to t to get

$$\int_0^t \left(y(s)e^{s^2}\right)'\,ds = \int_0^t e^{-s}e^{s^2}\,ds,$$

where we have changed the dummy integration variable from t to s. Then, using the fundamental theorem of calculus, we obtain

$$y(t)e^{t^2} - y(0)e^{0^2} = \int_0^t e^{-s}e^{s^2}\,ds,$$

or

$$y(t)e^{t^2} = 1 + \int_0^t e^{-s}e^{s^2}\,ds,$$

which gives $y(t)$.

Section 7.4.2

1. Substitute (7.20) into (7.19) and carry out the detailed (but tedious) verification.

3. (a) By the law of total probability (we note there are only two terms in the sum),

$$
\begin{aligned}
p_k(t+h) &= \Pr(N(t+h) = k) = \sum_{i=0}^{k} \Pr(N(t+h) = k | N(t) = i) \Pr(N(t) = i) \\
&= \Pr(N(t+h) = k | N(t) = k-1) \Pr(N(t) = k-1) \\
&\quad + \Pr(N(t+h) = k | N(t) = k) \Pr(N(t) = k) + o(h) \\
&= (\lambda h + o(h)) p_{k-1}(t)) + (1 - \lambda h + o(h)) p_k(t)) + o(h) \\
&= \lambda h (p_{k-1}(t) - p_k(t)) + p_k(t) + o(h).
\end{aligned}
$$

Then

$$
\frac{p_k(t+h) - p_k(t+h)}{h} = \lambda p_{k-1}(t) - \lambda p_k(t) + \frac{o(h)}{h}.
$$

Take the limit as $h \to 0$ to get

$$
p_k'(t) = -\lambda p_k(t) + \lambda p_{k-1}(t), \quad k = 1, 2, \dots
$$

(b) If $k = 0$, we go back to the previous derivation to get

$$
\begin{aligned}
p_0(t+h) &= \Pr(N(t+h) = 0 | N(t) = k) \Pr(N(t) = 0) \\
&= (1 - \lambda h + o(h)) p_0(t).
\end{aligned}
$$

Again writing the difference quotient and taking the limit gives

$$
p_0'(t) = -\lambda p_0(t).
$$

For initial conditions we have $p_0(0) = 1$ and $p_k(0) = 0$, $k \geq 1$. Notice that the equation for $p_0(t)$ is the usual decay equation with solution

$$
p_0(t) = e^{-\lambda t}.
$$

(c) For verification, we substitute the given expression for $p_k(t)$ into the differential equation in part (a). Taking the derivative using the product rule yields

$$
\begin{aligned}
\frac{d}{dt} p_k(t) &= \frac{d}{dt} \frac{1}{k!} (\lambda t)^k e^{-\lambda t} \\
&= \frac{1}{k!} (\lambda t)^k e^{-\lambda t} (-\lambda) + \frac{1}{k!} k (\lambda t)^{k-1} \lambda e^{-\lambda t} \\
&= -\lambda \frac{1}{k!} (\lambda t)^k e^{-\lambda t} + \lambda \frac{1}{(k-1)!} (\lambda t)^{k-1} e^{-\lambda t} \\
&= -\lambda p_k(t) + \lambda p_{k-1}(t).
\end{aligned}
$$

(d) The probability mass function in part (c) is that of a Poisson random variable average value λ replaced by λt. (See Section 5.4 on the Poisson distribution.) Thus, the expected value and variance of $N(t)$ are both λt.

Section 7.4.3

1. Simulate a Poisson process when $\lambda = 1$.

```
lambda=1;
n=linspace(1,50,50); t(1)=0;              % sets up a vector of length 50
for k=1:49
t(k+1)=t(k)-log(rand)/(lambda*n(k));      % computes random jump times
end
s=stairs(t,n)                             % plots the stair-step graph
```

Section 7.5.2

1. This is an exercise in partial differentiation. We have

$$u(x,t) = \frac{1}{\sqrt{4\pi Dt}} e^{-x^2/4Dt}.$$

Therefore, by the product rule

$$
\begin{aligned}
u_t &= \frac{1}{\sqrt{4\pi Dt}} e^{-x^2/4Dt} \frac{x^2}{4Dt^2} - e^{-x^2/4Dt} \frac{1}{2} \frac{4\pi D}{(4\pi Dt)^{3/2}} \\
&= \frac{1}{\sqrt{4\pi Dt}} e^{-x^2/4Dt} \left(\frac{x^2}{4Dt^2} - \frac{1}{2t} \right).
\end{aligned}
$$

Also,

$$u_x = -\frac{1}{\sqrt{4\pi Dt}} e^{-x^2/4Dt} \frac{x}{2Dt},$$

and by the product rule

$$
\begin{aligned}
u_{xx} &= -\frac{1}{\sqrt{4\pi Dt}} \left(e^{-x^2/4Dt} \frac{1}{2Dt} - \frac{x}{2Dt} e^{-x^2/4Dt} \frac{x}{2Dt} \right) \\
&= -\frac{1}{\sqrt{4\pi Dt}} e^{-x^2/4Dt} \left(\frac{1}{2Dt} - \frac{x^2}{4D^2t^2} \right).
\end{aligned}
$$

Now it is easy to see that $u_t = D u_{xx}$.

3. We compute the cdf of $W_1(t)$ defined by

$$
\begin{aligned}
\Pr(W_1(t) \le x) &= \Pr\left(W\left(\frac{t}{c^2}\right) \le \frac{x}{c} \right) \\
&= \int_{-\infty}^{x/c} \frac{1}{\sqrt{2\pi t/c^2}} e^{-y^2/(2t/c)} dy \\
&= \int_{-\infty}^{x} \frac{1}{\sqrt{2\pi t}} e^{-z^2/2t} dz,
\end{aligned}
$$

where in the last step we made the substitution $y = z/c$, $dy = dz/c$. So $W_1(t)$ has the same pdf as $W(t)$.

Section 7.5.3

1. Write the equation in its defined integral form and take the expected value of both sides of the equation to get

$$EX(t) = EX(0) + E \int_0^t X(s)\, ds + \frac{1}{2} E \int_0^t \sqrt{X(s)}\, dW(s).$$

By the result preceding this exercise, we have

$$EX(t) = EX(0) + \int_0^t EX(s)\, ds.$$

[In a later section we state the result that $E \int_0^t f(X(x), s)\, dW(s) = 0$ for reasonable functions f.] Using $X(0) = 1$, taking the derivative of both sides of the equation, and using the fundamental theorem of calculus, we get

$$\frac{d}{dt} EX(t) = EX(s).$$

This is the standard growth equation and the solution is

$$EX(t) = Ce^t,$$

where C is a constant. Because $X(0) = 1$, we have $EX(0) = 1 = C$, which gives C.

3. The SDE is $dX = X\, dt + \frac{1}{2}\sqrt{X}\, dW$, with $X(0) = 1$. Hence, $a(X, t) = X$, $b(X, t) = \frac{1}{2}\sqrt{X}$, and $\frac{\partial b}{\partial X} = 1/4\sqrt{X}$. Following is an m-file that implements Milstein's method.

```
x=1; xlist=x; dt=0.02; N=50;
for n=1:N
r1=randn; r2=randn;
x=x+x*dt+0.5*sqrt(x)*sqrt(dt)*r1+0.5*sqrt(x)*0.25./sqrt(x).*(dt*r2^ 2-dt);
xlist=[xlist,x];
end
t=0:0.02:1; plot(t,xlist)
```

Section 7.6

1. Square the matrix defining $\sqrt{C}$ and show that the result is the matrix C.

3. (a) To leading order, the expected value is

$$E(\Delta R) = \rho\lambda\Delta t + 0 \cdot (1 - \lambda\Delta t) = \rho\lambda\Delta t$$

and

$$E(\Delta R)^2 = \rho^2\lambda\Delta t + 0^2 \cdot (1 - \lambda\Delta t) = \rho^2\lambda\Delta t.$$

The variance is therefore

$$\text{var}(\Delta R) = E(\Delta R)^2 - (E(\Delta R))^2 = \rho^2\lambda\Delta t - (\rho\lambda\Delta t)^2 = \rho^2\lambda\Delta t.$$

Therefore,

$$\Delta R = \rho\lambda\Delta t + \rho\sqrt{\lambda\Delta t}\ Z,$$

where $Z \sim N(0, 1)$. In the limit as $\Delta t \to 0$,

$$dR = \rho\lambda\,dt + \rho\sqrt{\lambda}\,dW(t).$$

(b) Use the MATLAB code in Example 7.18.

5. (a) We write the equations as

$$dx = ax\,dt - bxy\,dt, \quad dy = cxy\,dt - my\,dt.$$

We interpret the right side of each equation as a birth rate minus a death rate. As usual, let $X(t)$ and $Y(t)$ denote the stochastic processes and $\mathbf{X}(t) = (X(t), Y(t))^{\mathsf{T}}$, with $\Delta\mathbf{X}(t) = (\Delta X(t), \Delta Y(t))^{\mathsf{T}}$. The quantity $\Delta X(t)$ can increase by 1, decrease by 1, or have no change; $\Delta Y(t)$ can do the same. All other changes are assumed to be small [order $(dt)^2$]. So there are nine possible changes to $\Delta\mathbf{X}(t)$. Only four of them are first order in Δt. We assemble the changes and probabilities, to first order, in the following table:

Change $\Delta\mathbf{X}(t)$	Probability
$(1, 0)^{\mathsf{T}}$	$aX\Delta t$
$(0, 1)^{\mathsf{T}}$	$cXY\Delta t$
$(-1, 0)^{\mathsf{T}}$	$bXY\Delta t$
$(0, -1)^{\mathsf{T}}$	$mY\Delta t$

For example, the change $\Delta\mathbf{X}(t) = (-1, 1)^{\mathsf{T}}$ means that $\Delta X(t) = -1$ and $\Delta Y(t) = 1$, with probabilities $bXY\Delta t$ and $cXY\Delta t$, respectively. The probability of both occurring is the product $(bXY\Delta t)(cXY\Delta t)$, which is o$(\Delta t)$. Now, the expected change is

$$\begin{aligned} E(\Delta\mathbf{X}(t)) &= (1, 0)^{\mathsf{T}}aX\Delta t + (0, 1)^{\mathsf{T}}cXY\Delta t + (-1, 0)^{\mathsf{T}}bXY\Delta t \\ &= +(0, -1)^{\mathsf{T}}mY\Delta t + (aX - bXY, cXY - mY)^{\mathsf{T}}\Delta t. \end{aligned}$$

To leading order, the covariance matrix is, to leading order,

$$
\begin{aligned}
C &= \operatorname{cov}\Delta\mathbf{X}(t) = \mathrm{E}(\Delta\mathbf{X}(t)\Delta\mathbf{X}(t)^{\mathrm{T}}) \\
&= \begin{pmatrix} 1 \\ 0 \end{pmatrix}(1,0)aX\Delta t + \begin{pmatrix} 0 \\ 1 \end{pmatrix}(0,1)cXY\Delta t \\
&\quad + \begin{pmatrix} 0 \\ -1 \end{pmatrix}(0,-1)mY\Delta t + \begin{pmatrix} -1 \\ 0 \end{pmatrix}(-1,0)bXY\Delta t \\
&= \begin{pmatrix} aX+bXY & 0 \\ 0 & cXY+mY \end{pmatrix}\Delta t.
\end{aligned}
$$

Therefore,

$$
\Delta\mathbf{X}(t) = \mathrm{E}(\Delta\mathbf{X}(t))\Delta t + B\mathbf{Z},
$$

where $B^2 = C$ and $\mathbf{Z} = (Z_1, Z_2)^{\mathrm{T}}$, where $Z_1, Z_2 \sim \mathrm{N}(0,1)$. Because C is a diagonal matrix, we have

$$
B = \begin{pmatrix} \sqrt{aX+bXY} & 0 \\ 0 & \sqrt{cXY+mY} \end{pmatrix}\sqrt{\Delta t}.
$$

Therefore,

$$
\begin{aligned}
\Delta\mathbf{X}(t) &= \mathrm{E}(\Delta\mathbf{X}(t))\Delta t + \begin{pmatrix} \sqrt{aX+bXY} & 0 \\ 0 & \sqrt{cXY+mY} \end{pmatrix}\sqrt{\Delta t}\,\mathbf{Z} \\
&= \begin{pmatrix} aX-bXY \\ cXY-mY \end{pmatrix}\Delta t + \begin{pmatrix} \sqrt{aX+bXY} & 0 \\ 0 & \sqrt{cXY+mY} \end{pmatrix}\Delta\mathbf{W},
\end{aligned}
$$

where $\Delta\mathbf{W} = (\Delta W_1, \Delta W_2)^{\mathrm{T}}$. Taking the limit as $\Delta t \to 0$ gives the pair of SDEs

$$
\begin{aligned}
dX &= (aX-bXY)dt + \sqrt{aX+bXY}\ dW_1, \\
dY &= (cXY-mY)dt + \sqrt{cXY+mY}\ dW_2.
\end{aligned}
$$

(b) The following MATLAB m-file produces a plot of the solution to the Lotka–Volterra equations using the Euler–Maruyama method:

```
x=120; y=40; xlist=x; ylist=y; dt=0.01; N=2000;
a=1; b=0.02; c=0.01; m=1;
for n=1:N
r1=randn; r2=randn;
u=x+(a*x-b*x.*y)*dt+sqrt(a*x+b*x.*y)*sqrt(dt)*r1;
v=y+(c*x.*y-m*y)*dt+sqrt(c*x.*y+m*y)*sqrt(dt)*r2;
xlist=[xlist,u];ylist=[ylist,v]; x=u; y=v;
end
t=0:0.01:20; plot(t,xlist,t,ylist)
```

Section 7.7

1. Take a partition of $[a, b]$ with N subintervals each of length $(b-a)/N$. Then the partition points are $t_n = a + n(b-a)/N$, $n = 0, 1, ..., N$, and $t_0 = a$ and $t_N = b$. Then

$$\sum_{n=0}^{N-1}[G(t_{n+1}, W(t_{n+1})) - G(t_{n+1}, W(t_{n+1}))] \;=\; G(t_N, W(t_N)) - G(t_0, W(t_0))$$
$$= G(b, W(b)) - G(a, W(a)),$$

because the series telescopes. Therefore,

$$\left(\sum_{n=0}^{N-1}[G(t_{n+1}, W(t_{n+1})) - G(t_{n+1}, W(t_{n+1}))] - [G(b, W(b)) - G(a, W(a))]\right)^2 = 0$$

So, the expected value of the left side is zero, and the limit as $N \to \infty$ is zero. Therefore by definition,

$$\int_a^b dG(t, W(t)) = G(b, W(b)) - G(a, W(a)).$$

3. We have $X = W$ and $dX = dW$; so $a(X, t) = 0$ and $b(X, t) = 1$. Take $Y = F(t, X) = f(t)X$. Then Itô's identity gives

$$dY = Xf'(t) dt + f(t) dW(t).$$

Integrating from 0 to t and using $Y(0) = f(0)X(0) = f(0)W(0) = 0$, we get

$$Y(t) = \int_0^t X(s)f'(s) ds + \int_0^t f(s) dW(s)$$

or

$$f(t)W(t) = \int_0^t W(s)f'(s) ds + \int_0^t f(s) dW(s).$$

5. Letting $Y = F(t, X) = \sigma^{-1} \ln(rX + \sigma)$, by Itô's identity (with $a = 0$ and $b = rX + \sigma$), we have

$$dY = -\frac{1}{2}\frac{r^2}{\sigma^2}dt + \frac{r}{\sigma}dW(t).$$

Integrating from 0 to t yields

$$Y(t) - Y(0) = -\frac{1}{2}\frac{r^2}{\sigma^2}t + \frac{r}{\sigma}\int_0^t dW(s).$$

Thus,

$$\sigma^{-1}\ln(rX + \sigma) - \sigma^{-1}\ln(rX_0 + \sigma) = -\frac{1}{2}\frac{r^2}{\sigma^2}t + \frac{r}{\sigma}W(t).$$

This gives

$$rX + \sigma = (rX_0 + \sigma)e^{-r^2t/2\sigma + rW(t)},$$

from which we can solve for X.

7. Let $Y = F(t, X) = X\exp\left(-\int_0^t a(s)\,ds\right)$. By Itô's identity, leaving the details to the reader,

$$dY = c(t)\exp\left(-\int_0^t a(s)ds\right)dW(t).$$

Integrating from 0 to t gives

$$Y(t) - Y(0) = \int_0^t c(s)\exp\left(-\int_0^s a(\tau)\,d\tau\right)dW(s),$$

or

$$X(t) = \left(X(0) + \int_0^t c(s)\exp\left(-\int_0^s a(\tau)\,d\tau\right)dW(s)\right)\exp\left(\int_0^t a(s)\,ds\right).$$

Section 7.8

1. For the particle to be at x at time $t + \Delta t$, either it was at x at time t and did not move (with probability $1 - p$), or it was at $x - \Delta x$ at time t and moved to x (with probability p). In symbols (from the law of total probability)

$$u(x, t + \Delta t) = u(x, t)(1 - p) + u(x - \Delta x, t)p.$$

Now expand in a Taylor series about (x, t) to obtain

$$u(x, t) + u_t(x, t)\Delta t + O(\Delta t^2) = u(x, t)(1 - p) + (u(x, t) - u_x(x, t)\Delta x + O(\Delta x^2))p.$$

Canceling $u(x, t)$, this formula becomes

$$u_t(x, t)\Delta t + O(\Delta t^2) = -pu_x(x, t)\Delta x + O(\Delta x^2).$$

Dividing by Δt yields

$$u_t(x, t) + O(\Delta t) = -pu_x(x, t)\frac{\Delta x}{\Delta t} + O\left(\frac{\Delta x^2}{\Delta t}\right).$$

Using the given limits as $\Delta t, \Delta x \to 0$, gives

$$u_t(x, t) = -pcu_x(x, t).$$

Index

PURE AND APPLIED MATHEMATICS
A Wiley-Interscience Series of Texts, Monographs, and Tracts

Founded by RICHARD COURANT
Editors Emeriti: MYRON B. ALLEN III, DAVID A. COX, PETER HILTON,
HARRY HOCHSTADT, PETER LAX, JOHN TOLAND

*Now available in a lower priced paperback edition in the Wiley Classics Library.
†Now available in paperback.

*Now available in a lower priced paperback edition in the Wiley Classics Library.
†Now available in paperback.

Printed in the United States
By Bookmasters